Computational Modeling of Genetic and Biochemical Networks

Computational Molecular Biology
Sorin Istrail, Pavel Pevzner, and Michael Waterman, editors

Computational Modeling of Genetic and Biochemical Networks
James M. Bower and Hamid Bolouri, editors, 2001

Computational Molecular Biology: An Algorithmic Approach
Pavel A. Pevzner, 2001

Computational Modeling of Genetic and Biochemical Networks

edited by James M. Bower and Hamid Bolouri

A Bradford Book
The MIT Press
Cambridge, Massachusetts
London, England

This book was set in Times Roman by Wellington Graphics, Westwood, Massachusetts.

Printed and bound in the United States of America.

Library of Congress Cataloging-in-Publication Data

Computational modeling of genetic and biochemical networks / edited by James M. Bower and Hamid Bolouri.
   p.cm. — (Computational molecular biology series)
   Includes bibliographical references and index.
   ISBN 0-262-02481-0 (hc : alk. paper)
   1. Biochemistry—Mathematical models. I. Bower, James M. II. Bolouri, Hamid. III. Series.

QP517.M3 C638   2000
572′.01′5118—dc21

                                                                                          00-024605

Contents

Foreword by John C. Wooley — vii

Acknowledgments — xi

Introduction: Understanding Living Systems — xiii

I   MODELING GENETIC NETWORKS

1   Modeling the Activity of Single Genes — 1
Michael Andrew Gibson and Eric Mjolsness

2   A Probabilistic Model of a Prokaryotic Gene and Its Regulation — 49
Michael Andrew Gibson and Jehoshua Bruck

3   A Logical Model of *cis*-Regulatory Control in a Eukaryotic System — 73
Chiou-Hwa Yuh, Hamid Bolouri, James M. Bower, and Eric H. Davidson

4   Trainable Gene Regulation Networks with Application to *Drosophila* Pattern Formation — 101
Eric Mjolsness

5   Genetic Network Inference in Computational Models and Applications to Large-Scale Gene Expression Data — 119
Roland Somogyi, Stefanie Fuhrman, and Xiling Wen

II   MODELING BIOCHEMICAL NETWORKS

6   Atomic-Level Simulation and Modeling of Biomacromolecules — 161
Nagarajan Vaidehi and William A. Goddard III

7   Diffusion — 189
Guy Bormann, Fons Brosens, and Erik De Schutter

8   Kinetic Models of Excitable Membranes and Synaptic Interactions — 225
Alain Destexhe

9   Stochastic Simulation of Cell Signaling Pathways — 263
Carl A. J. M. Firth and Dennis Bray

10   Analysis of Complex Dynamics in Cell Cycle Regulation — 287
John J. Tyson, Mark T. Borisuk, Kathy Chen, and Bela Novak

11   Simplifying and Reducing Complex Models — 307
Bard Ermentrout

Contributors — 325

Index — 327

Foreword
Computation, Modeling, and Biology in the Twenty-First Century

Over the past quarter century, extraordinary advances have been routine in the biological sciences and the computer sciences. For both, the popular press applies the term revolutionary advances and the academic world speaks of new paradigms. Our understanding of living systems and the development of computational science (along with its sibling, information science) have advanced to the point where each is ideally suited for the other, and the grandest advances in both fields will be made at the interface. While the nature of the progress seems likely to exceed even our dreams, press articles for the next quarter century will surely describe truly spectacular advances that arise from integrating computational and information technology with biological science research.

Living systems are very complex and their characterization will remain a challenge for the foreseeable future. To try to imagine what we may discover in the first few decades of the twenty-first century, consider where the biological sciences stood a few decades ago. Shocking to graduate students at the time, thirty-some years ago, a well-known, productive biologist, considered insightful, proclaimed that biology was now so well understood at the molecular level that no great new discoveries lay over the horizon and thus bright scientists should look elsewhere for challenges. Then came the deluge: the tools of molecular biology uncovered unexpected feature after feature of model organisms and began to lay open the vast complexity of metazoans. Today, the incredible diverse nature of living systems, often called biocomplexity, is clear to all of us.

The completion of the complete genome sequence for human and model organisms has given us the most powerful biological data set ever assembled, but even this information only tells us how finite our actual knowledge is. The richness of experimental observations and their scale require a data-driven approach. Biological scientists will need to embrace computational tools to uncover the meaning implicit in the DNA sequence. Large challenges for computational science lie ahead in extending knowledge representation and delivering data-driven computing, in contrast to numeric computing, which characterizes many physical science applications.

Throughout this same quarter century of remarkable progress, biologists have distrusted modeling, theory, and analytical and mathematical analysis. Researchers have focused on simple systems, the simplest possible to address their own area of interest, and whenever possible, asked questions believed to yield only binary answers, not quantitative ones. Completion and analysis of the DNA sequence of the first model organism, *H. influenza,* marked the end of our innocence; similarly, entry into the contemporary high throughput era, initiated by the human genome project, unambiguously demonstrated the limits of qualitative biology and introduced the vision of quantitative biology at a system level.

While we are just beginning to appreciate how we can approach quantitative analysis for biology, outlining how research can be refocused and simultaneously training a new generation of scientists is essential. The course developed at the California Institute of Technol-

ogy and represented in this book is a first step, much like the famous original moon
landing, a small step for each aspect but a large step for biology. The decision to deliver the
outcome of that course to a broader audience is very important and the effort should be
praised. Timing is said to be everything, and the timing of introducing a text to accelerate
progress at the frontier of computing and biology is perfect. About a decade ago, the National Science Foundation established the first government program directly and exclusively aimed at funding computational biology research and, simultaneously, a program
aimed at funding biological knowledge representation and analysis or what today is called
bioinformatics. At that time, experimentalists and even bureaucrats considered computational biology to be an oxymoron, or at least that living systems might be a great model for
computer scientists to emulate, but computers would never be significant research tools in
the biological sciences. The ever-pragmatic, hypothesis-driven, productive, experimental
biologist, seeking to be successful, to have an impact, and to be respected, avoided mathematical or analytical modeling and any experimental approach requiring quantitative insight. A famous biologist, on getting tenure, was told (in essence) by his chairman: well,
you have done great experimental work, the reference letters say so, but now that you have
been promoted, surely you will stay away from that theory and computation you were
(also) doing. I view this book as the answer to that challenge. Today, every biologist uses a
computer to search for a favorite gene and its attributes. Today, graduate students who exploit the access to the daily updates to the world's research findings leap ahead of those
plodding in their little corner of a lab. Bioinformatics is the "hottest commodity" for starting departments, obtaining jobs, or making headlines, and increasingly, a central component of academic strategic plans for growth. Information technology provides us with the
ability to create computational tools to manage and probe the vast data based on experimental discoveries, from sequence to transcription regulation, from protein structure to
function, and from metabolite to organ physiology. Today, just as the methods of molecular
biology transformed all of life sciences research, the methods of computational and information science will establish next generation biology. Tomorrow, I believe, *every* biologist
will use a computer to define their research strategy and specific aims, manage their experiments, collect their results, interpret their data, incorporate the findings of others, dissemi-
nate their observations, and extend their experimental observations—through exploratory
discovery and modeling—in directions completely unanticipated.

Finding patterns, insight, and experimental focus from the global literature, or discovery
science, provides a new approach for biological research. We now have the ability to create
models of the processes of living cells, models for how macromolecules interact and even
how cells actually work. For an example from this text, modeling differential gene expression as a function of time and space—specifically, the role of *cis*-regulatory control elements during sea urchin development—is an exemplar advance for what will become next-

generation biology. We have all been schooled in the regulation of the *lac* operon in the classic bacterium of choice and now must be awed by a first glimpse into the intricacies of eukaryotic gene regulation, incomplete as the model must be at this stage.

The limitations of experimental knowledge about biology and the limited power of computers led earlier generations of modelers to hit a brick wall seemingly infinite in height and thickness. Herein you will read numerous examples of cracks in the wall, which will tumble in the years ahead, in the era of completely sequenced genomes and high throughput biological science experiments. Despite the cliché nature of the expression, which we have all heard repeatedly for decades, biologists are now indeed working and living in exciting times, and computational biology already allows us to live/work "in the future." The extensive, in-depth contributions in this book, all truly exceptional in breadth, originality and insight, speak loudly, shout out, that biology is becoming a quantitative science, that biology can be approached systematically, that entire biological systems such as metabolism, signal transduction, and gene regulation, their interacting networks, or even higher levels of organization, can be studied and fully charactierzed through the combination of theory, simulation and experiment. The newly established funding of computational biology by private research foundations and by the National Institutes of Health is essential to provide the fuel through sustained, adequate funding. As the former government official who established those first programs for bringing the tools of computer and information science to bear on biological sciences research, I am myself honored to have been even a small part of this revolution and excited about what lies ahead.

Just read this book. Now. For once, read the entire book, which gives direct challenges for experimentalists and clear directions for current graduate students.

John C. Wooley
Associate Vice Chancellor, Research
University of California, San Diego

Acknowledgments

We thank The Caltech Computational Molecular Biology initiative (supported by a generous grant from the Burroughs-Wellcome foundation) and Scott Fraser in particular for generous support of our course at Caltech.

We are also grateful to the following Caltech biology faculty who kindly gave of their time and led modeling surgeries on their experimental systems: Pamela Bjorkman, Eric Davidson, Ray Deshaies, John Doyle, Scott Fraser, Elliot Meyerowitz, Vaidehi Nagaranjan, Ellen Rothenberg, Mel Simon, Paul Sternberg, and Kai Zinn; and to the students who took the course, for their keen interest and enthusiasm.

We would like to thank Michael Vanier for being much more than a teaching assistant for the course; Judy Macias for her resourceful help with course organization, and especially for finding the belly dancers who more succinctly than Bard Ermentrout ever could himself, demonstrated the wonders of coupled biological oscillations; and Michael Rutter of The MIT Press for his keenness and boundless energy.

J.M.B. thanks Carolina Livi for tolerating (and even supporting) this invasion into her own domain. You represent the future.

H.B. would like to thank everybody in his lab and STRC director Paul Kaye for their support of his work at Caltech.

Introduction: Understanding Living Systems

In 1982, at a symposium commemorating the tenth anniversary of the death of the biologist and modeler Aharon Katzir-Katchalsky, George Oster recalled Katchalsky joking that "Biologists can be divided into two classes: experimentalists who observe things that cannot be explained, and theoreticians who explain things that cannot be observed."[1] The joke is still funny because it sums up in one statement the traditionally awkward alliance between theory and experiment in most fields of biology as well as science in general. The statement that experimentalists "observe things that cannot be explained" mirrors the view of many traditional theorists that biological experiments are for the most part simply descriptive ("stamp collecting" as Rutherford may have put it) and that much of the resulting biological data are either of questionable functional relevance or can be safely ignored. Conversely, the statement that theorists "explain things that cannot be observed" can be taken to reflect the view of many experimentalists that theorists are too far removed from biological reality and therefore their theories are not of much immediate usefulness. In fact, of course, when presenting their data, most experimentalists do provide an interpretation and explanation for the results, and many theorists aim to answer questions of biological relevance. Thus, in principle, theorists want to be biologically relevant, and experimentalists want to understand the functional significance of their data. It is the premise of this book that connecting the two requires a new approach to both experiments and theory. We hope that this book will serve to introduce practicing theorists and biologists and especially graduate students and postdocs to a more coordinated computational approach to understanding biological systems.

Why Is Modeling Necessary?

Before considering the essential features of this new approach, it may be worthwhile to indicate why it as needed. Put as directly as possible, the sheer complexity of molecular and cellular interactions currently being studied will increasingly require modeling tools that can be used to properly design and interpret biological experiments. The advent of molecular manipulation techniques such as gene knockouts has made it very clear to most experimentalists that the systems they are studying are far more complex and dynamic than had often previously been assumed. Concurrently, the recent development of large-scale assay technologies such as cDNA arrays is already providing experimentalists with an almost overwhelming quantity of data that defy description or characterization by conventional means. As our understanding of the complexity of the systems we study grows, we will increasingly have to rely on something other than intuition to determine the next best experiment.

When the question of the role of modeling in biological investigations is raised, it is quite common for experimentalists to state that not enough is yet known to construct a model of their system. Of course, properly formulated, a model is often most useful as a way to decide what data are now necessary to advance understanding. As already stated, the complexity of biological systems makes it increasingly difficult to identify the next best experiment without such a tool. However, in reality, experimentalists already have and use models. Indeed, it is standard practice for experimentalists to begin and end research seminars with a "block and arrow" diagram. These diagrams constitute the experimentalists' understanding of how the components of their system interact. Accordingly, they are a kind of model. However, "models" in this form have no explicit form or structure but the diagram itself, with no associated formal specification. Without real parameters and mathematically defined relationships between the components, they cannot be falsified; they cannot make quantifiably testable predictions, or serve as a mechanism for conveying detailed information to other researchers on how the experimentalist is really thinking about the system under study.

From the point of view of a theorist, the enormous and growing amount of detailed information available about biological systems demands an integrative modeling approach. Atomic descriptions of molecular structure and function, as well as molecular explanations of cellular behavior, are now increasingly possible. Access to more and more sophisticated high-resolution in vivo and in vitro imaging technologies now provides the possibility of testing theoretical ideas at a fundamentally new level. At the atomic scale, technologies such as electron microscopy, X-ray crystallography, and nuclear magnetic resonance spectroscopy allow direct measurement of the way in which atomic interactions determine the three-dimensional geometry, the chemical and physical characteristics, and the function of biomolecules. At the molecular level, confocal microscopy, calcium imaging, and fluorescent tagging of proteins have made it possible to track the movement and reactions of molecules within single living cells. Combined with the new technologies for large-scale gene expression assays, these technologies are now yielding vast amounts of quantitative data on such cellular processes as developmental events, the cell cycle, and metabolic and signal transduction pathways. We now have the technology to track the expression pattern of thousands of genes during the lifetime of a cell, and to trace the interactions of many of the products of these genes. We are entering an era when our ability to collect data may no longer be the primary obstacle to understanding biology.

While our capacity for testing theoretical ideas is expanding dramatically, taking advantage of that capacity requires models that are directly linked to measurable biological structures and quantities. At the same time, the complexity of biological processes makes it less and less likely that simplified, abstract models will have the complexity necessary to capture biological reality. To be useful to experimentalists, models must be directly linked to biological data. For this reason, modelers interested in biology will have to develop new mathematical tools that can deal with the complexity of biological systems, rather than try to force those systems into a form accessible to the simpler tools now available.

What Type of Modeling Is Appropriate?

It is our belief that to make progress in understanding complex biological systems, experimentalists and theoreticians need each other. However, to make this connection, we need to rethink the role of models in modern biology and what types of models are most appropriate. Traditionally, biological models have often been constructed to demostrate a particular theorist's ideas of how a particular system or subsystem actually works. In this type of modeling, the functional idea exists before the model itself is constructed. While models of this sort can make experimentally testable predictions, it is usually easy enough to adjust the model's parameters to protect the original functional idea. When presented with surprising new results, often the theorist can adapt the model to fit the new data without having to reevaluate the basic premises of the model.

In our view, recent advances in biology require a more direct connection between modeling and experiment. Instead of being a means to demonstrate a particular preconceived functional idea, modeling should be seen as a way to organize and formalize existing data on experimentally derived relationships between the components of a particular biological system.[2] Instead of generating predictions that can be confirmed by their model, modelers should be more concerned with ensuring that the assumptions and structure of a particular model can be falsified with experimental data. As an aid in unraveling biological complexity, what a model cannot do is often more important than what it can do. It follows that models will have to generate experimentally measurable results and include experimentally verifiable parameters. Models of this kind should not be seen or presented as representations of the truth, but instead as a statement of our current knowledge of the phenomenon being studied. Used in this way, models become a means of discovering functionally relevant relationships rather than devices to prove the plausibility of a particular preexisting idea.

Many of these objectives can be more readily obtained if there is a close structural connection between biological data and the model itself. By structural we mean that the model itself reflects the structure of the system being studied. Such an arrangement makes it more likely that the modeler will learn something new (taken from the biology in some sense) from constructing the model. In fact, all modelers should be prepared to answer the question: What do you know now that you did not know before? If the answer is "that I was correct," it is best to look elsewhere.

Experimentally, models of this type almost always provide extremely important information on what types of experimental data are currently lacking. Thus, ironically, while we have said that the classical criticism of biological modeling is that we do not yet know enough to construct the model, model construction often makes it very clear that one knows even less than one thought about a particular system. When it is specific and experimentally addressable, This knowledge is extremely valuable in planning research. It follows that models should be evaluated by their ability to raise new experimental questions, not by their ability to account for what is already known or suspected. In consequence, understanding biological systems becomes an integrated, iterative process, with models feeding experimental design while experimental data in turn feed the model.

This type of interaction between models and experiment will inevitably alter the nature of both enterprises and the scientists that use them. If models are designed to directly integrate experimental data, then experimental investigations will also have to provide data for testing models. This in turn will require the development of new experimental techniques. The more closely models are linked to experimental data, the less obvious the distinction between a theorist and an experimentalist. Most of the authors in this book are avant garde in that they are theorists who collaborate closely with experimentalists. However, it is time to train a new generation of biologists who are equally comfortable with models and experiments.

What Is the Purpose of This Book?

This book arose from a graduate course of the same title that we organized and ran at the California Institute of Technology in the winter term of 1998. The objective of the course was to provide biology graduate students and postdoctoral scholars with an introduction to modeling techniques applied to a range of different molecular and cellular biological questions. The organization of the course and its pedagogy were informed by efforts over the past 15 years to introduce a similar approach to modeling neurobiology. As in our efforts to support the growth and extension of computational neuroscience,[2] this book reflects a conviction that modeling techniques must inevitably become a required feature of graduate training in biology. We hope that this book will serve as an introduction to those molecular and cellular biologists who are ready to start modeling their systems.

How Should this Book Be Used?

The book is aimed at the graduate and advanced-level student and is intended to provide instruction in the application of modeling techniques in molecular and cell biology. Since the book was originally used to support teaching at Caltech, we hope it will prove useful for others organizing similar courses. While we hope the book will be a useful instructional tool, we have also attempted to make it useful as a stand-alone introduction to the field for interested researchers. We expect the readers of this book to include cell biologists, developmental biologists, geneticists, structural biologists, and mathematical biologists, as well as the computational neuroscience community.

While each chapter starts with an overview of the biological system in question, no effort is made to provide a complete enough biological description to really understand the detailed objectives of the models described. Instead, attention is principally focused on the modeling approach itself. We regard our principal audience as biologists interested in modeling. However, we would hope that nonbiologists interested in making useful biological models would also find support here. Nonbiologists, however, are strongly encouraged to take advantage of the references at the end of each chapter to fill in their understanding of the biological systems being studied.

The span of disciplines contributing to this emerging field is vast. No reader will be fully conversant with all of them. Nor can this volume cover all aspects of them in sufficient depth and breadth. We therefore present this book, not as a comprehensive treatise, but as a sampler and a starting point for those with specific interests. For this reason, we have taken care to include extensive references to important work in each subject area. However, we anticipate explosive growth in this field, which will almost certainly require continued vigilant searching of the latest peer-reviewed literature for new applications of modeling techniques.

Organization of the Book

The book is organized in two parts: models of gene activity, i.e., genetic networks, and models of interactions among gene products, i.e., biochemical networks. Each part includes treatments of modeling at several different scales, from the smallest to the largest. Our hope is that in this way the reader will become aware of the advantages and disadvantages of modeling at finer scales, before considering more simplified models aiming at larger scales. It is our view that ultimately these levels of scale must be brought together and articulated into one large framework for understanding biology. For this reason, we have included a chapter by Bard Ermentrout that addresses the issue of scaling between models at different levels of description. The inclusion of this chapter also serves to emphasize the importance we place on the further development of mathematical tools to support biological modeling.

Part I.   Modeling Genetic Networks

The first part of the book deals with models of gene regulation, that is, protein–DNA and, indirectly, DNA–DNA interactions. The modeling approaches presented in these chapters range from models of individual genes, through models of interactions among a few genes, to techniques intended to help unravel the interactions among hundreds or thousands of genes. It is well understood that regulated gene expression underlies cellular form and function in both development and adult life. Changes in gene expression patterns generate the phenotypic variation necessary for evolutionary selection. An understanding of gene regulation is therefore fundamental to all aspects of biology. Yet, at present, knowledge of gene regulation, its evolution, and its control of cell function is patchy at best. The first part of this book reviews several tools and methods that are currently available for unraveling genetic regulatory networks at various levels of resolution and abstraction.

Chapter 1. Modeling the Activity of Single Genes How is the activity of individual genes regulated? This tutorial chapter begins by relating the regulation of gene activity to the basic principles of thermodynamics and physical chemistry. The authors then go on to discuss the need for different types of models to suit different model systems, the availability of experimental data, and the computational resources needed.

Chapter 2. A Probabilistic Model of a Prokaryotic Gene and Its Regulation This chapter provides a detailed investigation of an example model system in which the number of regulatory molecules determining gene activation is small and gene regulation is therefore stochastic. In addition to presenting a detailed model, it discusses the need for stochastic modeling, the advantages it provides in this case, and various examples of stochastic models applied to other systems.

Chapter 3. A Logical Model of *cis*-Regulatory Control in a Eukaryotic System The differential regulation of eukaryotic genes permits cell differentiation in space and time and is typically more complex than gene regulation in prokaryotes. Building on the theoretical framework presented in chapter 1, this chapter presents a detailed characterization of transcriptional regulation of a eukaryotic gene in which large numbers of regulatory factors interact in modular ways to control the transcription rate.

Chapter 4. Trainable Gene Regulation Networks with Application to *Drosophila* Pattern Formation Chapters 1 to 3 address the regulation of individual genes. However, gene regulation by its nature also involves interactions of whole systems of interacting genes. While we may someday be able to explore these interactions based on detailed models of individual genes, this chapter presents a framework for building phenomenological models that self-organize to replicate large-scale gene activity patterns. Such models can also be valuable aids in forming hypotheses and planning experiments.

Chapter 5. Genetic Network Inference in Computational Models and Applications to Large-Scale Gene Expression Data As already discussed, several emerging technologies now permit the simultaneous determination of the expression levels of thousands of genes in cell tissues. How can we unravel and reconstruct regulatory interactions among genes in such large-scale expression assays? This chapter reviews a number of potentially useful measures for uncovering correlated and potentially causally related events. It presents several studies that used these methods.

Part II. Modeling Biochemical Networks

The second part of this book considers interactions among the proteins produced by genetic regulation. It starts with modeling techniques that help us understand the interactions between single molecules, and then goes on to consider models aimed at understanding reactions and diffusion by large numbers of molecules. The later chapters emphasize the richness and complexity of the behavior that can arise in such networks and are a strong testament to the need for modeling in molecular biology.

Chapter 6. Atomic-Level Simulation and Modeling of Biomacromolecules Ultimately, we will want to build our understanding of biological systems from the first principles of the physics of atomic interactions. At present, however, the computational resources required to explain molecular function in terms of fundamental quantum mechanics limit our efforts to systems of at most a few hundred interacting atoms. This chapter presents a series of approximations to quantum mechanics, and demonstrates their use in modeling interactions of hundreds of thousands of atoms. Examples of predicting molecular structure and activity for drug design are presented. The chapter also provides a justification for the simpler, phenomenological models of molecular interaction used in the rest of this book by pointing out that atomic interaction models would be several orders of magnitude too computationally intensive.

Chapter 7. Diffusion Diffusion is the process by which random movement of individual molecules or ions causes an overall movement toward regions of lower concentration. Diffusion processes have a profound effect on biological systems; for example, they strongly affect the local rate of chemical reactions. Historically, diffusion has frequently been neglected in molecular simulations. However, as this chapter illustrates, diffusion often has a crucial impact on modeling results. The chapter describes a number of approximation methods for modeling diffusion, with particular reference to computational load. Example simulations are presented for diffusion and buffering in neural dendritic spines.

Chapter 8. Kinetic Models of Excitable Membranes and Synaptic Interactions
While chapter 7 looks at the free movement of molecules in a solvent, this chapter is concerned with selective molecular transport across cell membranes. It concentrates on models of inorganic ion transport channels. The stochastic mechanisms underlying voltage-dependent, ligand-gated, and second-messenger-gated channels are discussed. In each case, a range of representations, from biophysically detailed to highly simplified two-state (open/closed) models, are presented. Although such channels are structur- ally very different from protein transport channels, the modeling framework presented is applicable in all

cases. The chapter also shows how the same modeling formalism can be used to integrate the action of ion channels with intracellular biochemical processes.

Chapter 9. Stochastic Simulation of Cell Signaling Pathways   In Chapter 2, we see the need for stochastic modeling of gene regulation in a system with statistically few regulatory molecules. Here, we revisit stochastic modeling, this time in a much larger scale system and for a very different reason. The stochastic thermally driven flipping of proteins and protein complexes from one conformation to another turns out to be an essential ingredient of the signaling events considered. A detailed model of bacterial detection of attractants and repellents and an appropriate modeling framework and tool are presented.

Chapter 10. Analysis of Complex Dynamics in Cell Cycle Regulation   Ever since Goldbeter and Koshland's description[3] of "zero order ultra sensitivity," a central question in computational molecular biology has been the extent to which in vivo molecular interactions exhibit and rely on complex dynamic properties. This chapter uses numerical simulation and phase-plane analysis techniques to reveal the complex dynamics underlying the regulation of the cell cycle. The dynamics thus revealed are so complex that informal biochemical intuition could not have reliably deduced them.

Chapter 11. Simplifying and Reducing Complex Models   The final chapter of the book is a tutorial that provides the mathematical background for principled strategies to reduce the complexity of biologically based models. All chapters of this book—and any effort to model biological systems—must deal with the fact that biological modeling from first principles is not possible, given current limitations in our computational devices and analysis tools. In addition, it is often the case that a clear understanding of biological interactions can only be obtained when complex models are simplified. This chapter describes methods based on averaging over either instances or time which provide a solid basis for model simplification. It also emphasizes the importance of using formal techniques to connect models of differing complexity. This is an important and evolving area of research in the methods of biological modeling.

Notes

1. G. F. Oster, "Mechanochemistry and morphogenesis," in *Biological Structures and Coupled Flows-Proceedings of a Symposium Dedicated to Aharon Katzir-Katchalsky*, A. Oplatka and M. Balaban, eds. pp. 417–443. Academic Press, Orlando, Fla., 1983.

2. See for example, James M. Bower and David Beeman, *The Book of GENESIS: Exploring Realistic Models with the General Neural Simulation System*, Springer-Verlag, New York, 1995.

3. A. Goldbeter, and D. Koshland, "An amplified sensitivity arising from covalent modification in biological systems," *Proc. Natl. Acad. Sci. U.S.A.*, 78(11): 6840–6844, 1981.

I Modeling Genetic Networks

1 Modeling the Activity of Single Genes

Michael Andrew Gibson and Eric Mjolsness

1.1 Introduction

1.1.1 Motivation—The Challenge of Understanding Gene Regulation

The central dogma of molecular biology states that information is stored in DNA, transcribed to messenger RNA (mRNA), and then translated into proteins. This picture is significantly augmented when we consider the action of certain proteins in regulating transcription. These *transcription factors* provide a feedback pathway by which genes can regulate one another's expression as mRNA and then as protein.

To review: DNA, RNA, and proteins have different functions. DNA is the molecular storehouse of genetic information. When cells divide, the DNA is replicated, so that each daughter cell maintains the same genetic information as the mother cell. RNA acts as a go-between from DNA to proteins. Only a single copy of DNA is present, but multiple copies of the same piece of RNA may be present, allowing cells to make huge amounts of protein. In eukaryotes (organisms with a nucleus), DNA is found in the nucleus only. RNA is copied in the nucleus and then translocates (moves) outside the nucleus, where it is transcribed into proteins. Along the way, the RNA may be spliced, i.e., it may have pieces cut out. RNA then attaches to ribosomes and is translated to proteins. Proteins are the machinery of the cell—other than DNA and RNA, all the complex molecules of the cell are proteins. Proteins are specialized machines, each of which fulfills its own task, which may be transporting oxygen, catalyzing reactions, or responding to extracellular signals, just to name a few. One of the more interesting functions a protein may have is binding directly or indirectly to DNA to perform transcriptional regulation, thus forming a closed feedback loop of gene regulation.

The structure of DNA and the central dogma were understood in the 1950s; in the early 1980s it became possible to make arbitrary modifications to DNA and use cellular machinery to transcribe and translate the resulting genes. More recently, genomes (i.e., the complete DNA sequence) of many organisms have been sequenced. This large-scale sequencing began with simple organisms, viruses and bacteria, progressed to eukaryotes such as yeast, and more recently (1998) progressed to a multicellular animal, the nematode *Caenorhabditis elegans.* Sequencers have now moved on to the fruit fly *Drosophila melanogaster,* whose sequence was completed early in 2000. The Human Genome Project has completed a rough draft of the complete sequence of all 3 billion bases of human DNA. In the wake of genome-scale sequencing, further instrumentation is being developed to assay gene expression and function on a comparably large scale.

Much of the work in computational biology focuses on computational tools used in sequencing, finding genes that are related to a particular gene, finding which parts of the DNA code for proteins and which do not, understanding what proteins will be formed from a given length of DNA, predicting how the proteins will fold from a one-dimensional structure into a three-dimensional structure, and so on. Much less computational work has been done regarding the function of proteins. One reason for this is that different proteins function very differently, and so work on protein function is very specific to certain classes of proteins. There are, for example, proteins such as enzymes that catalyze various intracellular reactions, receptors that respond to extracellular signals, and ion channels that regulate the flow of charged particles into and out of the cell. In this chapter, we consider a particular class of proteins called *transcription factors* (TFs), which are responsible for regulating when a certain gene is expressed in a certain cell, which cells it is expressed in, and how much is expressed. Understanding these processes will involve developing a deeper understanding of transcription, translation, and the cellular activities that control those processes. All of these elements fall under the aegis of gene regulation or, more narrowly, transcriptional regulation.

Some of the key questions in gene regulation are: What genes are expressed in a certain cell at a certain time? How does gene expression differ from cell to cell in a multicellular organism? Which proteins act as transcription factors, i.e., are important in regulating gene expression? From questions like these, we hope to understand which genes are important for various macroscopic processes.

Nearly all of the cells of a multicellular organism contain the same DNA. Yet this same genetic information yields a large number of different cell types. The fundamental difference between a neuron and a liver cell, for example, is which genes are expressed. Thus, understanding gene regulation is an important step in understanding development (Gilbert 1977). Furthermore, understanding the usual genes that are expressed in cells may give important clues about various diseases. Some diseases, such as sickle cell anemia and cystic fibrosis, are caused by defects in single, nonregulatory genes; others, such as certain cancers, are caused when the cellular control circuitry malfunctions. An understanding of these diseases will involve pathways of multiple interacting gene products.

There are numerous challenges in the area of understanding and modeling gene regulation. First, biologists would like to develop a deeper understanding of the processes involved, including which genes and families of genes are important, how they interact, etc. From a computation point of view, embarrassingly little work has been done. In this chapter there are many areas in which we can pose meaningful, nontrivial computational questions, questions that have not been addressed. Some of these are purely computational. (What is a good algorithm for dealing with a model of type X?) Others are more mathematical. (Given a system with certain characteristics, what sort of model can one use? How does one find biochemical parameters from system-level behavior using as few experiments as possible?)

In addition to biological and algorithmic problems, there is also the ever-present issue of theoretical biology: What general principles can be derived from these systems? What can one do with models other than simulating time courses? What can be deduced about a class of systems without knowing all the details? The fundamental challenge to computationalists and theorists is to add value to the biology-to use models, modeling techniques, and algorithms to understand the biology in new ways.

1.2   Understanding the Biology

There are many processes involved in gene regulation. In this section, we discuss some of the most important and best-understood processes. In addition to biological processes, there are numerous physical processes that play a role in gene regulation in some systems. In addition, we discuss some of the key differences between prokaryotes and eukaryotes, which are important for including details in models and for determining which kind(s) of models are appropriate.

1.2.1   Biochemical Processes

The biological process of gene expression is a rich and complex set of events that leads from DNA through many intermediates to functioning proteins. The process starts with the DNA for a given gene. Recall that in eukaryotes the DNA is located in the nucleus, whereas prokaryotes have no nucleus, so the DNA may inhabit any part of the cell.

The Basics of Transcription   Transcription is a complicated set of events that leads from DNA to messenger RNA. Consider a given gene, as shown in figure 1.1(1). The gene consists of a coding region and a regulatory region. The coding region is the part of the gene that encodes a certain protein, i.e., this is the part that will be transcribed into mRNA and translated into a finished protein. The regulatory region is the part of the DNA that contributes to the control of the gene. In particular, it contains binding sites for transcription factors, which act by binding to the DNA (directly or with other transcription factors in a small complex) and affecting the initiation of transcription. In simple prokaryotes, the regulatory region is typically short (10-100 bases) and contains binding sites for a small number of TFs. In eukaryotes, the regulatory region can be very long (up to 10,000 or 100,000 bases), and contains binding sites for multiple TFs. TFs may act either positively or negatively; that is, an increase in the amount of transcription factor may lead to either more or less gene expression, respectively. Another input mechanism is phosphorylation or dephosphorylation of a bound TF by other proteins.

Transcription Factor
binding sites

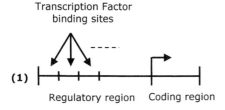

(1)

Regulatory region   Coding region

A single gene may be (in)activated by several
different protein combinations

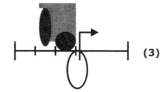

(3)

The TF complex enables RNAP binding

(2)

(In)activating Transcription Factors form
complexes attached to regulatory DNA

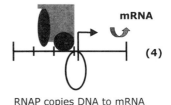

mRNA

(4)

RNAP copies DNA to mRNA

Figure 1.1
Processes involved in *gene regulation:* (1) transcription factor binding, (2) formation of transcriptional complex, (3) RNA polymerase binding, and (4) transcription initiation.

Typically, TFs do not bind singly, but in complexes as in figure 1.1(2). Once bound to the DNA, the TF complex allows RNA polymerase (RNAP) to bind to the DNA upstream of the coding region, as in figure 1.1(3). RNAP forms a transcriptional complex that separates the two strands of DNA, thus forming an open complex, then moves along one strand of the DNA, step by step, and transcribes the coding region into mRNA, as in figure 1.1(4). The rate of transcription varies according to experimental conditions (Davidson 1986, Watson et al. 1987). See von Hippel (1998) for a more detailed discussion of transcription.

A bit of terminology: TFs are sometimes called *trans-regulatory elements,* and the DNA sites where TFs bind are called *cis-regulatory elements.* The collection of *cis*-regulatory elements upstream of the coding region can be called the *promoter* (e.g., Small et al. 1992), although many authors (e.g., Alberts et al. 1994) reserve that term only for the sequence where RNA polymerase binds to DNA and initiates transcription. We will follow the former usage.

Cooperativity   There may be extensive cooperativity between binding sites, even in prokaryotes; for example, one dimer may bind at one site and interact with a second dimer

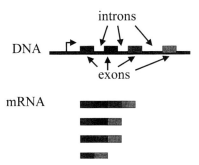

Figure 1.2
Eukaryotic DNA consists of exons and introns. Exons code for proteins, introns are spliced out of the mRNA. Alternative splicings are possible.

at a second site. If there were no cooperativity, the binding at the two sites would be independent; cooperativity tends to stabilize the doubly bound state. Competition is also possible, particularly for two different transcription factors binding at nearby sites.

Exons and Introns, Splicing   In prokaryotes, the coding region is contiguous, but in eukaryotes the coding region is typically split up into several parts. Each of these coding parts is called an *exon,* and the parts in between the exons are called *introns* (figure 1.2, top). Recall that in eukaryotes transcription occurs in the nucleus. For both types of organisms, translation occurs in the cytosol. Between transcription and translation, in eukaryotes, the mRNA must be moved physically from inside the nucleus to outside. As part of this process, the introns are edited out, which is called *splicing.* In some cases, there are alternative splicings, that is, the same stretch of DNA can be edited in different ways to form different proteins (figure 1.2, bottom). At the end of the splicing process, or directly after the transcription process in prokaryotes, the mRNA is in the cytosol and ready to be translated.

Translation   In the cytosol, mRNA binds to ribosomes, complex macromolecules whose function is to create proteins. A ribosome moves along the mRNA three bases at a time, and each three-base combination, or codon, is translated into one of the 20 amino acids (figure 1.3). As with transcription, the rate of translation varies according to experimental conditions (Davidson 1986).

Post-translational Modification   The function of the ribosome is to copy the one-dimensional structure of mRNA into a one-dimensional sequence of amino acids. As it does this, the one-dimensional sequence of amino acids folds up into a final three-dimensional protein structure. A protein may fold by itself or it may require the assistance of

AUGCUUGCUAAACUUGC

Figure 1.3
Translation consists of a ribosome moving along the mRNA one codon (three bases) at a time and translating each codon of the mRNA into an amino acid of the final protein.

other proteins, called *chaperones*. As previously mentioned, the process of protein folding is currently the subject of a large amount of computational work; we do not discuss it further here.

At this point, we have summarized the flow of information from DNA to a protein. Several more steps are possible. First, proteins may be modified after they are translated. As an example of this post-translational modification, certain amino acids near the N-terminus are frequently cleaved off (Varshavsky 1995). Even if there is no post-translational modification, proteins may agglomerate; for example, two copies of the same protein (monomers) may combine to form a homodimer. Multimers—complex protein complexes consisting of more than one protein—are also common. In particular, many TFs bind to DNA in a multimeric state, for example, as homodimers (two copies of the same monomer) or as heterodimers (two different monomers). These same proteins can exist as monomers, perhaps even stably, but only the multimer forms can bind DNA actively.

Degradation   DNA is a stable molecule, but mRNA and proteins are constantly being degraded by cellular machinery and recycled. Specifically, mRNA is degraded by a ribonuclease (RNase), which competes with ribosomes to bind to mRNA. If a ribosome binds, the mRNA will be translated; if the RNase binds, the mRNA will be degraded.

Proteins are degraded by cellular machinery, including proteasomes signaled by ubiquitin tagging. Protein degradation is regulated by a variety of more specific enzymes (which may differ from one protein target to another). For multimers, the monomer and multimer forms may be degraded at different rates.

Other Mechanisms   Eukaryotic DNA is packaged by complexing with histones and other chromosomal proteins into chromatin. The structure of chromatin includes multiple levels of physical organization such as DNA winding around nucleosomes consisting of histone octamers. Transcriptionally active DNA may require important alterations to its physical organization, such as selective uncoiling. Appropriately incorporating this kind of organization, and other complications we have omitted, will pose further challenges to the modeling of gene regulation networks.

1.2.2 Biophysical Processes

In addition to the chemical processes described above, numerous physical processes can be important in gene regulatory systems. One may need to include a detailed model of the physical processes in some systems; in others, these processes may have only minor effects and can be ignored without a significant degradation of model performance.

Most of the models that follow deal with the amount of protein in a cell. In the simplest case, we may be able to assume that the cell is well mixed, i.e., that the amount of protein is uniform across the cell. For more complicated systems, in particular for large systems, that is not a good approximation, and we must consider explicitly the effect of diffusion or of transport. Diffusion will be the most important physical effect in the models we consider, but in other systems active transport could be as important or more important.

Diffusion is a passive spreading out of molecules as a result of thermal effects. The distance a molecule can be expected to diffuse depends on the square root of time, so molecules can be expected to move small distances relatively quickly (this is why we can ignore this effect in small systems), but will take much longer to diffuse longer distances. The diffusion distance also depends on a constant specific to the molecule and to the solvent. Larger molecules tend to diffuse more slowly than smaller ones, and all molecules diffuse more slowly in solvents that are more viscous.

In addition to the passive diffusion process, cells have active machinery that moves proteins from one part to another. In humans, for example, some neurons can be up to a meter long, yet the protein synthesis machinery is concentrated in the cell body, the part of the cell where the nucleus is located. Cells have developed an elaborate system of active transport from the site of protein synthesis to their most distal parts. It may be necessary to model active transport more completely in some systems than has been done so far.

Depending on the type of model and the time scale, cell growth may be an important effect. The rate of a chemical reaction of second or higher order (discussed in section 1.3.1) depends on the volume in which the reaction occurs. This consistent change of volume, due to growth, may be an effect that it is important to include in models of some systems.

DNA and most DNA-binding proteins are electrically charged. Some cells, especially neurons, but also muscle cells, heart cells, and many others, change their electrical properties over time. This has no significant effect on gene regulation in any system we will discuss, but that is not to say that it will not have an effect in *any* gene regulatory system.

1.2.3 Prokaryotes vs. Eukaryotes

The key difference between prokaryotes and eukaryotes is that eukaryotes have a nucleus and prokaryotes do not. That difference belies the difference in complexity between these two types of organisms. Fundamentally, prokaryotes are much simpler organisms—the lack of nucleus is just one example of that. Other differences are in the complexity of the transcription complex, mRNA splicing, and the role of chromatin.

Transcription   The eukaryotic transcription complex is much more complicated than the prokaryotic one. The latter consists of a small number of proteins that have been isolated, and this small number is sufficient for transcription. For that reason, in vitro measurements of prokaryotes are thought to be more related to in vivo processes than are the corresponding measurements in eukaryotes.

Eukaryotic promoters may have large numbers of binding sites occurring in more or less clustered ways, whereas prokaryotes typically have a much smaller number of binding sites. For N binding sites, a full equilibrium statistical mechanics treatment (possibly oversimplified) will have at least 2^N terms in the partition function (discussed later), one for each combination of bound and unbound conditions at all binding sites. The most advantageous way to simplify this partition function is not known because there are many possible interactions between elements of the transcription complex (some of which bind directly to DNA, some of which bind to each other). In the absence of all such interactions, the partition function could be a simple product of N independent two-term factors, or perhaps one such sum for each global active and inactive state.

The specific transcription factors are proteins that bind to DNA and interact with one another in poorly understood ways to regulate transcription of specific, large sets of genes. These protein-protein interactions inside the transcription complex really cloud the subject of building models for eukaryotic gene expression. An example of transcription factor interaction in a budding yeast is shown in figure 1.4.

A further complication is found in the "specific" transcription factors such as TFIID that assemble at the TATA sequence of eukaryotic transcription complexes, building a complex with RNA polymerase II, which permits the latter to associate with DNA and start transcribing the coding sequence. Finally, signal transduction (e.g., by MAP kinase cascades (Madhani and Fink 1998) may act by phosphorylating constitutively bound transcription factors, converting a repressive transcription factor into an enhancing one.

Modules   Transcription factors work by binding to the DNA and affecting the rate of transcription initiation. However, severe complications ensue when interactions with other transcription factors in a large transcription complex become important. For simple prokaryotes, it is sometimes possible to write out all possible binding states of the DNA, and to measure binding constants for each such state. For more complicated eukaryotes, it is not. It has been hypothesized that transcription factors have three main functions. Some are active in certain cells and not in others and provide positional control; others are active at certain times and provide temporal control; still others are present in response to certain extracellular signals (figure 1.5).

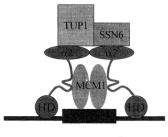

asg operator

Figure 1.4
Transcriptional regulation at a specific gene (asg) operator in budding yeast, for α and a/α cells. Note multiple protein–protein and protein–DNA interactions in a complex leading to activation of the transcriptional apparatus. DNA is at the bottom and includes the asg operator. Redrawn from Johnson (1995).

Many binding sites occur in spatial and functional clusters called *enhancers, promoter elements,* or *regulatory modules.* For example, the 480 base pair *even-skipped (eve)* stripe 2 "minimal stripe element" in *Drosophila* (Small et al. 1992) has five activating binding sites for the *bicoid (bcd)* transcription factor and one for *hunchback (hb).* It also has three repressive binding sites for each of *giant (gt)* and *Kruppel (Kr).* The minimal stripe element acts as a "module" and suffices to produce the expression of *eve* in stripe 2 (out of seven *eve* stripes in the developing *Drosophila* embryo). Similar modules for stripes 3 and 7, if they can be properly defined (Small et al. 1996), would be less tightly clustered. These promoter regions or modules suggest a hierarchical or modular style of modeling the transcription complex and hence single gene expression, such as that provided by Yuh et al. (1998) for *Endo16* in sea urchin, or the hierarchical cooperative activation (HCA) model suggested in section 4.3.1 of chapter 4 in this volume.

A different way to think about these binding site interactions is provided by Gray et al. (1995), who hypothesize three main forms of negative interaction between sites:

• *competitive binding,* in which steric constraints between neighboring binding sites prevent both from being occupied at once

• *quenching,* in which binding sites within about 50 base pairs of each other compete

• *silencer regions,* promoter regions that shut down the whole promoter when cooperatively activated

Given these observations, we can see that the biological understanding of eukaryotic *cis*-acting transcriptional regulation is perhaps "embryonic."

Translation   Eukaryotic genes are organized as introns and exons (figure 1.2), and splicing is the process by which the introns are removed and the mRNA edited so as to produce

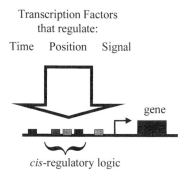

Transcription Factors
    that regulate:

Time   Position   Signal

gene

cis-regulatory logic

Figure 1.5
Some transcription factors act at certain times, some in certain cells, and others in response to signals.

the correct proteins. There can be alternative splicings, and control elements can be located in introns. There may be a nontrivial role for chromatin state at scales of several lengths. These complications will not be considered further in any of the models discussed in this chapter.

1.2.4   Feedback and Gene Circuits

We have considered some of the mechanisms by which transcription of a single gene is regulated. A key point we have avoided is feedback. Simply stated, the TFs are themselves subject to regulation. This leads to interconnected systems that are more difficult to analyze than feedforward systems. For single-variable systems, there are two major kinds of feedback—positive and negative; for multivariable systems, feedback is more complicated.

Negative feedback is the way a thermostat works: when a room gets too hot, the cooling system kicks in and cools it down; when a room gets too cool, the heater kicks in and warms it up. This leads to stabilization about a fixed point. More complicated negative feedback is also possible, which leads to better control. The field of control, both of single-input systems and of multiple-input systems, is well beyond the scope of this chapter.

Positive feedback can create amplification, decisions, and memory. Suppose your thermostat were wired backward, in the sense that if the room got too hot, the heater would turn on. This would make it even hotter, so the heater would turn on even more, etc., and soon your room would be an oven. On the other hand, if your room got too cold, the air conditioner would kick in, and cool it down even more. Thus, positive feedback would amplify the initial conditions—a small hot temperature would lead to maximum heat, a small cold temperature would lead to maximum cooling. This results in two stable fixed points as final states—very hot and very cold—and a decision between them. Roughly, this is how it is possible for a cell to pick one of several alternative fates and to remember its decision amidst thermal noise and stochastic environmental input. Several models of multiple-gene feedback circuits or networks will be presented in the remainder of this book.

1.3   Modeling Basics

Having described the processes involved in gene regulation, we turn to the question of modeling. This section and the next will show how to create a predictive model from biochemical details, spelled out textually and in pictures. We split this process into two parts. First, this section develops the concept of a calculation-independent model, that is, a formal, precise, and quantitative representation of the processes described in section 1.2. Section 1.4 describes how to start with a calculation-independent model, do calculations, and make predictions about the behavior of the system. There are numerous ways to do the calculation, depending on the assumptions one makes; each set of assumptions has its pros and cons.

There are two reasons why having a calculation-independent model is useful. First, biologists can develop a model—a precise representation of the processes involved in gene regulation—without regard to the computational problems involved. For instance, the section 1.4 will show certain areas where the computational theory has not been worked out fully; a precise description of biological processes should not be held hostage to computational problems. Rather, a precise model should be made, and when computations are to be done, additional assumptions and constraints can be added, which are understood to be computational assumptions and constraints, not biological. The second use of calculation-independent models is that theorists can develop the tools—both computational and mathematical—to deal with all the models that fit into this calculation-independent framework, rather than ad hoc methods that apply only to a particular biological system.

The rest of this section introduces the notation of chemical reactions and describes some fundamental ideas underlying physical chemistry: kinetics, equilibrium, and the connection to thermodynamics.

1.3.1   Chemical Reactions

Chemical reactions are the lingua franca of biological modeling. They provide a unifying notation by which to express arbitrarily complex chemical processes, either qualitatively or quantitatively. Specifying chemical reactions is so fundamental that the same set of chemical reactions can lead to different computational models, e.g., a detailed differential equations model or a detailed stochastic model. In this sense, representing processes by chemical equations is more basic than using either differential equations or stochastic processes to run calculations to make predictions.

A generic chemical reaction, such as

$$n_a A + n_b B \xrightarrow{\ k\ } n_c C + n_d D$$

states that some molecules of type A react with some of type B to form molecules of types C and D. The terms to the left of the arrow are called the *reactants;* those on the right are called the *products.* There can be an arbitrary number of reactants and an arbitrary number of products, not always just two, and the number of reactants and products do not have to match.

In the reaction given, n_a molecules of A react with n_b molecules of B to give n_c molecules of C and n_d of D. The n terms are called *stoichiometric coefficients* and are small integers. For example, the reaction

$$A \xrightarrow{\ k\ } A'$$

has one reactant, A, and one product, A'; n_a is 1 and $n_{a'}$ is also 1. In other words, this reaction says that one molecule of type A reacts to give one molecule of type A'. Note that, as in this example, stoichiometric coefficients of 1 are frequently omitted.

In these two examples, the value k on the reaction arrow is a rate constant. Chemical reactions do not occur instantaneously, but rather take some time to occur. The value k is a way of specifying the amount of time a reaction takes. The rate constant depends on temperature. At a fixed temperature, suppose n_a molecules of A collide with n_b molecules of B; the rate constant measures the probability that this collision will occur with sufficient energy for the molecules to react and give the products.

Example: The process illustrated in figure 1.6 is a simple example of TFs binding to DNA (as discussed in section 1.2 and shown in figure 1.1(2)). In this particular example, two different proteins, P and Q, can bind to DNA. Using the notation P·DNA to mean "P bound to DNA," etc., the chemical equations for the process are:

$$1P + 1DNA_{free} \xrightarrow{\ k_{01}\ } 1P \cdot DNA$$

$$1Q + 1DNA_{free} \xrightarrow{\ k_{02}\ } 1Q \cdot DNA$$

$$1P + 1Q \cdot DNA \xrightarrow{\ k_{23}\ } 1P \cdot Q \cdot DNA$$

$$1Q + 1P \cdot DNA \xrightarrow{\ k_{13}\ } 1P \cdot Q \cdot DNA$$

$$1P \cdot DNA \xrightarrow{\ k_{10}\ } 1P + 1DNA_{free}$$

$$1Q \cdot DNA \xrightarrow{\ k_{20}\ } 1Q + 1DNA_{free}$$

$$1P \cdot Q \cdot DNA \xrightarrow{\ k_{32}\ } 1P + 1Q \cdot DNA$$

$$1P \cdot Q \cdot DNA \xrightarrow{\ k_{31}\ } 1Q + 1P \cdot DNA.$$

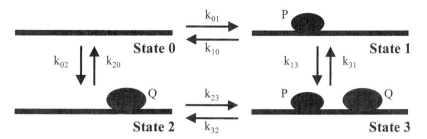

Figure 1.6
Kinetic model of two proteins, P and Q, binding to DNA.

1.3.2   Physical Chemistry

Concept of State   The *state* of a system is a snapshot of the system at a given time that contains enough information to predict the behavior of the system for all future times. Intuitively, the state of the system is the set of variables that must be kept track of in a model. Different models of gene regulation have different representations of the state:

• In Boolean models, the state is a list, for each gene involved, of whether the gene is expressed ("1") or not expressed ("0").

• In differential equations models, the state is a list of the concentrations of each chemical type.

• In stochastic models, a configuration is a list of the actual number of molecules of each type. The state is either the configuration or the current probability distribution on configurations, depending on the particular variant of stochastic model in question.

• In a molecular dynamics model of gene regulation, the state is a list of the positions and momenta of each molecule.

Although the state in each of these models is very different, there are some fundamental similarities.

• Each model defines what it means by the state of the system.

• For each model, given the current state and no other information, the model predicts which state or states can occur next.

• Some states are equilibrium states in the sense that once in that state, the system stays in that state.

Physical chemistry deals with two problems: kinetics, that is, changes of state, and equilibria, which states are equilibrium states. Amazingly, the latter question can be answered by thermodynamics, just by knowing the energy differences of the different possible states of the system, without knowing the initial state or anything about the kinetics. We will consider in turn kinetics, equilibria, and thermodynamics as they apply to gene regulation.

Kinetics-Changes of State   The chemical equation

$$A \xrightarrow{\;\;k\;\;} A'$$

deals with two chemical species, A and A'. According to the equation, A is transformed to A' at a rate of k. In other words, the chemical equation specifies how the state of the system changes, and how fast that change occurs.

In the differential equations approach, this equation specifies that state changes in the following way: the concentration of A decreases and the concentration of A' increases correspondingly. For a small time, dt, the amount of the change is given by $k \times [A] \times dt$ where $[A]$ is the concentration of A.

In the stochastic approach, this equation specifies that the state changes in a different way: a single molecule of A is converted into a molecule of A'; the total number of molecules of A decreases by one and the total number of molecules of A' increases by one. The probability that this event occurs in a small time dt is given by $k \times \{\#A\} \times dt$, where $\{\#A\}$ is the number of molecules of A present.

Equilibrium   A system is said to be at equilibrium when its state ceases to change. Consider the previous example:

$$A \xrightarrow{\;\;k\;\;} A'$$

We saw that, as long as A is available, the state will change at a rate determined by k. Thus, when this system reaches equilibrium, all of the A will be gone, and only A' will remain. At the equilibrium point, once all the A has been used up, there can be no more changes of state.

It is not in general true that equilibrium only occurs when one of the components has been used up. Consider, for example, a certain DNA binding protein X, whose binding to and unbinding from DNA follow simple and complementary chemical kinetics:

$$X + DNA \xrightarrow{\;\;k_1\;\;} X \cdot DNA \tag{1.1}$$

$$X \cdot DNA \xrightarrow{\;\;k_{-1}\;\;} X + DNA. \tag{1.2}$$

This set of reactions involves chemicals of three types: X, DNA and X·DNA. Equilibrium will occur when the rate at which X and DNA are converted to X·DNA [according to equation (1.1)] exactly equals the rate at which X·DNA is converted to X and DNA [according to equation (1.2)]. Note that in this equilibrium, there are still changes: both reactions still occur constantly. However, there are no net changes. And since the state of the system is a list of (for example) concentrations of all chemical types involved, the state does not change.

The concepts of kinetics and equilibrium are general—they appear in one form or another in all the different models in this chapter. At this point, however, we present a detailed example in one particular framework, that of differential equations for reaction kinetics, to illustrate key points. This example will illustrate concepts and ideas that are true, not just for differential equations, but also in general.

Example (equilibrium in the differential equations framework):   The two chemical equations above lead to three differential equations, one for the concentration of X (denoted [X]), one for [DNA] and one for [X·DNA]. A later section will explain how to derive the differential equations from chemical equations and in particular will show that the differential equation for [X·DNA] is

$$\frac{d[X \cdot DNA]}{dt} = k_1[X][DNA] - k_{-1}[X \cdot DNA].$$

Note that the right-hand side has two terms: the first one is production of new X·DNA due to equation (1.1), the second is degradation of existing X·DNA due to equation (1.2). With the corresponding equations for [X] and [DNA], plus the initial concentrations of each of the three, we can solve for [X·DNA] as a function of time. Doing so would be the kinetics approach to the problem.

Instead, assume that the two competing processes have reached equilibrium, that is, there are no further net changes. Thus,

$$0 = \frac{d[X \cdot DNA]}{dt} = k_1[X][DNA] - k_{-1}[X \cdot DNA]. \tag{1.3}$$

This leads to the following equation, which is valid for equilibrium concentrations:

$$\frac{[X \cdot DNA]}{[X][DNA]} = \frac{k_1}{k_{-1}} \equiv K_{eq}. \tag{1.4}$$

Here K_{eq} is called the *equilibrium constant* of this reaction. Notice that applying the equilibrium condition in (1.3) reduced a differential equation to an algebraic equation. Applying the equilibrium condition in an arbitrary framework removes the time dependence of the kinetic equations, and the resulting equations are algebraic. Now, in addition to the rate constants of the previous subsection, chemical reactions may have an equilibrium constant, which is a property only of the system, not the computational framework of the system. For example, in a stochastic framework, it is possible to define an equilibrium constant just as in (1.4). Since the stochastic framework typically uses the number of molecules, rather than the concentration of molecules, the stochastic equilibrium constant is typically reported in different units.

Real physical systems tend toward equilibrium unless energy is continually added. So another interpretation of equilibrium is the state of the system as time→∞, in the absence of energy inputs. To simplify models (and experiments), reactions that are fast (compared with the main reactions of interest) are often assumed to be at equilibrium. Given this assumption, for example, the pair of equations (1.1) and (1.2) can be abbreviated as

$$X + DNA \xleftarrow{\quad K_{eq} \quad} X \cdot DNA. \tag{1.5}$$

It is possible to go directly from the chemical equation (1.5) to the algebraic equation (1.4) by multiplying all the concentrations of the products (in this case only [X·DNA]) and dividing by the product of the concentrations of the reactants (in this case [X][DNA]).

Consider equation (1.5) and the additional equilibrium equation:

$$X \cdot DNA \xleftarrow{\quad K_{eq2} \quad} X' \cdot DNA. \tag{1.6}$$

For example, this could mean that X undergoes a conformational change to X' while bound to DNA. Returning to the differential equations framework, the equilibrium equation is

$$\frac{[X' \cdot DNA]}{[X \cdot DNA]} = K_{eq2}. \tag{1.7}$$

Notice that

$$\frac{[X' \cdot DNA]}{[X][DNA]} = \frac{[X \cdot DNA][X' \cdot DNA]}{[X][DNA][X \cdot DNA]} = K_{eq} K_{eq2}.$$

This is an important property of equilibria: if A and B are in equilibrium, and B and C are in equilibrium, then A and C are in equilibrium, and the resulting A-C equilibrium constant is simply the product of the A-B and B-C equilibrium constants.

Thermodynamics   Two ways of determining equilibrium constants have been presented thus far: calculating equilibrium constants from the forward and reverse rate constants, and calculating them from the equilibrium concentrations of the chemicals. There is a third way—equilibrium constants can be calculated from thermodynamics, using only the energy difference between the products and reactants. From this energy difference alone, one can predict the final state of the system, but not the time course of the state from initial to final.

For chemical reactions, the Gibbs free energy, G, is defined by

G = (total internal energy) − (absolute temperature) × (entropy) + (pressure) × (volume)

$\quad$ = Σ(chemical potential) × (particle number).

Typically, values of ΔG are reported instead of the values of K_{eq}. From thermodynamics in dilute solutions (Hill 1985, Atkins 1998), it follows that for reactants and products with the free energy difference ΔG,

$$K_{eq} = e^{\frac{-\Delta G}{RT}},$$

where R is the ideal gas constant, namely Boltzmann's constant times Avogadro's number, and T is the absolute temperature.

To deal with multiple states in equilibrium with each other, one uses partition functions. In general, the fraction of the system in a certain configuration c (e.g., the fraction of the DNA bound to protein P) is given by

$$\text{frac}_c = \frac{\exp(-\Delta G_c / RT)[\text{species}_1]^{\text{power}_1} \cdots [\text{species}_n]^{\text{power}_n}}{\sum_i \exp(-\Delta G_i / RT)[\text{species}_1]^{\text{power}_1} \cdots [\text{species}_n]^{\text{power}_n}}.$$

The power of chemical species x in configuration c is simply the number of molecules of type x present in configuration c. The denominator of this equation is the partition function.

Example (partition functions):   Consider the example in figure 1.6, but this time assume that equilibrium has been reached. Then

$$\frac{[\text{DNA}]_1}{[\text{DNA}]_{\text{total}}} = \frac{K_1[\text{P}]}{Z},$$

$$\frac{[\text{DNA}]_2}{[\text{DNA}]_{\text{total}}} = \frac{K_2[\text{Q}]}{Z}, \quad \text{and}$$

$$\frac{[\text{DNA}]_3}{[\text{DNA}]_{\text{total}}} = \frac{K_3[\text{P}][\text{Q}]}{Z},$$

where each of the Ks is defined to be the equilibrium constant between a given state and state 0, and $Z \equiv 1 + K_1[\text{P}] + K_2[\text{Q}] + K_3[\text{P}][\text{Q}]$. Here Z is the partition function.

Derivation: Let $[\text{DNA}]_0$ be the concentration of DNA in state 0, and let $[\text{P}]$ and $[\text{Q}]$ be the concentrations of free (i.e., unbound) proteins P and Q, respectively. Then there are equilibrium constants K_1, K_2 and K_3, so that at equilibrium

$$\frac{[\text{DNA}]_1}{[\text{P}][\text{DNA}]_0} = K_1,$$

$$\frac{[DNA]_2}{[Q][DNA]_0} = K_2, \quad \text{and}$$

$$\frac{[DNA]_3}{[P][Q][DNA]_0} = K_3.$$

The total concentration of DNA is given by

$$[DNA]_{total} = [DNA]_0 + [DNA]_1 + [DNA]_2 + [DNA]_3,$$

which leads to

$$[DNA]_{total} = [DNA]_0 + K_1[P][DNA]_0 + K_2[Q][DNA]_0 + K_3[P][Q][DNA]_0$$

$$= [DNA]_0 Z.$$

The result of the previous example follows directly.

1.4 Generating Predictions from Biochemical Models

This section will assume that a computationally independent model of a biological process has been created. Such a model, as described in section 1.3, may consist of chemical equations, parameter values, and physical constraints such as diffusion and growth. The model is a formal, precise way to describe the biological process, and writing the model is a biological problem. Once the model has been created, as assumed in this section, the remaining computational problem is how to generate predictions from the model.

This section discusses several different modeling methods. Although this selection is by no means complete, it should serve as a good comparison of different types of models and as a jumping-off point for further investigation. There are three main questions to keep in mind when evaluating the different types of models:

• What is the state in each model, that is, which variables does one consider?

• How does the state change over time, that is, what are the kinetic or dynamic properties of the model?

• What are the equilibrium states of the model, that is, which states are stable?

There are also several practical questions to keep in mind, which relate to the applicability of the model:

• What assumptions does the model make? Which are biological and which are strictly computational?

Note that all models make some computational assumptions, so the mere existence of assumptions or approximations should not be grounds for rejecting a model. The specific assumptions made, however, may be.

• For what time scale is the model valid?

At very short time scales (seconds or less), the low-level details of binding and unbinding and protein conformational changes have to be modeled. At longer time scales, these can be considered to be in equilibrium, and certain average values can be used (this point will be discussed in more detail later). At very long time scales (e.g., days), processes such as cell division, which can be ignored at shorter time scales, may be very important.

• How many molecules are present in the biological system?

If the number of molecules is very small (i.e., in the tens or low hundreds), stochastic models may need to be used. However, once the number of molecules becomes very large, differential equations become the method of choice.

• How complex is the computation resulting from this model?

The more detail and the longer time span one wants to model a process for, the higher the complexity, and hence the longer it takes computationally.

• How much data are available?

For many systems, there are a lot of high-level qualitative data, but less quantitative, detailed data. This is particularly true of complex eukaryotes. The nature of the data required by a model in order to make predictions, an important practical property of the model, may depend on the power of the data-fitting algorithms available.

• What can the model be expected to predict? What can it not predict?

To begin modeling, one must focus on what types of predictions are sought. For simple predictions, simple models are sufficient. For complex predictions, complex models may be needed.

1.4.1   Overview of Models

Rather than advocating a single, definitive model of gene regulation, this section describes a variety of modeling approaches that have different strengths, weaknesses, and domains of applicability in the context of the foregoing questions. Note that improvements in modeling precision typically carry a cost; for example, they may require more data or more precise data. First we outline the basic modeling approaches, then give a detailed explanation of each, with examples where appropriate.

   At one end of the modeling spectrum are *Boolean network* models, in which each gene is either fully expressed or not expressed at all. This coarse representation of the state has certain advantages, in that the next state, given a certain state, is a simple Boolean function; also, the state space is finite, so it is possible to do brute-force calculations that would be intractable in an infinite state space. These models are typically used to obtain a first representation of a complex system with many components until such time as more detailed data become available.

One step along the spectrum are *kinetic logic* models. These models represent the state of each gene as a discrete value: "not expressed," "expressed at a low level," "expressed at a medium level," or "fully expressed," for example, thus providing more granularity than in Boolean networks. In addition, these models attempt to deal with the rates at which the system changes from one state to another. Rather than assuming that all genes change state at the same time (as in Boolean networks), these models allow genes to change state at independent rates. The functions describing the changes of state are more complicated in these models than in Boolean networks, so more data are required to find the functions. In the end, however, the predictions of this type of model are more precise and tie more closely to the biology.

Next in complexity come the *continuous logical* models, which lies between kinetic logic and differential equations. These models still contain discrete states, but the transition from one state to another is governed by linear differential equations with constant coefficients. These differential equations allow more modeling precision, but again, require more detailed data.

Differential equation models provide a general framework in which to consider gene regulation processes. By making certain assumptions, one can transform essentially any system of chemical reactions and physical constraints into a system of nonlinear ordinary differential equations (ODEs), whose variables are concentrations of proteins, mRNA, etc. One way to do this is by reaction kinetics (or enzyme kinetics), considering the transition rates between all microscopic states. When the number of states becomes too large or poorly understood, as for example in protein complexes, coarser and more phenomenological ODE systems may be postulated. Either way, the equations can be solved numerically for their trajectories, and the trajectories analyzed for their qualitative dependence on input parameters. There are also many different ways to approximate the nonlinear differential equations, some of which are advantageous for parameter estimation, for example, as discussed in the next section.

Not all systems can be modeled with differential equations. Specifically, differential equations assume that changes of state are continuous and deterministic. Discontinuous transitions in deterministic systems can be modeled with various hybrids between discrete and continuous dynamics, including continuous-time logic, special kinds of grammars, and discrete event systems. Nondeterministic systems, systems where the same state can lead to different possible outcomes, generally must be modeled in a different framework. One such framework, borrowed from physics, is the *Langevin* approach. Here, one writes a system of differential equations as before, then adds a noise term to each. The noise term is a random function. For a particular noise function, one may solve the differential equations as before. Given the statistics of the noise function, one may make statements about the statistics of a large number of systems; for example, which outcomes occur with which probabilities.

Another approach to overcoming the limitations of differential equations is the Fokker–Planck method. Here, one starts from a probabilistic framework and writes a full set of equations that describes the change in probability distribution as a function of time. One still assumes continuity, that is, that the probability distribution is a continuous function of concentration. This leads to a certain partial differential equation of the probability distribution (partial in time and in concentration). One can solve this numerically—although partial differential equations are notoriously harder to solve than ordinary differential equations—to get full knowledge of the probability distribution as a function of time.

Both the Langevin and Fokker–Planck equations are approximations of the fully stochastic model to be considered later. Under certain conditions, the approximations are very good, that is, the difference between approximation and exact solution is much smaller than the variance of the approximation. Under other conditions, the difference may be on the order of the variance, in which case these models do not make precise predictions.

A more general formalism, van Kampen's $1/\Omega$ expansion, phrases the fully stochastic problem as a Taylor expansion in powers of a parameter Ω, e.g., the volume. Collecting terms of the same Ω order, one first gets the deterministic differential equation, then the Fokker–Planck equation (or equivalently, the Langevin equation), then higher-order terms.

Fully stochastic models consider the individual molecules involved in gene regulation, rather than using concentrations and making the continuity assumption of differential equations. The probability that the next state consists of a certain number of molecules, given the current state, can be expressed in a straightforward way. Typically, it is computationally intensive to deal with this framework, so various computational methods have been developed.

Section 1.4.2 considers three main views of gene regulation—fully Boolean, fully differential equations, or fully stochastic—in greater detail. Each of these can be stated relatively simply. Section 1.4.3 discusses all the other methods, which will be viewed as combinations or hybrids of these three main methods.

1.4.2 "Pure" Methods

4.2.1 Boolean   There are numerous Boolean network models based on Boolean logic (Kauffman 1993). Each gene is assumed to be in one of two states: "expressed" or "not expressed." For simplicity, the states will be denoted "1" and "0," respectively. The state of the entire model is simply a list of which genes are expressed and which are not.

Example (Boolean state):   Consider three genes, X, Y, and Z. If X and Y are expressed, and Z is not, the system state is 110. From a given state, the system moves deterministically to a next state. There are two ways to express this: First, one may write out a truth table, which specifies what the next state is for each current state. If there are n genes, there are 2^n states, each of which has exactly one next state. If one considers a reasonably small number of genes, say five, one can then write out the entire truth table exhaustively.

Example (truth table): A possible truth table for the three genes in the previous example is

Current state	Next state
000	000
001	001
010	010
011	010
100	011
101	011
110	111
111	111

Boolean functions provide an alternative representation of truth tables. Each variable is written as a function of the others, using the operators AND, which is 1 if all of its inputs are 1; OR, which is 1 if any of its inputs are 1; and NOT, which is 1 if its single input is 0. Truth tables and Boolean functions are equivalent in the sense that one can convert from one to the other using standard techniques. However, Boolean functions for the next state are typically relatively simple; although arbitrarily complex truth tables can exist, most of them are not possible for gene regulation logic.

Example (Boolean functions): The functions that lead to the truth table in the previous example are

X(next) = X(current) AND Y(current)

Y(next) = X(current) OR Y(current)

Z(next) = X(current) OR {[NOT Y(current)] AND Z(current)}.

A simple description of a biological system might be expressed as Boolean functions; for example, "C will be expressed if A AND B are currently expressed." Translating from this statement to a formal Boolean network model is straightforward if it is known how the system behaves for all possible states. Typically, however, one does not know all states, and a Boolean network is one way to explicitly list states and find previously unknown interactions.

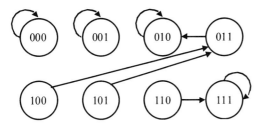

Figure 1.7
Finite-state machine for the Boolean network example in the text.

Another equivalent way to represent changes of state is as a finite-state machine. Here, one draws and labels as many circles as there are states, then draws arrows from one state to the next state, as in figure 1.7.

Equilibrium states can be found exhaustively. In the above example, the truth table shows that 000, 001, 010, and 111 are equilibrium states. In the finite-state machine representation, the equilibrium states are easily recognizable as the states whose transition is to themselves.

Differential Equations (Reaction Kinetics)   In the differential equations approach to modeling gene regulation, the state is a list of the concentrations of each chemical species. These concentrations are assumed to be continuous; they change over time according to differential equations. It is possible to write differential equations for arbitrary chemical equations.

Consider a chemical equation, such as

$$A \xrightarrow{\ k\ } A'.$$

This denotes a reaction that occurs at a rate of $k[A]$, for example, as [A] increases, there is a linear increase in the rate at which A′ is created. As the reaction occurs, [A] decreases and [A′] increases. In particular,

$$\frac{d[\mathrm{A}]}{dt} = -k[A] \quad \text{and}$$

$$\frac{d[\mathrm{A}']}{dt} = k[\mathrm{A}].$$

Note that the rate of reaction depends only on the reactants, not on the products. For reactions with more than one reactant, the rate of reaction is the rate constant k times the product of the concentrations of each reactant.

Example: The chemical equation

$$A + B \xrightarrow{\ k\ } AB$$

leads to the differential equations

$$\frac{d[A]}{dt} = -k[A][B],$$

$$\frac{d[B]}{dt} = -k[A][B], \quad \text{and}$$

$$\frac{d[AB]}{dt} = k[A][B].$$

For stoichiometric coefficients other than 1, one must raise the corresponding concentration to the power of the coefficient. Also, the rate of change of such a concentration will be equal to the stoichiometric coefficient times the rate of the reaction.

Example: The chemical equation

$$2A \xrightarrow{\ k\ } A_2$$

leads to

$$\text{rate} = k[A]^2,$$

$$\frac{d[A_2]}{dt} = k[A]^2, \quad \text{and}$$

$$\frac{d[A]}{dt} = -2k[A]^2.$$

When more than one chemical equation is being considered, the differential equation for a certain chemical is simply the sum of the contributions of each reaction. For example, a previous section asserted that the chemical equations

$$X + DNA \xrightarrow{\ k_1\ } X \cdot DNA$$

$$X \cdot DNA \xrightarrow{\ k_{-1}\ } X + DNA$$

lead to the differential equation

$$\frac{d[X \cdot DNA]}{dt} = k_1[X][DNA] - k_{-1}[X \cdot DNA].$$

This transformation is simply the sum of the rates of the two reactions.

Notation: One typically writes a vector, **v,** consisting of all the concentrations, that is, the entire state. The change of state can then be expressed as $(d/dt)\mathbf{v} = f(\mathbf{v})$, where f is a known but often nonlinear function.

To solve a system of differential equations, one needs not only the equations themselves but also a set of initial conditions—the values of the concentrations at time 0, for example. Typically, writing the equations and finding the correct values of the rate constants and of the initial values are the conceptually interesting parts—in most actual gene regulatory systems, the equations become too complicated to solve analytically and so must be solved numerically. Many standard tools exist for solving and analyzing systems of differential equations.

As stated in the previous section, the equilibrium states in the differential equations framework are found by setting all derivatives to 0, that is, by imposing the condition that there is no further change in state. Finding the equilibrium states is then only an algebraic problem: find **v** so that $0 = f(\mathbf{v})$.

Example: Ackers et al. (1982, later extended in Shea and Ackers 1985) modeled a developmental switch in λ phage. The model considers the regulation of two phage proteins, repressor and cro. Regulation occurs in two ways: production and degradation. Degradation of the proteins follows the chemical equation

$$P \xrightarrow{k} \text{nothing.}$$

Production is more complicated—a detailed model is provided of the binding of these two proteins and RNAP at three DNA binding sites—OR_1, OR_2, and OR_3. The binding is rapid compared with protein production, and hence is assumed to have reached equilibrium. Recall that the fraction of the DNA in each possible binding configuration, c, is

$$\text{frac}_c = \frac{\exp(-\Delta G_c / RT)[\text{species}_1]^{\text{power}1} \cdots [\text{species}_n]^{\text{power}n}}{\sum_i \exp(-\Delta G_i / RT)[\text{species}_1]^{\text{power}1} \cdots [\text{species}_n]^{\text{power}n}}.$$

In this model, each configuration also corresponds to a rate of production of cro and of repressor. The total average rate of production is the sum of the rates of each configuration, weighted by the fraction of DNA in each configuration. Thus the differential equation for the concentration of repressor is

$$\frac{d[\text{rep}]}{dt} = \frac{\sum_i \{\text{rate}_i(\text{rep})\}\{\exp(-\Delta G_i / RT)[\text{rep}]^p [\text{cro}]^q [\text{RNAP}]^r\}}{\sum_i \exp(-\Delta G_i / RT)[\text{rep}]^p [\text{cro}]^q [\text{RNAP}]^r} - k_{rep}[\text{rep}].$$

There is a corresponding equation for [cro]. The key biochemical work is in determining the configurations (c) of the binding site, the free energies (ΔG_i) of these configurations, the rate constants of protein production—rate_i (rep) and rate_i (cro)—in each configuration, and the degradation constants (k_{rep} and k_{cro}). The original 1982 model considered only 8 configurations and did not consider RNAP. The more complete 1985 model considered 40 configurations and also RNAP.

A generic problem with reaction kinetics models (and some others as well) is exponential complexity: the number of equations, hence unknown constants, can increase exponentially with the number of interdependent state variables. For example, k phosphorylation sites on a protein or k DNA binding sites in a promoter region would in principle yield 2^k states eligible to participate in reactions. This problem is one reason for moving to coarser, more phenomenological models as in section 1.5. Much of the practical work in gene regulation revolves around finding approximations to f that have certain properties. For example, linear models assume that f is linear in each of the concentrations, and neural network models assume that f consists of a weighted linear sum of the concentrations, followed by a saturating nonlinearity. These sorts of approximations are useful in analyzing systems and in estimating parameters, that is, in taking some incomplete knowledge of the biochemistry plus experimental knowledge of the change in the system state, and using it to develop better estimates of the rate and equilibrium constants. These topics are introduced in section 1.5.

Stochastic Models   The stochastic framework considers the exact number of molecules present, a discrete quantity. The state indicates how many molecules of each type are present in the system. The state changes discretely, but which change occurs and when it occurs is probabilistic. In particular, the rate constants specify the probability per unit time of a discrete event happening.

Example: A simple chemical equation, such as

$$A \xrightarrow{k} A'$$

is interpreted as "One molecule of A is transformed into A'; the probability of this event happening to a given molecule of A in a given time dt is given by $k\,dt$." So, the probability of some molecule of A being transformed in a small time is $k\,\{\#A\}\,dt$, where $\{\#A\}$ indicates the number of molecules of A present. For reactions with more than one reactant, the probability per unit time is given by the rate constant times the number of molecules of each reactant; this is completely analogous to the differential equations case.

An aside: There is a subtlety in reactions with multiple copies of the same reactant, which will be discussed in the next chapter.

If there are multiple possible reactions, the state will change when any one of the reactions occurs. For example, consider the chemical reactions

$$X + DNA \xrightarrow{k_1} X \cdot DNA$$

$$X \cdot DNA \xrightarrow{k_{-1}} X + DNA.$$

A short time later, the state will change from state ({#X}, {#DNA}, {#X · DNA}) to state ({#X} − 1, {#DNA} − 1, {#X · DNA} + 1) with the probability k_1 {#X} {#DNA} dt, to state ({#X} + 1, {#DNA} + 1, {#X · DNA} − 1) with the probability k_{-1} {#X · DNA} dt, or will stay in the same state with the probability $1 − k_1$ {#X} {#DNA} $dt − k_{-1}$ {#X · DNA} dt.

This framework is very different from the other two considered thus far in that changes of the state of the system are probabilistic, not deterministic. From the same initial state, it is possible to get to multiple possible successor states. One way to deal with this is to use a Monte Carlo simulation, that is, to use a random number generator to simulate the changes of state. For any given set of random numbers picked, the change of state is deterministic. By making multiple trajectories through state space, that is, by picking different sets of random numbers, one generates statistics of the process. There are efficient algorithms available to do such Monte Carlo simulations (see Gillespie 1977, McAdams and Arkin 1998, and chapter 2). For general systems, this numerical Monte Carlo approach is the best known way to deal with the problem.

For small systems, that is, systems where there are a small number of possible states, there is another technique: One deals with probabilities rather than numbers of molecules. In other words, the list of the numbers of molecules is a *configuration*, and the state is the probability distribution over all configurations. If the number of configurations is small, the number of probabilities one must deal with is also small. The probabilities obey a master equation, a linear differential equation with constant coefficients and certain other properties, whose solution is particularly easy if the number of states is finite and (typically) small (McQuarrie 1967).

Example (stochastic binding and unbinding): Consider the detailed diagram of transitions for protein-DNA binding in figure 1.6. Let $P(0,t)$ be the probability that a single molecule of DNA is in configuration 0 at time t. Assume there are p molecules of protein P present and q of protein Q. The probability of being in each state at time $t + \Delta t$ is given by

$$\mathbf{P}(t + \Delta t) \equiv \begin{bmatrix} P(0,t + \Delta t) \\ P(1,t + \Delta t) \\ P(2,t + \Delta t) \\ P(3,t + \Delta t) \end{bmatrix} = A\mathbf{P}(t),$$

where

$$A = \begin{bmatrix} 1 - pk_{01}\Delta t - qk_{02}\Delta t & k_{10}\Delta t & k_{20}\Delta t & 0 \\ pk_{01}\Delta t & 1 - k_{10}\Delta t - qk_{13}\Delta t & 0 & k_{31}\Delta t \\ qk_{02}\Delta t & 0 & 1 - k_{20}\Delta t - pk_{23}\Delta t & k_{32}\Delta t \\ 0 & qk_{13}\Delta t & pk_{23}\Delta t & 1 - k_{31}\Delta t - k_{32}\Delta t \end{bmatrix}.$$

Taking the limit as $\Delta t \to 0$ leads to a differential equation in the probabilities,

$$\frac{d\mathbf{P}}{dt} = \lim_{\Delta t \to 0} \frac{\mathbf{P}(t + \Delta t) - \mathbf{P}(t)}{\Delta t} = B\mathbf{P}(t),$$

where

$$B = \begin{bmatrix} -pk_{01} - qk_{02} & k_{10} & k_{20} & 0 \\ pk_{01} & -k_{10} - qk_{13} & 0 & k_{31} \\ qk_{02} & 0 & -k_{20} - pk_{23} & k_{32} \\ 0 & qk_{13} & pk_{23} & -k_{31} - k_{32} \end{bmatrix}.$$

This type of system is called a *continuous-time Markov chain* because the probability of the next transition depends on the current state only, not on the history of states (Feller 1966, van Kampen 1992). Given initial conditions, one can solve this system of differential equations using standard methods. This example is a relatively simple case in that there is precisely one molecule of DNA. In a different process, with two substances that could each be present in multiple copies, the approach would be similar, but the number of states would grow quickly.

In the stochastic framework, it is not possible to define an equilibrium configuration in terms of number of molecules since any reaction that occurs changes the number of molecules. Recall that chemical equilibria in the differential equations framework are dynamic-chemicals are constantly changing from one form to another-but changes balance each other, and so there is no net change. This concept of equilibrium suffices for differential equations models, but not for stochastic models. For stochastic models, there is an equilibrium probability distribution, a set of probabilities that the system has certain numbers of molecules, and even though the system the changes number of molecules, the changes and their probabilities balance out exactly.

Example (equilibrium of stochastic binding and unbinding): The equilibrium distribution for the previous example can be found by applying the equilibrium condition to the differential equation for the probabilities, namely:

$$0 = \frac{d\mathbf{P}}{dt} = B\mathbf{P}(t).$$

Thus, the equilibrium distribution is simply the normalized eigenvector of B that corresponds to the eigenvalue 0. A different formulation of stochastic models for molecular simulations is provided by stochastic Petri nets as described by Goss and Peccoud (1998).

An aside (*formal basis of the relationship between low-level models and higher ones*): In principle, one could write out the full molecular dynamics—the positions and momenta of each particle. At an even lower level, one could create a quantum mechanical description of a system. Since either of these approaches would be computationally intractable for gene regulation and would not provide insight into relevant system behavior, one makes certain assumptions and reduces the model to the stochastic framework, the differential equations framework, or the Boolean framework. The assumptions made should be noted. The precise definition and rigorous justification of the conditions—number of molecules, time scale, etc.—under which each of the modeling frameworks is appropriate remains an open problem.

To use a stochastic model rather than molecular dynamics, one assumes the solution is well mixed, at least locally; that is, that a given molecule is equally likely to be anywhere in the solution, or equivalently, that the rate of molecular collisions is much greater than the rate of reactions.

Example: The diffusion rate of green fluorescent protein in *Escherichia coli* has been measured to be on the order of 1–10 μm^2/s (Elowitz et al. 1999). *E. coli* has dimensions on the order of 1 μm, so for gene regulation on the order of seconds or tens of seconds, this assumption is fine.

Two additional assumptions are required in order to use the differential equations framework: The first is that the number of molecules is sufficiently high that discrete changes of a single molecule can be approximated as continuous changes in concentration. The second is that the fluctuations about the mean are small compared with the mean itself.

It is straightforward to check whether the first condition is met. The second often appeals for its justification to the central limit theorem of probability theory (Feller 1966), which states that the sum of independent, identically distributed random variables with finite moments tends to a Gaussian distribution whose variance grows as the square root of the mean. Typically, one assumes this $1/\sqrt{N}$ noise is sufficiently small for $N \geq 100 - 1000$.

Example: Consider a large number of molecules, each of which can degrade or not degrade independently. At a fixed time, the number remaining is simply the sum of the random variable q, defined for a single molecule to be 1 if the molecule is present and 0 if it has degraded. This sum meets the criteria of the mean value theorem, so the differential equations approach is probably valid.

Example: Consider the process of transcriptional elongation, that is, DNA is transcribed into mRNA nucleotide by nucleotide. One may model each nucleotide step as taking an exponentially distributed amount of time, so the total time to move several steps down the DNA is a random variable that meets the criteria of the central limit theorem.

For more complex processes, it is not so simple to apply the central limit theorem; doing so may not even be legitimate. Under what precise conditions one can and cannot do so remains an open problem.

Fortunately we can also appeal to experimental data, rather than further modeling, to settle the question of model applicability for a particular system. From quantitative immunofluorescence measurements of *hunchback* and *Kruppel* protein expression levels in *Drosophila* syncytial blastoderms (Kosman et al. 1998), variations in fluorescence measured between nuclei occupying similar positions on the anterior-posterior axis of a single blastoderm seem to be about 10% of the average value for an "on" signal and 50–100% of the signal average when it is very low, or "off." These values seem to be consistent with the requirements of differential equation modeling, and indeed differential equation models have high predictive value in this system.

To go from differential equations to Boolean models, one assumes that the function f in the differential equation $(d/dt)\mathbf{v} = f(\mathbf{v})$ has saturating nonlinearities; that is, that when $\mathbf{v}$ is very small or very large, $f(\mathbf{v})$ tends toward a limit, rather than growing without bounds. Further, one assumes that the nonsaturated region between the two extremes is transient and can be ignored. The lower and upper limits become "0" and "1," respectively.

1.4.3   Intermediate and Hybrid Methods

The three main views of gene regulation have been described. The remaining models can be considered intermediates or hybrids, models that combine aspects of the three approaches described earlier.

Between Boolean Networks and Differential Equations   There have been numerous attempts to use the basic ideas of Boolean networks—that the state of gene expression can be represented by discrete values, and the change in state follows simple rules—yet make models that are more related to the detailed biology than are Boolean networks themselves. Two examples are kinetic logic and the continuous logical networks of Glass and Kaufmann (1973). Others models combine grammars rather than Boolean networks with differential equations.

Kinetic Logic   The kinetic logic formalism of Thomas and D'Ari (1990) and Thomas et al. (1995) is more complex than Boolean networks, but has greater predictive value. The state is still discrete, but instead of each gene being "not expressed" or "expressed" only, as in the Boolean model, this formalism considers levels 0, 1, 2, 3, etc., which might correspond to "no expression," "low level expression," "medium expression," and "high expression." Different genes may have a different granularity; one may only have states 0 and 1, while another may have multiple intermediate levels.

The rules for change of state are somewhat complex. One can write a desired next state for each possible current state. However, the actual next state is not necessarily the desired next state; one must first apply two constraints: continuity and asynchronicity. Continuity says that if a gene's current state is 0 and its desired next state is 3, for example, then its actual next state will be 1; that is, it takes one step at a time toward its final goal. Biologically, this means a gene that is "not expressed" will be "expressed at a low level" before becoming "fully expressed." Asynchronicity means that the next state is found by letting a single gene change its state. This is in contrast to Boolean network models, where all genes change their states at the same time (synchronously). For a given state, it is possible that there can be more than one next state, each one consisting of a single gene changing its state.

Example: Consider two genes, each of which has three possible states: "not expressed," "low level expression," and "fully expressed." The following truth table, which states the desired next state in terms of the current state, has an extra column with possible next states, which are found by applying the conditions of continuity and asynchronicity.

Current state(XY)	Desired next state	Possible next states
00	20	10
01	02	02
02	01	01
10	20	20
11	00	10, 01
12	00	02, 11
20	00	10
21	00	11, 20
22	00	12, 21

As was true for Boolean networks, one can represent this truth table as a finite-state machine, as in figure 1.8. Note that some states have multiple possible next states, indicated by multiple arrows out of those states.

It is useful to write logical equations for the desired next state rather than listing the truth table explicitly. As with Boolean networks, the functions that are possible are typically a subset of all possible functions. For multivalued logic, however, it is somewhat more complicated to express the next state. Toward that end, we define a Boolean threshold function, defined by

$$X_T = \begin{cases} 0 & \text{if } X < T \\ 1 & \text{if } X \ge T \end{cases}.$$

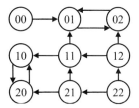

Figure 1.8
Finite-state machine for the kinetic logic example in the text.

The next-state logic can be defined in terms of such threshold functions. The biological meaning of these functions is simply that below a certain threshold level (T), gene X does not affect gene Y, but above that level, it does.

Example: The desired next state function in the truth table was generated by the functions:

$$X_{next} = \begin{cases} 0 & \text{if } -2X_2 - 4Y_1 < -1 \\ 2 & \text{if } 1 < -2X_2 - 4Y_1 \end{cases}$$

$$Y_{next} = \begin{cases} 0 & \text{if } \quad -4X_1 + 4Y_1 - 2Y_2 < 1 \\ 1 & \text{if } \quad 1 < -4X_1 + 4Y_1 - 2Y_2 < 3 \\ 2 & \text{if } \quad 3 < -4X_1 + 4Y_1 - 2Y_2 \end{cases}$$

This corresponds to the textual statements: "At low concentrations, X represses expression of Y"; "at low concentrations, Y represses expression of X and enhances its own expression"; and "at high concentrations, each gene represses further expression of itself."

An interesting question arises when searching for equilibrium states. In the example, it appears there are no equilibrium states: The system will eventually go to the pair of states {01, 02} or to {10, 20}. What this really means is that the equilibrium points are on the thresholds between states. By defining a threshold state T_{12}, the model now reaches one of two equilibria, $0T_{12}$ or $T_{12}0$. In addition to the formalism, Thomas et al. (1995) have worked out an analysis to find such equilibria, which relies on the notion of the characteristic state of a feedback loop.

Continuous Logical Networks   Another formalism between Boolean and differential equations is the work of Mestl et al. (1995, 1996). Here the state consists of continuous concentrations, as in differential equations, but the change of state is simpler. Specifically, the formalism uses a simple linear differential equation for protein production, namely,

$$\frac{d[X]}{dt} = k_1 - k_2[X].$$

Here k_1 and k_2 are functions of the concentrations of [X], [Y], [Z] and of any other proteins present in the system, but those functions are piecewise constant in the discrete ranges of "low," "medium," etc. (Boolean-valued regions are a special case). So within a certain region of state space, the coefficients k_1 and k_2 are constant, and one can solve the simple linear differential equations. By piecing together these simple solutions, one gets the complete behavior of the system.

This formalism allows one to calculate steady states and more complete trajectories than are possible in the strictly algebraic approach (Boolean networks or kinetic logic), but it may require more detailed data to determine the dynamics from rate parameters.

Other Hybrid Systems   A more expressive modeling formalism than logical models or ODEs alone is provided by hybrid models such as developmental "grammars" that incorporate subsets of differential equations whose applicability is modulated by grammatical "rules" in a discrete, discontinuous manner (Mjolsness et al. 1991, Prusinkiewicz et al. 1993, Fleisher 1995). A related formalism is provided by discrete event systems.

Between Differential and Stochastic Equations   The theoretical biology community has done very little work between fully averaged formalisms, such as deterministic differential equations, and fully stochastic ones, such as master equations and transition probability matrices. There is, however, a useful level of modeling that has been used successfully by the physics and chemistry communities. This level is described equivalently by either a differential equation in concentrations with a stochastic noise term added (the Langevin equation) or a deterministic differential equation for the dynamics of a probability distribution (the Fokker–Planck equation). More generally, one may expand the master equation and drop higher-ordered terms as in van Kampen (1992). For example, a stochastic process may be approximately Gaussian at any fixed time point, with a mean and variance that follow differential equations. These formalisms and some conditions for their validity are introduced, for example, in Risken (1989) and van Kampen (1992). They hold some promise for modeling the dynamics of chemical species (including many gene products) where the number of molecules involved is perhaps 10-1000, in-between "a few" and "large numbers." We outline these approaches here, with the warning that the literature on these topics is more mathematical than many of the other references for which a good biological introduction is available.

Langevin   Recall that in the differential equations formalism, one writes the state as a vector **v** of concentrations of different chemicals, which changes in time according to a differential equation:

$$\frac{d}{dt}\mathbf{v} = f(\mathbf{v}).$$

Here f is derived from the chemical equations of the system. In the simplest Langevin approach, one extends this function to be

$$\frac{d}{dt}\mathbf{v} = f(\mathbf{v}) + n(t). \tag{1.8}$$

The function n is a (multivariable) function of time only; it does not depend on the current state. The function n is a noise term; it induces stochastic fluctuations of the state $\mathbf{v}$ about its deterministic value.

For a particular function n, one can solve the equation (1.8) using the standard techniques of differential equations. The key, however, is that n is a random function, so one deals with the expected value of the state $\mathbf{v}$, denoted <v>, and with fluctuations about that state. Specifically, one assumes that the noise term has a zero mean and is uncorrelated with itself over time:

$$\langle n(t) \rangle = 0 \quad and$$

$$\langle n(t)n(\tau) \rangle = \Gamma \times \delta(t - \tau). \tag{1.9}$$

The latter equation means that the correlation is zero for two different times t and τ, and is Γ for the same time (i.e., the variance is Γ).

The general approach to using the Langevin equation is

- Write the deterministic set of differential equations.
- Add a random term $n(t)$ with the properties (1.9).
- Adjust Γ to give the correct equilibrium distribution, as determined by thermodynamics.
- Take expectations of (1.8) to get the state and the variance as functions of time.

An aside (applicability of Langevin approach):   Consider the last step of this approach, taking the expectations of (1.8):

$$\left\langle \frac{d}{dt}\mathbf{v} \right\rangle = \langle f(\mathbf{v}) \rangle + \langle n(t) \rangle.$$

Because expectation is a linear operator, the right-hand side may legitimately be split into two parts. This equation would be a deterministic differential equation with the variable <v> if it were

$$\frac{d}{dt}\langle \mathbf{v} \rangle = f(\langle \mathbf{v} \rangle) + 0.$$

The first part, changing the order of expectation and the derivative, is always legitimate. The last part, the zero, follows from (1.9). The middle part, switching the order of f and of the expectation, is strictly valid only if f is linear. Alternatively, if f is nonlinear, but can be approximated as linear, this is also valid. In physics, the Langevin approach has been applied successfully to linear cases. If, however, f has large nonlinearities on the order of the fluctuation, this approach breaks down. Specifically, two different linear approximations may produce very different final results. A more complicated version of the Langevin approach consists of adding a noise term that can depend on the state. In particular, one replaces (1.8) with

$$\frac{d}{dt}\mathbf{v} = f(\mathbf{v}) + g(\mathbf{v})n(t).\qquad(1.10)$$

Here $g(\mathbf{v})$ is a deterministic function of $\mathbf{v}$ only, which determines the scaling of the noise term. Unfortunately, this simple addition complicates the mathematics significantly, because $n(t)$ is a singular function consisting of δ functions, as in (1.9). We will not go into the details of this problem, but only say that there are two different interpretations of (1.10), the Itô interpretation and the Stratonovich interpretation. See van Kampen (1992) or Risken (1989) for a more detailed discussion.

Fokker–Planck   One way of characterizing the Langevin approach would be to start with the differential equations approach and make it more like the stochastic approach. In a complementary way, the Fokker–Planck approach starts with the fully stochastic model and makes it more like the differential equations model.

Recall that the state of the fully stochastic model consists of the number of molecules present, a discrete quantity. Equivalently, the probability distribution P(number of molecules, t) is a multivariable function that depends on both the continuous variable t and the discrete variables, number of molecules. This function is hard to deal with explicitly unless the number of discrete states is small, in which case one can handle it exhaustively as above. The Fokker–Planck approach deals with P, but assumes that it depends continuously on the number of molecules. By letting the vector $\mathbf{v}$, the number of each kind of molecule, consist of real numbers and not just integers, the function $P(\mathbf{v}, t)$ will be continuous in all its variables. Given this approximation, the Fokker–Planck approach writes a partial differential equation for P, namely,

$$\frac{\partial}{\partial t}P(\mathbf{v},t) = -\frac{\partial}{\partial v}A(\mathbf{v})P(\mathbf{v},t) + \frac{1}{2}\frac{\partial^2}{\partial v^2}B(\mathbf{v})P(\mathbf{v},t).\qquad(1.11)$$

From the particular form of this equation, it follows that

$$\frac{\partial}{\partial t}\langle\mathbf{v}\rangle = \langle A(\mathbf{v})\rangle \quad \text{and}$$

$$P_{eq}(\mathbf{v}) = \frac{const}{B(\mathbf{v})} \exp\left[2\int_0^v \frac{A(\mathbf{v}')}{B(\mathbf{v}')} d\mathbf{v}'\right]. \tag{1.12}$$

Note that P_{eq} is an equilibrium probability distribution, as in the fully stochastic case, not an equilibrium state, as in the differential equations case.

As with the Langevin approach, the Fokker–Planck equation (1.11) is most useful when the dynamics are linear. (Here "linear" means A is a linear function of $\mathbf{v}$ and B is constant.) For nonlinear equations, the same caveats apply as in the Langevin method. For linear functions,

$$\frac{\partial}{\partial t}\langle \mathbf{v}\rangle = A(\langle \mathbf{v}\rangle) \tag{1.13}$$

and one may use the following algorithm to apply the Fokker–Planck formalism:

1. Let A be the (linear) function that describes the differential equations version of the dynamics, as in (1.13).

2. From thermodynamics, find the equilibrium probability distribution, P_{eq}.

3. Use this equilibrium probability distribution, A, and equation (1.12) to find B.

4. From A and B, solve the partial differential equation (1.11) to find the complete dynamics as a function of time.

A generalization of the Fokker–Planck approach, the Kramers–Moyal expansion, views (1.11) as the first term in a Taylor expansion for the actual function P. The Kramers–Moyal expansion consists of terms of all orders. Cutting off after the first-order term leads to equation (1.11). More generally, one can cut off the Kramers–Moyal expansion of the master equation after any number of terms and use that approximation.

Note: The Fokker–Planck formalism is provably equivalent to the Langevin approach.

1/Ω Expansion (van Kampen)   There are other, more general, approaches to dealing with a stochastic system where one can assume the fluctuations are small compared with the average. One such approach, van Kampen's $1/\Omega$ expansion, assumes the fluctuations are on the order of the square root of the average and writes the actual number of molecules, n, in terms of the average concentration, the fluctuations, and a parameter, Ω, say, the volume:

$$n(t) = \Omega c(t) + \sqrt{\Omega}f(t). \tag{1.14}$$

One now rewrites the master equation for the dynamics of the system (see chapter 2) in terms of $c(t)$ and $f(t)$, not $n(t)$, and expands in powers of Ω. Collecting terms with the same power of Ω gives, in order:

1. The macroscopic deterministic equation.

2. The Fokker–Planck equation (equivalent to a Langevin approach).

3. Higher-order terms.

The details of this expansion are beyond this chapter. We refer the reader to van Kampen (1992) for a detailed explanation.

1.5   Parameter Estimation, Data Fitting, Phenomenology

1.5.1   Introduction

Section 1.4 described how to use complete models to generate predictions. This section considers the inverse problem—how to use observed system dynamics to generate or improve models. This can be considered as two subproblems:

• Write equations with unknown parameters.

• Find the values of those unknown parameters that lead to predictions that best match the dynamics.

If all the chemical equations are known, and only the parameters are unknown, the first problem is solved, and one need only consider the second. However, coarser and more phenomenological alternative models may be required as a practical matter when the relevant states and transition rates are unknown or are known to a very limited extent, as is likely for large protein complexes such as subcomplexes of the eukaryotic transcription complex.

The general problem of estimating parameters is hugely important across different scientific disciplines. Different communities have different names for parts of the process, different problems with which they are concerned, and different techniques. The linear systems community, for example, considers parameter estimation in connection with linear models; the parameter estimation is called *system identification.* The speech recognition community has considered many problems related to estimating parameters of Markov chains, in particular, hidden Markov models. The computational learning theory community refers to this type of estimation as *learning,* and considers which structures (such as neural networks) provide a good functional basis for estimating arbitrary nonlinear functions, how much data are necessary to make an estimate with a particular degree of certainty, which algorithms (such as simulated annealing, backpropagation, etc.) can be used to do estimation, and many other problems.

A major impetus for the parameter-fitting point of view is that systems with many interacting state variables, such as medium-sized or large protein complexes, have exponentially many states; therefore detailed state-by-state modeling becomes prohibitive to use with real data. More approximate, phenomenological models become attractive as targets of parameter fitting. We introduce several such models in the following discussion.

1.5.2 Boolean Models

There are several ways to learn Boolean models from data. If one can make a complete truth table, there are standard techniques that are taught in every introductory computer engineering class. Typically, this exhaustive enumeration of states is not feasible in biological systems, and one uses a more refined technique. The computational learning theory community (see Kearns and Vazirini 1994) has developed techniques for learning Boolean functions from observations of the inputs and outputs. For example, Jackson (1994) and Somogyi (chapter 5 this volume), have developed a technique based on information theory that has been applied to biological systems and will be covered in detail in chapter 5. Finally, some computational learning theory work has focused on uncovering the structure of a finite-state machine without being able to see the individual states, only the final states and the inputs required to get to that state (the algorithm is covered in Kearns and Vazirini 1994). Although this has not been applied to biology, it could provide an interesting way to deduce regulatory details that one cannot observe directly.

1.5.3 Methods for ODE Models

The training methods for models consisting of systems of ordinary differential equations are so far the most developed. Loosely, one can consider two problems: techniques for approximating the full dynamics to allow easier training, and techniques for training, given the approximated dynamics.

Approximating Dynamics Here the problem is to find a family of functions that can approximate the function f in the dynamics,

$$\frac{d}{dt}\mathbf{v} = f(\mathbf{v}) \tag{1.15}$$

given only time-course or related behavioral data. Below we list series of formulations for f that have the universal approximation property that given enough unobserved state variables, they can approximate the updated dynamics of "most" other dynamic systems. The linear approximation is an exception, but is sometimes convenient. Each of these ODE formulations generalizes immediately from one gene to many interacting genes in a feedback circuit.

Linear As a first approach, one may approximate f as a linear function, using a matrix L:

$$\frac{d}{dt}\mathbf{v} = L\mathbf{v} \tag{1.16}$$

or in component notation,

$$\frac{dv_i}{dt} = \sum_j L_{ij} v_j.$$ (1.17)

The techniques for learning the matrix L are well studied under the name *system identification* (see Ljung 1987).

Recurrent Artificial Neural Networks Another approach, which has been applied to real gene expression data in several cases as described in chapter 4 of this volume, is to use analog-valued recurrent artificial neural network (ANN) dynamics with learnable parameters. One version of such network dynamics, in use for gene regulation modeling, is given by Mjolsness et al. (1991):

$$\tau_i \frac{dv_i}{dt} = g\left(\sum_j T_{ij} v_j + h_i \right) - \lambda_i v_i.$$ (1.18)

Here the state variables v are indexed by i, and therefore connection strengths T between them take two such indices, i and j. The function g is a saturating nonlinearity, or sigmoid. The parameters T, τ, λ, and h have been successfully "trained" from spatiotemporal patterns of gene expression data (Reinitz et al. 1992).

Diffusion and cell-cell signaling interactions have been incorporated in this model (Mjolsness et al. 1991, Marnellos 1997). Controlled degradation could perhaps be handled as well using a degradation network $\hat{T}_{ij}$:

$$\tau_i \frac{dv_i}{dt} = g\left(\sum_j T_{ij} v_j + h_i \right) - \lambda_i v_i g\left(\sum_j \hat{T}_{ij} v_j + \hat{h}_i \right),$$ (1.19)

though no computer parameter-fitting experiments have been performed on such a model yet.

Sigma-Pi Neural Networks Two further models can be mentioned as attempts to incorporate promoter-level substructure into gene regulation networks that are otherwise similar to the ANN approach. In chapter 4 of this volume, sigma-pi units (Rumelhart et al. 1986a, b) or "higher-order neurons" are introduced to describe promoter-level substructure: regulatory modules in sea urchin *Endo16* promoter. A typical feedforward sigma-pi neural network is described by the relationships between variables v and weights T at layer network l and variables v at network layer $l+1$:

$$v_i^{l+1} = g\left(\sum_{jk} T_{ijk}^l v_j^l v_k^l + \sum_j T_{jk}^l v_j^l + h_j^l \right).$$ (1.20)

A novelty here is the third-order (three-index) connections T_{ijk} in which two input variables indexed by j and k jointly influence the activity of the ith output variable. This equation can hold at each of a number of layers in a feedforward architecture. If the three-index connection parameters T_{ijk} are all zero, the system specializes to a conventional feedforward layered neural network with second-order (two-way) connections T_{ij} at each layer. Yet higher-order connections may also be included, but can be transformed into multiple third-order interactions with the addition of extra variables v to compute intermediate products. (To be exact, this transformation requires that some gs be linear.) The generalization to continuous-time recurrent ODEs as in previous subsection is straightforward.

Hierarchical Model   In chapter 4, section 4.3.1, partition functions for promoter regulatory regions, dimerization, and competitive binding are proposed as a way to expand a single-node description of selected genes in a regulatory circuit into a subcircuit of partial activation states within a eukaryotic transcription complex. A simplified version of the HCA model, omitting heterodimers and competitive binding, and restricting inputs to two levels of hierarchy, is

$$\tau_i \frac{dv_i}{dt} = \frac{Ju_i}{1 + Ju_i} - \lambda_i v_i \tag{1.21}$$

$$u_i = \prod_{\alpha \in i} \left(\frac{1 + J_\alpha \tilde{v}_\alpha}{1 + \hat{J}_\alpha \tilde{v}_\alpha} \right) \tag{1.22}$$

$$\tilde{v}_\alpha = \frac{\tilde{K}_\alpha \tilde{u}_\alpha}{1 + \tilde{K}_\alpha \tilde{u}_\alpha} \tag{1.23}$$

$$\tilde{u}_\alpha = \prod_{b \in \alpha} \left(\frac{1 + K_b v_{j(b)}^{n(b)}}{1 + \hat{K}_b v_{j(b)}^{n(b)}} \right), \tag{1.24}$$

where i and j are transcription factors, α indicates promoter modules, b indicates binding sites, and the function $j(b)$ determines which transcription factor j binds at site b.

Generalized Mass Action   Another example of a phenomenological model for gene regulation is the proposal by Savageau (1998) to use "generalized mass action" (GMA) with nonintegral exponents C for modeling transcriptional regulation. In the above notation,

$$\tau_i \frac{dv_i}{dt} = \sum_a k_{ia}^+ \prod_j v_j^{C_{iaj}^+} - \sum_a k_{ia}^- \prod_j v_j^{C_{iaj}^-}, \tag{1.25}$$

which generalizes enzyme kinetics equations to nonintegral stoichiometric coefficients, and is also related to sigma-pi neural networks. The state variables v_i remain positive if $C_{iai}^- > 0$, as they should if they are to be interpreted as concentrations. A more computationally and analytically tractable specialization of GMA is to "S-systems" (Savageau and Voit 1987):

$$\tau_i \frac{dv_i}{dt} = k_i^+ \prod_j v_j^{C_{ij}^+} - k_i^- \prod_j v_j^{C_{ij}^-}, \qquad (1.26)$$

which have only one positive and one negative summand (interaction) per gene v_i. This form retains universal approximation capability.

The simplified enzyme kinetics model of Furusawa and Kaneko (1998), a specialization of GMA with diffusion added, is also worth mentioning as a possible target for parameter fitting.

Training Parameters   Several techniques exist for training parameters from trajectory (behavioral) data. System identification has already been mentioned in the context of linear models. For the nonlinear approximation schemes discussed earlier, several different techniques are available.

Backpropagation   For feedforward neural networks with two-way connections or higher-order ones, one can develop efficient gradient descent algorithms for training the network to match a training set of input/output patterns. The best known of these algorithms is the batch mode version of backpropagation (see Rumelhart et al. 1986a, b) in which an error signal propagates backward through the network layers and alters each connection. In practice, a simpler, incremental version of the algorithm, which only follows the gradient when averaged over many time steps, is often more effective. The incremental algorithm is in effect a form of stochastic gradient descent in which minor local minima can sometimes be escaped.

An easy generalization of the backpropagation algorithm applies to discrete-time recurrent neural networks such as fixed-timestep approximate discretizations of continuous-time recurrent neural networks. This backpropagation through time (BPTT) algorithm involves unrolling the network into a deep, many-layer feedforward neural net in which each layer shares connection strength parameters with the others. However, it is not necessary to make this approximation; a directly applicable gradient-descent learning algorithm for continuous-time recurrent neural networks was investigated by Pearlmutter (1989). These and other variations of backpropagation training are surveyed in Hertz et al. (1991).

Simulated Annealing   Another form of stochastic gradient descent is simulated annealing. This technique is typically much less efficient, but may still work when other data-fitting optimization procedures become ineffective due to large numbers of local minima. A particular variant of simulated annealing (Lam and Delosme 1988a, b) has proven effective at inferring small analog neural networks from trajectory data (Reinitz et al. 1992,

Reinitz and Sharp 1995). This version adaptively chooses discrete step sizes for adjusting analog parameters such as connection strengths, and also adjusts the temperature schedule so as to maintain a constant ratio of accepted to rejected annealing steps. A relatively gently saturating sigmoid function g is used to help smooth out the objective function for simulated annealing optimization.

An attractive possibility is to combine recurrent backpropagation with simulated annealing, perhaps by using the former as an inner loop in a simulated annealing optimization. Erb and Michaels (1999) have provided a sensitivity analysis of the fitting problem for ANN models, illustrating some of the computational difficulties encountered.

Genetic Algorithms   Genetic algorithms are another class of stochastic optimization methods that sometimes work relatively well on difficult optimization problems. Here, single-parameter update steps such as those of simulated annealing are augmented by crossover operations in which two complete configurations of the unknown parameters, chosen out of a population of N partially optimized ones, are combined to produce "progeny" subject to selection as part of the optimization procedure. This kind of algorithm has been investigated for inferring analog neural network models of gene regulation by Marnellos (1997), where it outperformed simulated annealing on artificial test problems, but not on the more realistic biological problems tried. It is possible that further work could substantially improve the situation.

Boolean Network Inference   The REVEAL algorithm of Liang et al. (1998) provides a starting point for inferring Boolean networks of low input order and 50 network elements. Building on this work, much larger networks were identified by Akutsu et al. (1999). It is possible that inference of Boolean networks will prove to be essentially more tractable than inference of continuous-time recurrent neural networks in cases where a parsimonious Boolean network approximation of the target trajectory data actually exists.

1.5.4   Stochastic Models

A future challenge will be to develop efficient algorithms to train parameters in the stochastic framework. The speech recognition community and the bioinformatics community have done a significant amount of work using hidden Markov models (Rabiner 1989). This work includes estimating parameters when the state (and maybe even the model) is not uniquely determined. The ion-channel community has done work on estimating parameters when better knowledge of the state and the biology are available (Colquhoun and Hawkes 1982). To our knowledge, these techniques have not been applied to gene regulation systems.

1.6   Summary and Road Map

1.6.1   Summary

One way to approach this chapter is to ask how to use the techniques covered here to aid in understanding a biological system. The method outlined in this chapter is:

• Create a calculation-independent model, a system of chemical and physical equations summarizing what is known or suspected about the system.

• Make some additional, computational assumptions, and use one of the model frameworks in the chapter. Chose a modeling framework whose assumptions are reasonable for the biological system.

• If biochemical constants are available, use them. If some or all of the biochemical constants are not available, use the parameter estimation techniques covered to find values of the missing constants.

• Use the model to make predictions.

• Compare the predictions of the model with experimental data to improve the model and to increase understanding of the system.

  Another view of the chapter is as a starting point for further research. There are several major areas:

• Develop faster or better algorithms to deal with the frameworks available. In particular, develop tools that are easy for the nonspecialist to use.

• Fill in gaps in this process. There are numerous places in the chapter where a point is missing because no research has been done in that area. Examples include but are not limited to applying the Fokker–Planck or Langevin formalisms to gene regulation systems, developing better ways to parameterize complex eukaryotic systems, and developing estimation techniques for the fully stochastic formalism.

• Use modeling and theory to add value to the biology—there is more do with a model than just simulate time courses. Some useful analysis techniques have been worked out for the differential equations formalism; techniques for the other formalisms are incomplete or nonexistent.

1.6.2   Where to Go from Here

The remainder of this book contains a number of chapters related to gene regulation. Chapter 2 presents a simple model of prokaryotic gene regulation in the fully stochastic formalism. Chapter 3 presents a model of a eukaryotic system, and includes elements of the differential equations formalism and some discrete interactions as well. Chapter 4 presents a neural network model of *Drosophila melanogaster* and illustrates, not only the differential equations framework, but also the usefulness of phenomenological models and the techniques of parameter estimation. Chapter 5 presents an extensive example of the Boolean network framework and parameter estimation in that framework.

In addition to the chapters in this book, several other references are available. The list that follows is by no means complete, but should provide a starting point for further study of gene regulation and mathematical modeling.

Acknowledgement

1. This work was supported in part by Office of Naval Research grants N00014-97-1-0293 and N00014-97-1-0422, the National Aeronautics and Space Administration's, Advanced Concepts program, and by a Jet Propulsion Laboratory, Center for Integrated Space Microsystems (JPS-CISM) grant.

References

Ackers, G. K., Johnson, A. D., and Shea, M. A. (1982). Quantitative model for gene regulation by lamba repressor. *Proc. Natl. Acad. Sci. U.S.A.* 79: 1129-1133.

Akutsu, T., Miyano, S., and Kuhara, S. (1999). Identification of genetic networks from a small number of gene expression patterns under the Boolean network model. In *Pacific Symposium on Biocomputing '99,* R. B. Altman, A. K. Dunker, L. Hunter, T. E. Klein, and K. Lauderdale, eds. World Scientific, Singapore.

Alberts, B., Bray, D., Lewis, J., Raff, M., Roberts, K. and Watson, J. D. (1994). *Molecular Biology of the Cell,* 3rd ed. Garland Publishing, New York.

Atkins, P. W. (1998). *Physical Chemistry*, 6th ed. Freeman, New York.

Colquhoun, D. and Hawkes, A. G. (1982). On the stochastic properties of bursts of single ion channel openings and of clusters of bursts. *Phil. Trans. R. Soc. Lond.* B. 8000: 1-59.

Davidson, E. H. (1986). *Gene Activity in Early Development.* Academic Press, Orlando, Fla.

Elowitz, M. B., Surette, M. G., Wolf, P. E., Stock, J. B. and Leibler, S. (1999). Protein mobility in the cytoplasm of *Escherichia coli. J. Bacteriol.* 181(1): 197-203.

Erb, R. S., and Michaels, G. S. (1999). Sensitivity of biological models to errors in parameter estimates. In *Pacific Symposium on Biocomputing '99* R. B. Altman, A. K. Dunker, L. Hunter, T. E. Klein, and K. Lauderdale, eds. World Scientific, Singapore.

Feller, W. (1966). *An Introduction to Probability Theory and Its Applications.* Wiley, New York.

Fleischer, K. (1995). A Multiple-Mechanism Developmental Model for Defining Self-Organizing Geometric Structures. Ph.D. thesis, California Institute of Technology, Pasadena, Calif.

Furusawa, C., and Kaneko, K. (1998). Emergence of multicellular organisms with dynamic differentiation and spatial pattern. In *Artificial Life VI*, C. Adami, R. K. Belew, H. Kitano, and C. E. Taylor, eds. MIT Press, Cambridge, Mass.

Gilbert, S. F. (1997). *Developmental Biology.* Sinauer Associates, Sunderland, Mass.

Gillespie, D. T. (1977). Exact stochastic simulation of coupled chemical reactions. *J. Phys. Chem.* 81(25): 2340-2361.

Glass, L., and Kauffman, S. A. (1973). The logical analysis of continuous, non-linear biochemical control networks. *J. Theor. Biol.* 39: 103-129.

Goss, P. J. E., and Peccoud, J. (1998). Quantitative modeling of stochastic systems in molecular biology by using stochastic Petri nets. *Proc. Natl. Acal. Sci. U.S.A.* 95: 6750-6755.

Gray, S., Cai, H., Barolo, S. and Levine, M. (1995). Transcriptional repression in the *Drosophila* embryo. *Phil. Trans. R. Soc. Lond.* B 349: 257-262.

Hertz, J., Krogh, A., and Palmer R. G., (1991). *Introduction to the Theory of Neural Computation.* Addison-Wesley, Redwood City, Calif.

Hill, T. L. (1985). *Cooperativity Theory in Biochemistry: Steady-State and Equilibrium Systems.* Springer Series in Molecular Biology, Springer-Verlag, New York. See especially pp. 6-8, 35-36, and 79-81.

Jackson, J. (1994). An efficient membership-query algorithm for learning DNF with respect to the uniform distribution. *IEEE Symp. Found. Comp. Sci.* 35: 42-53.

Johnson, A. (1995). Molecular mechanisms of cell-type determination in budding yeast. *Curr. Opinion in Genet. Devel.* 5: 552-558.

Kauffman, S. A. (1993). *The Origins of Order.* Oxford University Press, New York. See also *J. Theor. Biol.* 22: 437, 1969.

Kearns, M. J., and Vazirani, U. V. (1994). *An Introduction to Computational Learning Theory.* MIT Press, Cambridge, Mass.

Kosman, D., Reinitz, J., and Sharp, D. H. (1998). Automated assay of gene expression at cellular resolution. In *Pacific Symposium on Biocomputing '98*, R. Altman, A. K. Dunker, L. Hunter, and T. E. Klein, eds. World Scientific, Singapore.

Lam, J., and Delosme, J. M. (1988a). An Efficient Simulated Annealing Schedule: Derivation. Technical Report 8816, Yale University Electrical Engineering Department, New Haven, Conn.

Lam, J., and Delosme, J. M. (1988b). An Efficient Simulated Annealing Schedule: Implementation and Evaluation. Technical Report 8817, Yale University Electrical Engineering Department, New Haven, Conn.

Liang, S., Fuhrman, S., and Somogyi, R. (1998). REVEAL, a general purpose reverse engineering algorithm for inference of genetic network architectures. In *Pacific Symposium on Biocomputing '98,* R. B. Altman, A. K. Dunker, L. Hunter and T. E. Klein, eds. World Scientific, Singapore.

Ljung, L. (1987). *System Identification: Theory for the User.* Prentice-Hall, Englewood Cliffs, NJ.

Madhani, H. D., and Fink, G. R. (1998). The riddle of MAP kinase signaling specificity. *Trends Genet.* 14: 4.

Marnellos, G. (1997). Gene Network Models Applied to Questions in Development and Evolution. Ph.D. diss. Yale University, New Haven, Conn.

McAdams, H. H., and Arkin, A. (1998). Simulation of prokaryotic genetic circuits. *Ann. Rev. Biophys. Biomol. Struct.* 27: 199-224.

McQuarrie, D. A. (1967). Stochastic approach to chemical kinetics. *J. Appl. Prob.* 4: 413-478.

Mestl, T., Plohte, E., and Omholt, S. W. (1995). A mathematical framework for describing and analyzing gene regulatory networks. *J. Theor. Biol.* 176: 291-300.

Mestl, T., Lemay, C., and Glass, L. (1996). Chaos in high-dimensional neural and gene networks. *Physica* D 98: 33.

Mjolsness, E., Sharp, D. H., and Reinitz, J. (1991). A connectionist model of development. *J. Theor. Biol.* 152: 429-453.

Pearlmutter, B. A. (1989). Learning state space trajectories in recurrent neural networks. *Neural Comp.* 1: 263-269.

Prusinkiewicz, P., Hammel, M. S., and Mjolsness, E. (1993). Animation of plant development. In *SIGGRAPH '93 Conference Proceedings.* Association for Computing Machinery, Anaheim.

Rabiner, L. R. (1989). A tutorial on hidden Markov models and selected applications in speech recognition. *Proc. IEEE* 77(2): 257-286.

Reinitz, J. and Sharp, D. H. (1995). Mechanism of *eve* stripe formation. *Mechanisms Devel.* 49: 133-158.

Reinitz, J., Mjolsness, E., and Sharp, D. H. (1992). Model for Cooperative Control of Positional Information in *Drosophila* by Bicoid and Maternal Hunchback. Technical Report LAUR-92-2942, Los Alamos National Laboratory, Los Alamos, N.M. See also *Exper. Zoo.* 271: 47-56, 1995.

Risken, H. (1989). *The Fokker–Planck Equation: Methods of Solution and Applications,* 2nd ed. Springer-Verlag, New York.

Rumelhart, D. E., Hinton, G. E., and McClelland, J. L. (1986a). A general framework for parallel distributed processing. In *Parallel Distributed Processing*, Vol. 1, chapter 2, D. E. Rumelhart and J. L. McClelland, eds. MIT Press, Cambridge, Mass.

Rumelhart, D. E., Hinton, G. E., and Williams, R. J. (1986b). Learning internal representations by error propagation. In *Parallel Distributed Processing,* Vol. 1, chapter 8, D. E. Rumelhart and J. L. McClelland, eds. MIT Press, Cambridge, Mass.

Savageau, M. A. (1998). Rules for the evolution of gene circuitry. In *Pacific Symposium on Biocomputing '98*, R. Altman, A. K. Dunker, L. Hunter, and T. E. Klein, eds. World Scientific, Singapore.

Savageau, M. A., and Voit, E. O. (1987). Recasting nonlinear differential equations as S-systems: a canonical nonlinear form. *Math. Biosci.* 87: 83-115.

Shea, M. A., and Ackers, G. K. (1985). The OR control system of bacteriophage lambda. A physical-chemical model for gene regulation. *J. Mol. Biol.* 181: 211-230.

Small, S., Blair, A., and Levine, M. (1992). Regulation of *even-skipped* stripe 2 in the *Drosophila* embryo. *EMBO J.* 11(11): 4047-4057.

Small, S., Blair, A., and Levine, M. (1996). Regulation of two pair-rule stripes by a single enhancer in the *Drosophila* embryo. *Devel. Biol.* 175: 314-324.

Thomas, R., and D'Ari, R. (1990). *Biological Feedback*. CRC Press, Boca Raton, Fla.

Thomas, R., Thieffry, D. and Kaufman, M. (1995). Dynamical behaviour of biological regulatory networks-I. Biological role of feedback loops and practical use of the concept of the loop-characteristic state. *Bull. Math. Biol.* 57(2): 257-276.

van Kampen, N. G. (1992) *Stochastic Processes in Physics and Chemistry.* North-Holland, Amsterdam.

Varshavsky, A. (1995). The N-end rule. *Cold Spring Harb. Symp. Quant. Biol.* 60: 461-478.

von Hippel, P. H. (1998). An integrated model of the transcription complex in elongation, termination, and editing. *Science* 281: 660-665.

Watson, J. D., Hopkins, N. H., Roberts, J. W., Argetsinger Steitz, J., and Weiner, A. M. (1987). *Molecular Biology of the Gene*. Benjamin/Cummings, Menlo Park, Calif.

Yuh, C. H., Bolouri, H., and Davidson, E. H. (1998). Genomic *cis*-regulatory logic: experimental and computational analysis of a sea urchin gene. *Science* 279: 1896-1902.

2 A Probabilistic Model of a Prokaryotic Gene and Its Regulation

Michael Andrew Gibson and Jehoshua Bruck

2.1 Introduction

Chapter 1 described several different kinds of models for gene regulation. This chapter focuses exclusively on stochastic models (van Kampen 1992, McQuarrie 1967). As mentioned in chapter 1, the stochastic framework is important when the number of molecules involved is small, the time scale is short, or both. This chapter will not rehash the theoretical underpinnings already covered, but rather will provide an extensive example that illustrates many of the biological processes involved in (prokaryotic) gene regulation and many of the stochastic processes used to model these biological processes. The system to be modeled is part of the regulatory circuitry of λ phage and is a small subset of the complete model in Arkin et al. (1998).

2.2 λ Phage Biology

Lambda phage (Ptashne 1992) is a virus that infects *Escherichia coli* cells. It is called a *temperate* phage, because it has two possible developmental pathways (figure 2.1). It can either replicate and lyse (dissolve) the host cell, thus releasing about 100 progeny or it can integrate its DNA into the bacterial DNA and form a lysogen. In the latter case, the virus will replicate passively whenever the bacterium replicates. A lysogen also has immunity from subsequent lambda infections; this protects the lysogen from being destroyed should another phage infect the host cell. Under the right conditions (e.g., exposure to UV light), a lysogen can be induced, that is, the viral DNA excises itself from the bacterial DNA and undergoes normal replication and lysis (see Ptashne 1992).

Lambda has been studied extensively because it is one of the simplest developmental switches—systems with multiple possible end states. Its complete genome (≈50,000 base pairs) was sequenced long before the era of large-scale genome sequencing. The details of which genes are expressed, when, and in what quantity have been studied extensively, as has lambda's host, *E. coli*. As a result, lambda is one of the best-understood organisms in existence. It is also an organism that exhibits probabilistic or stochastic behavior; some infected cells end up in lysis, others in lysogeny. Hence it is a perfect system for demonstrating the stochastic framework for modeling gene regulation.

In this chapter, we focus on three key regulatory proteins in lambda: N, Cro, and the lambda repressor. There are a handful of other proteins that are equally important, but we have chosen only these three to illustrate the stochastic framework here.

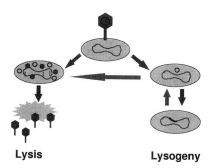

Lysis   **Lysogeny**

Figure 2.1
The life cycle of λ phage. A phage injects its DNA into the *E. coli* host, then either (left) replicates and destroys
the host cell—lysis— or (right) integrates its DNA into the host DNA—lysogeny.

• Repressor is present at high levels in lysogens and represses the expression of all other
genes.

• Cro (control of repressor and others) is important in the lytic pathway; it inhibits the production of repressor and controls the production of the proteins that cause viral shell formation, cell lysis, replication of λ DNA, etc.

• Repressor and Cro are mutually inhibitory.

• N is a temporal regulator. Once lambda injects its DNA into the host, N is produced; thus
N is produced very early in the life cycle. N facilitates the expression of early genes, genes
necessary for both lysis and lysogeny. Late in the life cycle, once the phage has decided between one of the two fates (by expressing Repressor and not Cro, or vice versa), N production is halted and other specialized proteins are produced, some for lysis, some for
lysogeny.

Figure 2.2 illustrates the concentrations of each of these proteins as a function of time for
typical cells—one that ends in lysis and one that ends in lysogeny.

Understanding the entire lambda decision process would require a model much more
detailed than the one in this chapter. For now, we will focus almost entirely on N and how
it is regulated. The *N* gene lies downstream of lambda's left promoter, P_L (figure 2.3). (A
bit of notation: a gene is written in italics, for example, *N*; its protein product is in roman
text, e.g., N.) In the absence of transcription factors, RNA polymerase (RNAP) can bind to
P_L and initiate transcription. (Note that this is one of the key differences between
prokaryotes and eukaryotes—in prokaryotes, DNA and RNAP are sufficient for transcription; in eukaryotes, more factors are required.) Thus, when lambda first injects its DNA
into a host, N can be produced. Later, as Repressor and Cro accumulate, these two transcription factors bind to P_L and inhibit further production of N. We will be modeling production of N as a function of the concentrations of Repressor and Cro.

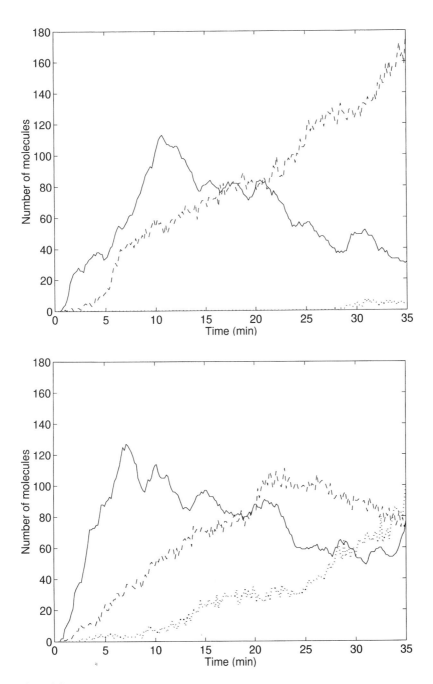

Figure 2.2
A typical plot of number of proteins as a function of time for the complete lambda model (not presented in this chapter). Solid line, N; dashed line, Cro_2; dotted line, $Repressor_2$. (Top) Lysis, (bottom) Lysogeny.

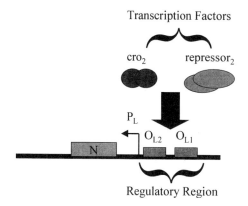

Figure 2.3
Regulatory logic for gene *N*, which lies downstream of promoter P_L. The transcription factors Cro_2 and $Repressor_2$ bind upstream of P_L at operator sites O_{L1} and O_{L2}. Either transcription factor can bind to either site.

2.2.1   Why a Probabilistic Model Is Useful

Lambda DNA is inserted into its *E. coli* host cell. At that time, there are no lambda proteins or mRNA present. Yet, despite this, sometimes the host will go to lysis; other times it will go to lysogeny. Modeling probabilistic outcomes requires a probabilistic model. A deterministic model would predict that the same initial conditions always lead to the same final conditions.

Figure 2.4 shows data from a more complete model of lambda, which includes the full interactions of five genes and their protein products. We have simulated protein production and have plotted the final concentration of repressor and of cro at the time when the host cell divides. Each dot represents one phage. Each phage is modeled with the same chemical reactions, the same parameters for those reactions, and the same initial conditions. The only difference between one dot and another is the set of random numbers chosen in our simulation. This is the key point: *In a deterministic model, there would be only one dot on this figure.*

Since the stochastic model and the deterministic model make different predictions, they are not equally good for modeling this system. Because the final lysis or lysogeny decision is stochastic (some phages go to lysis, some go to lysogeny), a deterministic model would be inappropriate.

Origins of Stochastic Behavior   Where does this probabilistic or stochastic behavior come from? As mentioned in chapter 1, there are tradeoffs to consider in modeling. With low numbers of molecules and short time steps, systems behave stochastically, while with high numbers of molecules and long time steps, averaging occurs and there is a deterministic outcome.

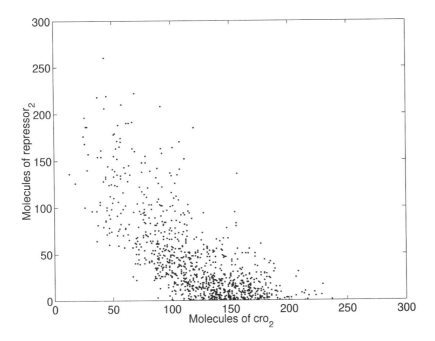

Figure 2.4
Number of molecules of Cro$_2$ and Repressor$_2$ at the end of the simulation. Each dot represents one phage.

An *E. coli* cell is a rod-shaped bacterium 2 μm long with a diameter of 1 μm (Watson et al. 1987). The volume of an *E. coli* cell is

$$V = \pi r^2 l = \pi / 2 \times 10^{-15} \text{ liters.}$$

As will be explained later, significant differences in the amount of binding occur in the range 10^{-9} M to 10^{-7} M. The number of molecules that corresponds to, say, 10^{-8} M is ($\pi/2$ × 10^{-15} liters)(10^{-8} mol/liter)(6×10^{23} molecules/mole) ≈ 10 molecules. The lambda model deals with one molecule of DNA, a few molecules of mRNA, and tens to hundreds of molecules of proteins.

Differential equations with concentrations as variables assume that the concentrations vary continuously and deterministically, or at least that the fluctuation around the average value of concentration is small relative to the concentration. Figure 2.4 showed that these assumptions lead to incorrect predictions.

An aside (molecular dynamics):   At a lower level, one could (in principle) write out the molecular dynamics equations for each of the molecules involved, solve these deterministic equations, and average over all possible initial conditions. Note that the variables for this approach are positions and momenta of atoms, not concentrations. Such an approach would be computationally intractable—first in the sheer number of atoms involved (DNA, multiple copies of each protein and mRNA, water molecules, other cellular machinery), second in the time scale (tens of minutes), and finally in the uncertainty of initial conditions because many of the cellular components are not even known.

The molecular dynamics approach to modeling is sometimes called *microscopic* chemistry; differential equations with concentrations as variables are called *macroscopic* chemistry; between these two lies the stochastic approach, or *mesoscopic* chemistry (van Kampen 1992). As in the microscopic approach, one considers individual molecules—proteins and mRNA, for example—but ignores many molecules, such as water and nonregulated parts of the cellular machinery. Rather than considering the positions and momenta of the molecules modeled, one considers the statistics of which reactions occur and how often. The end result is that the statistics (e.g., the probability that there are n molecules at time t) follow simple laws, as described in chapter 1.

2.3   Outline of a Model

Chapter 1 mentioned that gene regulation consists of many processes: transcription, splicing, translation, post-translational modifications, degradation, diffusion, cell growth, and many others. Fortunately, many of these processes are not present in general in prokaryotes or in particular in lambda. Our model of the production of N will consider only transcription, translation, and degradation. The full lambda model (Arkin et al. 1998) contains cell growth as well, but we will argue that it can be neglected in this chapter.

The model consists of two parts: TF–DNA binding and N production. The former is modeled as an equilibrium process according to the principles of statistical thermodynamics explained in chapter 1. The latter process consists of several individual subprocesses, each of which is modeled as a simple chemical reaction. The complete set of chemical reactions is found in table 2.1, and the parameters for those reaction are found in table 2.2. Note that these are mesoscopic rate constants, that is, they are the stochastic rate constants that apply to individual molecules. The relationship between these rate constants and the usual macroscopic rate constants, which apply to concentrations consisting of large numbers of molecules, will be discussed in a later section.

2.3.1   Transcription

Transcription involves two process: initiation and elongation. Initiation involves all of the biological details necessary for RNAP to create a transcription complex and start transcription. For this model, these processes will be lumped into reaction (2.1) in table 2.1, with rate constant k_1. The value of k_1 is determined from an equilibrium model, which will be described in detail in a later section.

Table 2.1
Chemical reactions involved in gene expression in λ phage

$\text{RNAP} \cdot \text{DNA}_{\text{closed}} \xrightarrow{k_1} \text{RNAP} \cdot \text{DNA}_{\text{open},0}$	(2.1)
$\text{RNAP} \cdot \text{DNA}_{\text{open},n} \xrightarrow{k_2} \text{RNAP} \cdot \text{DNA}_{\text{open},n+1}$	(2.2)
$\text{RNAP} \cdot \text{DNA}_{\text{open,MAX}} \xrightarrow{k_3} \text{RNAP}_{\text{free}} + \text{DNA}_{\text{free}} + \text{mRNA}_{\text{free}}$	(2.3)
$\text{Ribosome} + \text{mRNA}_{\text{free}} \xrightarrow{k_4} \text{Ribosome} \cdot \text{mRNA}_0$	(2.4)
$\text{RNase} + \text{mRNA}_{\text{free}} \xrightarrow{k_5} \text{RNase} \cdot \text{mRNA}$	(2.5)
$\text{RNase} \cdot \text{mRNA} \xrightarrow{k_6} \text{RNase}$	(2.6)
$\text{Ribosome} \cdot \text{mRNA}_n \xrightarrow{k_7} \text{Ribosome} \cdot \text{mRNA}_{n+1}$	(2.7)
$\text{Ribosome} \cdot \text{mRNA}_{\text{MAX}} \xrightarrow{k_8} \text{Ribosome}_{\text{free}} + \text{mRNA}_{\text{free}} + \text{protein}$	(2.8)
$\text{protein} \xrightarrow{k_9} \text{no protein}$	(2.9)

Source: Arkin et al. (1998).

Elongation is the process, described by reactions (2.2) and (2.3) in table 2.1, by which RNAP transcribes DNA into mRNA nucleotide by nucleotide. Reaction (2.2) in the table describes most of the elongation process, and reaction (2.3) describes the final step of elongation, namely, that the RNAP reaches the final base of the DNA it is transcribing, releases mRNA, and separates from the DNA.

2.3.2 Translation

Translation is mathematically similar to transcription: It consists of an initiation step and an elongation step. The translation initiation rate is not regulated by any transcription factors, so it is simply a constant. The translational elongation process is entirely analogous to the transcriptional elongation process. The reactions for translation are (2.4), (2.7), and (2.8) in table 2.1.

2.3.3 Degradation

Transcription and translation end in the production of new mRNA and proteins. If these were the only processes present, the amount of mRNA and of protein would grow without limit. Instead, both mRNA and proteins are degraded by various cellular processes. Degradation of mRNA occurs when a ribonuclease (rather than a ribosome) binds to the mRNA and then degrades the mRNA [see reactions (2.5) and (2.6) in table 2.1]. This competitive binding will be explored further in section 2.6.

Degradation of a protein occurs when various cellular machinery, such as proteosomes, binds to the protein and degrades it. Reaction (2.9) in table 2.1 shows the pseudo-first-

Table 2.2
Values of mesoscopic rate constants in this model

Constant	Value
k_1	Calculated via binding model (Shea and Ackers, 1985)
k_2	30 s^{-1}
k_3	30 s^{-1}
$k_4 \times$ [ribosome]	0.3 s^{-1}
$k_5 \times$ [RNase]	0.03 s^{-1}
k_6	Fast compared with k_4 and k_5
k_7	100 s^{-1}
k_8	100 s^{-1}
k_9	0.00231 s^{-1} (half-life = 7 min)
Transcription length	550
Translation length	320

Source: Arkin et al. (1998).

order version of that process; that is, we assume the rate depends only on the amount of protein N present, not on the concentration of the degradation machinery. This point is discussed further below.

2.3.4 Host Cell Growth

Over the course of about 35 min, an *E. coli* host cell will double in size and divide. In general, this is a very important nonlinear effect; namely, concentrations are the number of molecules per unit volume, so if the volume change is ignored, calculations that involve concentrations will be incorrect. Essentially all reaction rates will involve a rate constant and some power of the volume. However, first-order reactions (and pseudo-first-order reactions) involve the 0th power of the volume, and thus do not depend on the volume. Thus, reactions (2.2) through (2.9) in table 2.1 will not be affected by the change in volume.

As for reaction (2.1) in table 2.1, we will assume that the concentrations of the transcription factors, Cro and Repressor, in their active (dimer) forms are constant, rather than the number of Cro and Repressor molecules being constant. Thus reaction (2.1) is not affected by changes in volume, either. In summary, the host cell growth will not affect our model, so we will ignore it.

An aside (pseudo-first-order degradation):   As mentioned earlier, reaction (2.9) in table 2.1 shows the pseudo-first-order version of the protein degradation process—that is, the rate depends only on the amount of protein N present, not on the concentration of the degradation machinery. In other words, the protein degradation machinery is present in constant concentration, regardless of what happens to N, Cro, and Repressor. A more complete model would include some of the cellular machinery and treat degradation as a second-order effect. In λ phage, for example, saturation of the machinery that degrades

Table 2.3
Free energies used to calculate k_1 from a transcription factor–DNA binding model

State s	O_L1	O_L2	ΔG_s (kcal/mol)	i	j	k	$k_1(s)$ (1/s)
1	—	—	0	0	0	0	0
2	Cro$_2$	—	−10.9	1	0	0	0
3	—	Cro$_2$	−12.1	1	0	0	0
4	Repressor$_2$	—	−11.7	0	1	0	0
5	—	Repressor$_2$	−10.1	0	1	0	0
6	—	RNAP	−12.5	0	0	1	0.011
7	Cro$_2$	Cro$_2$	−22.9	2	0	0	0
8	Cro$_2$	Repressor$_2$	−20.9	1	1	0	0
9	Repressor$_2$	Cro$_2$	−22.8	1	1	0	0
10	Repressor$_2$	Repressor$_2$	−23.7	0	2	0	0

Source: Arkin et al. (1998).

protein CII is thought to be one of the important ways in which lambda decides between lysis and lysogeny (Ptashne 1992, Arkin et al. 1998). To model this second-order effect, we must explicitly include the degradation machinery in the model. This chapter will use the pseudo-first-order model only. We mention the full second-order model only for completeness.

2.4   Details of Model

This section spells out the details of the model outlined in section 2.3.

2.4.1   TF–DNA Binding

The promoter for N has two binding sites, O_L1 and O_L2 (see figure 2.3). Each can bind either Cro$_2$ or Repressor$_2$. The possible states of the promoter are shown in table 2.3 (Arkin et al. 1998). For each state we have listed:

• A state number s, which is used to discuss the states only; it has no physical meaning

• the configuration of the two binding sites

• the difference in free energy from the ground state (state 1)

• three numbers, i, j, and k, which are the number of molecules of Cro$_2$, Repressor$_2$, and RNAP, respectively

and finally,

• $k_1(s)$, the rate of transcription initiation, given that the DNA is in state s

Chapter 1 mentioned the possibility of measuring rate constants for transitions among the various states of such a model. We will assume that the transitions have reached equilibrium, and so we ignore kinetics and use equilibrium thermodynamics to find the probability that the promoter is in each of the 10 possible states. Applying the theory from chapter 1, those probabilities are of the form

$$P_s = \frac{[DNA]_s}{[DNA]_{total}} = \frac{e^{\frac{-\Delta G_s}{RT}} [Cro_2]^i [Repressor_2]^j [RNAP]^k}{Z}, \qquad (2.1)$$

where the partition function Z is given by

$$Z = \sum_{s,i,j,k} e^{\frac{-\Delta G_s}{RT}} [Cro_2]^i [Repressor_2]^j [RNAP]^k.$$

As mentioned earlier, we assume that $[Cro_2]$ and $[Repressor_2]$ are given. We will use a temperature of 37°C and an RNAP concentration of 30 nM for all subsequent calculations.

Example: Consider a simple example: $[Cro_2] = [Repressor_2] = 0$. Then nearly all of the states will have a probability of 0, according to equation (2.1). Only states 1 and 6 have a nonzero probability. Using $T = 37°C = 310$ K, $[RNAP] = 30$ nM, $R = 8.314$ J/(K mole), and 1 kcal = 4184 J:

$$P_1 = \frac{1}{1+19.5} = 0.049$$

$$P_6 = \frac{19.5}{1+19.5} = 0.951.$$

Example: Consider a more complex example: $[Cro_2] = 50$ nM, $[Repressor_2] = 10$ nM. The partition function Z, evaluated at these conditions, is

$$Z = 1 + 2.41 + 17.0 + 1.77 + 0.13 + 19.5 + 34.9 + 0.27 + 5.9 + 5.1 = 88.0$$

The probabilities of the 10 states are 0.011, 0.028, 0.193, 0.020, 0.002, 0.222, 0.397, 0.003, 0.067, and 0.058, respectively.

Using the TF–DNA binding model, we can calculate k_1. For a given $[Cro_2]$ and $[Repressor_2]$, we calculate the probability P_s of each state s. Given the probabilities of finding the DNA in each of the 10 states, it is a simple matter to calculate k_1:

$$k_1 = \sum_s k_1(s) P_s.$$

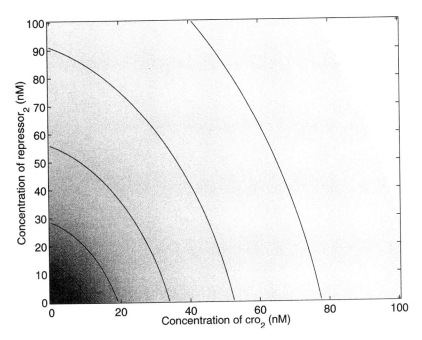

Figure 2.5
Rate constant k_1 as a function of [Cro$_2$] and [Repressor$_2$]. Contours represent (from bottom left to top right) $k_1 =$ 0.0075, 0.005, 0.0025, 0.00125, and 0.000625.

Example: For the two examples above, $k_1(s)$ is 0 except for $s = 6$. Thus, k_1 is simply $0.011 \times P_6$. Numerically, k_1 is 0.0105 s^{-1} for the first example and 0.0024 s^{-1} for the second. Figure 2.5 shows k_1 as a function of [Cro$_2$] and [Repressor$_2$].

2.4.2   Doing the Calculation

Given [Cro$_2$] and [Repressor$_2$], we can calculate k_1 as in the previous section. Tables 2.1 and 2.2, together with the value of k_1 constitute a system of chemical equations with constant coefficients. In this section, we will look at how to deal with such a system computationally in a stochastic framework. We explain the basic ideas of the calculation only; the detailed algorithms can be found in the references.

2.4.3 Probability of Reaction per Unit Time

To model a system in a stochastic framework, assume the system is in a given state, that is, that there are a specific number of molecules of each kind. Then the probability that a particular reaction will occur per unit time is constant. The symbol μ will denote this probability per unit time. In particular, μ is equal to the rate constant, k, times the product of the number of molecules of each reactant.

Example: Consider the protein degradation reaction [(2.9) in table 2.1]. For protein N, this reaction can be written as

$$N \xrightarrow{k_9} \text{no N}.$$

From this,

$$\mu_9 = k_9 \times (\text{number of molecules of N}).$$

Table 2.2 gave first-order and pseudo-first-order rate constants only. First-order rate constants have the property that the mesoscopic (i.e., stochastic) rate constant is the same as the macroscopic (i.e., deterministic) rate constant. Second- and higher-order rate constants do not have this property.

Example: A second-order mesoscopic rate constant has units of 1/(molecules $\times$ seconds). A second-order macroscopic rate constant has units of 1/(molar $\times$ seconds). Dimensional analysis shows that:

$$k_{\text{macroscopic}} = k_{\text{mesoscopic}} \times V \times A \times C$$

for second-order rate constants, where V is volume, A is Avogadro's number, and C is a dimensionless constant. In all the second-order reactions in table 2.1, C is 1; in other reactions it may be 1/(2!), 1/(3!), etc. For a more detailed discussion, see Gillespie (1977).

2.5 Chemical Master Equation

Chapter 1 gave an example of using the chemical master equation (see, for example, McQuarrie 1967). Essentially, the approach is this:

• Write out all possible states, that is, all possible combinations of numbers of molecules.

• Use as variables the probabilities of being in each of the states. These variables are functions of time.

• Write a system of linear differential equations with constant coefficients (the probabilities of reaction per unit time, or μs) that express how the variables (the probabilities) change as a function of time.

• Solve this system of linear equations explicitly.

This approach works only if the number of possible states is small. In the example in Chapter 1, the number of possible states was four. For larger systems, it is sometimes possible to use mathematical techniques to lessen the number of states, but this is not possible for arbitrary systems. Thus, for arbitrarily large systems, the chemical master equation approach is not feasible.

Example: Consider reaction (2.7) of table 2.1. Suppose there is a single molecule of mRNA. Then, since the subscript n in reaction 2.7 can range from 0 to max ($= 320$), there are max $+ 1$ possible states, namely

$$(n = 0), (n = 1), K \ (n = \max).$$

Suppose there are two molecules of mRNA. Since they behave independently, there are now about $\frac{1}{2}\max^2$ states. For m molecules of mRNA, there are about $\max^m \div m!$ states just for mRNA! Because of this rapid growth in the number of states, the chemical master equation approach quickly becomes unfeasible.

2.5.1   Monte Carlo Algorithms

In the chemical master equation approach, one used as variables the probabilities of being in each state. These variables were continuous and followed simple differential equations. However, the approach became computationally intractable because the number of variables grew quickly for complex systems. Monte Carlo algorithms provide a way around this growth of the number of variables:

• Use as variables the number of molecules of each type. This gets around the combinatorial explosion in the number of variables—the number of variables is simply the number of types of molecules.

• Simulate one trajectory through state space:
From each state, there are several possible next states. Pick one next state randomly, using the appropriate probability distribution.
Using the new state, pick a new next state.
Go to the second step and repeat until some final time.

• Simulate many trajectories to get the statistics of the system.

The general idea is that rather than calculating probability distributions explicitly, one uses a computer's random number generator to take samples of those distributions, and by sampling repeatedly one gets a good approximation of the actual distributions. The outline given here does not specify how to deal with time, or how exactly to pick the next state. There are different algorithms that solve these problems in different ways.

Simple Approach   A simple algorithm is to use a fixed timestep Δt and step forward through time, using a random number generator to calculate whether any reactions occur during the timestep. This algorithm has the advantage that the running time is a simple function of Δt. It has the disadvantage of approximating continuous time by a discrete step. If the timestep is small enough, this approximation does not cause any problems, but if the timestep is too large, multiple reactions that occur very close together may be missed. There is a tradeoff between accuracy of computation and computation time, which comes down to choosing the timestep. In particular, the optimal value of Δt is a function of the reaction constants and may change over time.

More complicated schemes, with adaptive timesteps, are possible. However, there is another approach, which we now consider, that circumvents this problem and never trades off accuracy.

Gillespie Algorithm   A different, and in many ways better approach is that of Gillespie (1977). His exact simulation algorithm never approximates continuous time by a discrete value; rather, it asks: What is the next reaction that occurs, and when does it occur? The first part, which reaction occurs, is a discrete question. The appropriate probability distribution from which to draw the reaction can be calculated efficiently. The second part of the question deals with a time random variable, which is picked from the (exact) distribution, without any discrete approximation of time. In this way, no discrete approximations are made. By using properties of the chemical reactions, one can solve some of the mathematics off-line and use those results in the algorithm to make these calculations efficient.

Gibson–Bruck Algorithm   The present authors have developed several improvements to the Gillespie algorithm that make it run faster for gene regulatory systems (Gibson and Bruck 2000).

2.6   Modeling Results

Using an appropriate algorithm, one may calculate the probability that there are n protein copies at time t. We start with the simplest case, $[\text{Cro}_2] = [\text{Repressor}_2] = 0$. We have previously shown that under these conditions, k_1 is 0.0105 s^{-1}. Using the Gibson–Bruck algorithm, generating 5000 trajectories took just over a minute on a Pentium 2. More complicated simulations involving feedback and high degrees of coupling take longer.

Figure 2.6 shows a typical trajectory (solid line) and the average trajectory (dotted line). Note that the typical trajectory fluctuates wildly around the average and that the fluctuations are on the same order as the average; they are emphatically *not* small compared with the average. Note also that the amount of N continues to increase up to and including the final time because we are not including feedback regulation, in which increasing amounts of Repressor and Cro cause N production to shut down.

Figure 2.7 shows the probability that the system has n molecules of protein N at time t, for each n and t. The axes are t and n, and the color (see plate 1) represents the probability.

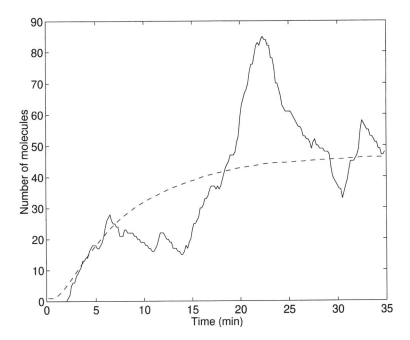

Figure 2.6
Number of molecules of protein N as a function of time, for the model presented in this chapter with [Cro$_2$] = [Repressor$_2$] = 0. Solid line, a typical trajectory; dotted line, average trajectory.

2.6.1   Plots—Inhibition by Cro, Repressor

The proteins Cro and Repressor inhibit expression of N. Consider [Cro$_2$] = 50 nM, [Repressor$_2$] = 10 nM. In a previous example, we showed that under these conditions, k_1 is 0.0024 s^{-1}. For this value, generating 5000 trajectories takes under a minute because on average there are fewer proteins and hence fewer reaction events to deal with. Figures 2.8 and 2.9 (see also plate 2) correspond to figures 2.6 and 2.7 and use the new k_1 value. These show that Cro and Repressor do indeed inhibit production of N. This is perhaps easier to see in figure 2.9; increasing the concentration of Cro and Repressor shifts the probability distribution toward 0 molecules; that is, the probability of ever having 10, 20 or more molecules of N drops. Also, the probability distribution shifts to the right; with inhibition it is likely to take longer to accumulate a fixed number of molecules.

2.6.2   Adding Intuition

One way to deal with the stochastic framework is to take the chemical reactions, plug them into numerical software that simulates within the stochastic framework, and look at the re-

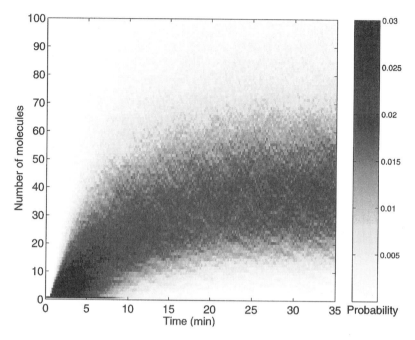

Figure 2.7
Probability of *n* molecules of protein N at time *t*, for the model presented in this chapter with $[Cro_2] = [Repressor_2] = 0$. Color represents probability (see plate 1).

sults. For some systems, that may be the only thing to do. This section will take another tactic and analyze key subsystems in order to gain experience and intuition with the stochastic framework.

There are several ways reactions can depend on each other. Chapter 1 considered reversible reactions in equilibrium. Here we consider reactions that compete with each other and reactions that follow each other in succession.

Competition—Concurrent Reactions   Consider reactions (2.4) and (2.5) in table 2.1. A single molecule of mRNA can do one of two things: It can bind a ribosome and begin translation, or it can bind ribonuclease (RNase) and be degraded (see figure 2.10). We now ignore the rest of the model and focus solely on the subsystem. In this special case of a single molecule, we can explicitly write out all the states and use the chemical master equation approach directly.

We start by defining the states 0, 1, and 2 as in figure 2.10. Let $\mathbf{P}(t)$ be the probability vector $[P(0), P(1), P(2)]^T$. Then, using Markov chains as in chapter 1,

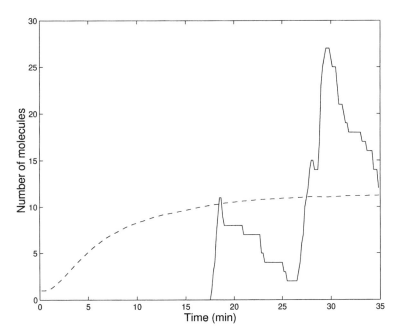

Figure 2.8
Number of molecules of protein N as a function of time, for the model presented in this chapter with [Cro$_2$] = 50 nM, [Repressor$_2$] = 10 nM. Solid line, a typical trajectory; dotted line, average trajectory.

$$\frac{d\mathbf{P}}{dt} = \frac{d}{dt}\begin{bmatrix} P(0) \\ P(1) \\ P(2) \end{bmatrix} = \begin{bmatrix} -k_4 - k_5 & 0 & 0 \\ k_4 & 0 & 0 \\ k_5 & 0 & 0 \end{bmatrix} \mathbf{P}(t). \tag{2.2}$$

Using standard techniques, we can solve this as a function of time. Assuming we always start in state 0, we have:

$$\mathbf{P}(t) = \begin{bmatrix} P(0) \\ P(1) \\ P(2) \end{bmatrix} = \begin{bmatrix} \exp(-(k_4 + k_5)t) \\ k_4 / (k_4 + k_5) \times [1 - \exp(-(k_4 + k_5)t)] \\ k_5 / (k_4 + k_5) \times [1 - \exp(-(k_4 + k_5)t)] \end{bmatrix}. \tag{2.3}$$

Question: What is the probability that the given mRNA will start translation rather than be degraded, and vice versa?

From (2.3), it is easy to calculate the probability of eventually winding up in state 1 or 2, respectively; we simply take the limit as $t \to \infty$. Thus, starting at state 0, an mRNA winds up in state 1 with the probability $k_4/(k_4 + k_5)$ and in state 2 with the probability $k_5/(k_4 + k_5)$.

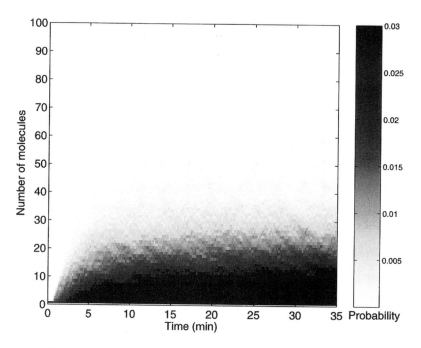

Figure 2.9
Probability of *n* molecules of protein N at time *t*, for the model presented in this chapter with $[Cro_2] = 50$ nM, $[Repressor_2] = 10$ nM. Color represents probability (see plate 2).

There are easier ways to calculate this than solving equation (2.2) for all possible times; for the sake of this example, we have chosen to show the time solution as well.

Question: What is the probability that n proteins will be produced from one mRNA molecule before it degrades?

The previous example showed that translation occurs with the probability $p = k_4/(k_4 + k_5)$ and mRNA degradation occurs with the probability $1 - p = k_5/(k_4 + k_5)$. Using this definition of p, the answer is simply:

$$P(n) = p^n(1 - p). \tag{2.4}$$

In particular, this is the probability that translation occurs n times and then degradation occurs. This distribution (2.4) is called a *geometric distribution*. The fact that the values of p and $1 - p$ came from our previous analysis of a Markov chain is irrelevant for answering this question.

Question: What is the average number of proteins produced per mRNA transcript?

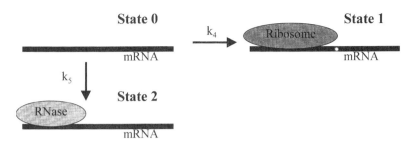

Figure 2.10
Competitive binding of ribosomes and RNase to mRNA, leading to transcription and degradation, respectively.

It can easily be shown (or looked up) that the expected value of a geometric distribution is $p/(1 - p)$. For the numbers given, $p = 0.3/(0.3 + 0.03) = 10/11$ and $1 - p = 1/11$. Thus, the average number of proteins produced per mRNA is 10. [It might seem surprising that this worked out so cleanly, but in fact, the values of k_4 and k_5 were chosen to make the average number of proteins 10 (Arkin et al. 1998).]

Simplification—Sequential Reactions Consider once again reaction (2.7) in table 2.1. There are max steps of the same type, each of which takes an exponential writing time (this is a property of time-invariant Markov processes). In the master equation approach, this process consists of max simple (exponential) reactions. In algorithms that deal only with simple reactions, such as Gillespie's (1977) exact simulation algorithm, there is no good way to simplify the calculation based on the observation that the interesting parts of reaction (2.7) are the first step, translation initiation, and the last step, translation termination; the intermediate states are not interesting.

The entire process of translational elongation constitutes a Poisson process, that is, a process consisting of sequential steps, with the time of each step exponentially distributed. The number of steps made up to a fixed time follows a Poisson distribution, while the time to make a fixed number of steps obeys a gamma distribution. The probability of finishing translational elongation at time t, $P_e(t)$ is given by

$$P_e(t) = \frac{t^{\max-1}}{(k_7)^{\max}(\max-1)!} \exp(-k_7 \times t).$$

Thus a single equation combines max separate steps. In this way, we can build up the entire simulation out of a series of more complicated steps, such as the gamma distribution, instead of a much larger series of simple steps, as in the strict interpretation of Gillespie's algorithm. One advantage that the authors' algorithm has over Gillespie's is the ability to deal with arbitrary stochastic processes, not just exponentials. It should be mentioned, however, that this only works because of the loose coupling between reactions. If the reactions were highly coupled, it might be important to know which of the intermediate states the ribosome was in, and these simplifications would not be appropriate. See Gibson and Bruck (2000) for a more complete discussion.

2.6.3 A More Complete Model

The model presented here lacks several subtleties that are necessary for a complete model. Among them are feedback and second-order reactions, notably degradation, which are affected by changes in volume.

Feedback   The model presented is, however, complicated, in the sense that it contains lots of parts, each of which must be understood individually and in relation to the others. In another sense, though, we have simply shown (in a very detailed way) how to write the probability of having n molecules of protein N as a function of time, of [Cro_2] and of [$Repressor_2$], where these latter two quantities are constant. In other words, this is a feedforward model; the concentrations of Cro_2 and $Repressor_2$ feed into the regulatory circuitry for N (figure 2.3), but the concentration of N does not feed back and affect them.

   To get a complete model, it is necessary to write an analogous model for the regulation of Cro_2 and $Repressor_2$. The tricky part of the complete model is that the regulatory inputs to each of the genes are not constant, but depend on each other. Systems that involve feedback are in general more complex than the open-loop system we have modeled, for this reason. Subsequent chapters will discuss modeling feedback systems in a deterministic framework; for work on modeling such systems in a probabilistic framework, see Arkin et al. (1998).

Second-Order Reactions   This chapter has ignored changes in volume by treating all reactions as first order or pseudo-first order. The probability per unit time of higher-order reactions depends on volume, so the growth of the host cell may be an important effect. One cannot always ignore growth, although the magnitude of this effect compared with others is hard to quantify. In the lambda genetic switch, there is strong positive feedback of Repressor onto itself. Since positive feedback can cause amplification, a small error (due to ignoring volume change or otherwise) could, in principle, be magnified to create a large calculation error.

   Degradation provides another set of second-order reactions. For a first approximation, we assumed that the processes modeled do not regulate the protein degradation machinery and used a pseudo-first-order model of degradation. In the Markov chain approach discussed in Chapter 1, this amounts to saying the following: suppose there are n copies of protein N at time t. Then, at time $t + dt$,

$$P(n-1, t+dt \mid n, t) = n \times k_9 \times dt \quad \text{and}$$

$$P(n, t+dt \mid n, t) = 1 - n \times k_9 \times dt.$$

We may explicitly include the degradation machinery. In the simplest case, consider n copies of protein N at time t and m copies of the degradation machinery. At time $t+dt$,

$$P(n-1, m-2, t+dt|n, m, t) = n \times m \times k_9' \times dt \quad \text{and}$$

$$P(n, m, t+dt|n, m, t) = 1 - n \times m \times k_9' \times dt.$$

Here k_9' is the true second-order rate constant for reaction (2.9) in table 2.1. It has units of $1/(\text{molecules} \times \text{seconds})$. Notice that the pseudo-first-order approximation consists of treating m as constant and letting $k_9 = m\, k_9'$.

In general, and for lambda in particular, second-order, nonlinear reactions are important. For example, in lambda, saturation of a second-order degradation reaction, the degradation of protein CII, biases the genetic switch toward lysogeny.

2.7 Conclusion

We have presented an example of the stochastic framework for modeling gene regulation. Once elucidated, the basic ideas are simple, but there are a lot of details. Regulatory proteins Cro and Repressor bind to DNA and achieve a fast equilibrium. Those proteins affect the rate of transcription initiation. Once transcription initiates, RNA polymerase moves step by step down the DNA and transcribes gene N into mRNA. The mRNA can then be degraded by ribonuclease, or it can bind to a ribosome and begin translation. Once translation is initiated, the ribosome proceeds step by step along the mRNA and produces a protein. Proteins in solution can be degraded by a pseudo-first-order process or a second-order process, depending on the particular protein and the assumptions made.

All of these details can be written out as chemical reactions, as in table 2.1. Depending on the number, complexity, and coupling of such reactions, different techniques can be used to solve them. If there are only a very few reactions, or the reactions are uncoupled, it may be possible to get an analytical solution. If there are more reactions, but they are still uncoupled, it may be possible to use the sort of simplifications mentioned here to simplify the calculation. In the general case of many, coupled reactions, the best approach seems to be a Monte Carlo simulation of the type in Gillespie (1977) or that of Gibson and Bruck (1999).

There are numerous open areas of research. For larger systems, computational simplifications will be needed. There has not been much systematic work on such simplifications in feedforward systems, let alone in feedback systems. One important point to be noted is that there is no need to reinvent the wheel to analyze each new regulatory system. There is already a significant amount of computer software devoted to solving chemical kinetics in the deterministic limit. Some programs have been developed to do stochastic simulations as well. Ideally, certain conventions would be established for specifying the biological details of a gene regulation system in a calculation-independent way, and then different programs could do the analysis in different ways.

It is rare to have a system for which so much experimental data are readily available. Even in this system, certain assumptions were made; for example, the promoter data were taken from a related promoter for which experimental data are available. There are several strategies for dealing with systems that have missing data:

• Use some sort of statistical technique to estimate (or "fit" or "learn") the missing parameters.

• Measure the missing parameters.

• Use theoretical methods to determine which parameters affect the performance of the model the most; measure those experimentally and estimate the rest.

• Only model systems for which all parameters are available.

Each of these approaches has its strengths and weaknesses. The experimental measurement procedures are still too slow and labor intensive to provide all the data theorists would like. The parameter estimation procedure is well worked out for deterministic models, but much less so for stochastic models of this sort. We are not aware of any systematic work involving estimation of stochastic parameters for gene regulation systems.

Finally, there are many open theoretical questions. For example, what can one do with a model, other than just simulate trajectories? This question has been considered in detail for deterministic models, but not for stochastic models. The challenge in theoretical biology is to add value—to use mathematics and computation to gain insight into systems that would be impossible (or very difficult) to gain with simpler, more intuitive methods. Stochastic systems defy much conventional intuition. The field is wide open for theoretical advances that help us to reason about systems in greater detail and with greater precision.

Acknowledgment

This work has been supported in part by Office of Naval Research grant N00014-97-1-0293, by a Jet Propulsion Laboratory Center for Integrated Space Microsystems grant, by National Science Foundation Young Investigator Award CCR-9457811, and by a Sloan Research Fellowship.

References

Arkin, A., Ross, J., and McAdams, H. H. (1998). Stochastic kinetic analysis of developmental pathway bifurcation in phage lambda-infected *Escherichia coli* cells, *Genetics* 149: 1633–1648.

Gibson, M. A., and Bruck, J. (2000). Efficient exact stochastic simulation of chemical systems with many species and many channels. *J. Phys. Chem. A* 104: 1876–1889.

Gillespie, D. T. (1977). Exact stochastic simulation of coupled chemical reactions, *J. Phys. Chem.* 81: 2340–2361.

McQuarrie, D. A. (1967). Stochastic approach to chemical kinetics, *J. Appl. Prob.* 4: 413–478.

Ptashne, M. (1992). *A Genetic Switch: Phage Lambda and Higher Organisms.* Cell Press and Blackwell Scientific Publications, Oxford.

Shea, M. A., and Ackers, G. K. (1985). The OR control system of bacteriophage lambda. A physical-chemical model for gene regulation, *J. Mol. Biol.* 181: 211–230.

van Kampen, N. G. (1992). *Stochastic Processes in Physics and Chemistry.* Elsevier, Amsterdam.

Watson, J. D., Hopkins, N. H., Roberts, J. W., Steitz, J. A., and Weiner, A. M. (1987). *Molecular Biology of the Gene.* Benjamin/Cummings Publishing, Menlo Park, Calif.

3 A Logical Model of *cis*-Regulatory Control in a Eukaryotic System

Chiou-Hwa Yuh, Hamid Bolouri, James M. Bower, and Eric H. Davidson

3.1 Introduction

The regulated expression of thousands of genes controls the mechanisms by which morphological form and cell functions are spatially organized during development. We refer to the set of transcription factors required for any given state of gene expression as the *trans*-regulatory system. The organized set of transcription factor target sites in the genomic DNA sequence that is sufficient to cause the properly regulated expression of the gene is referred to as the *cis*-regulatory system. The sequences of the transcription factor target sites in the *cis*-regulatory DNA of each gene determine the inputs to which the gene will respond (Arnone and Davidson 1997). Reorganization of developmental *cis*-regulatory systems and the networks in which they are linked must have played a major role in metazoan evolution because differences in the genetically controlled developmental process underlie the particular morphologies and functional characteristics of diverse animals. Thus, understanding embryonic development and evolution will require an understanding of genomic *cis*-regulatory systems. As described in chapters 1 and 2, quite detailed models of the regulatory systems of prokaryotic genes have been developed and successfully demonstrated. However, the large size and additional complexity of eukaryotic genes has in the past not permitted a similar level of analysis. We present here an experimental analysis of the multiple functions of a well-defined *cis*-regulatory element that controls the expression of a gene during the development of the sea urchin embryo. The outcome is a computational model of the element, in which the logical functions mediated through its DNA target site sequences are explicitly represented. The regulatory DNA sequences of the genome may specify thousands of such information-processing devices.

3.2 Modeling Considerations

3.2.1 The Problem

cis-Regulatory domains specify the conditions (e.g., spatial location, or time sequence) under which a given gene is transcribed. Many post-transcriptional regulatory mechanisms can also control both the function of the gene product and the rate at which it is synthesized. In many cases, such additional regulatory stages complicate the relationship between the transcription rate of a gene and the concentration of the protein that it encodes, thus requiring much detailed knowledge for quantitative analysis of the *cis*-regulatory system. In addition, proteins are frequently involved in complex interactions with other proteins so that the concentration of a gene product during some experimentally measured period may not reflect its rate of synthesis (see chapter 10 for an example).

Thus, the first problem in attempting to understand *cis*-regulation is the identification of experimental systems of gene expression that are amenable to direct measurement. An example is the expression of the *Endo16* gene. To understand why *Endo16* offers an ideal opportunity for the study of *cis*-regulation, we need first to look at the sea urchin and its embryonic development.

3.2.2   Sea Urchin Development

Sea urchin species are members of the echinoderm phylum, which also includes sea stars and the sea cucumber. Figure 3.1 shows adult sea urchins of the species studied here: *Strongylocentrotus purpuratus,* commonly known as the California purple sea urchin. The biology of the adult sea urchin is beyond the scope of this chapter, but the embryonic development of its larva is particularly pertinent. Sea urchin larvae develop in a very different manner from those of insects and vertebrates. The larvae are very simple compared with the adult because in sea urchins the larva is merely a host within which the adult body is grown. Indeed, the larva produced from a fertilized egg bears no morphological resemblance to the adult (compare figures 3.1 and 3.2). The adult body grows through a separate postembryonic developmental process from special populations of cells set aside during larval development (Ransick and Davidson 1995). When the embryonic growth of the larva is complete (about 3 days after fertilization for the species studied here), it starts feeding. Formation of the adult body begins shortly afterward and takes about 11 weeks for this species. At metamorphosis, the emerging juvenile simply jettisons the larval body.

The fully grown *S. purpuratus* larva has approximately 1500 cells, about 100 times fewer cells than a newly formed adult. This relative simplicity makes the larva an attractive model system. Furthermore, the fate of many embryonic cells can be traced back to events during cleavage. So the number of processes that translate maternal asymmetries into differentiated cell function must be relatively small. This makes sea urchin embryos an ideal material in which to study the mechanisms underlying spatial and temporal regulation of gene activity. Figure 3.2 shows the embryo at various stages of development. The drawings are by Hörstadius (1973) and were done in the early part of this century.

In *S. purpuratus,* the egg undergoes nine stereotyped and synchronous cell cleavages, resulting in a hollow spherical body (blastula) one cell thick and consisting of about 400 cells (corresponding to figure 3.2, d). Asynchronous cell divisions after the ninth cleavage increase the number of cells to approximately 600 before gastrulation and the onset of morphogenesis (figure 3.2, e). From the onset of gastrulation to a fully formed pluteus larva (figure 3.2, g), the total number of larva cells increases to about 1500 and then remains constant throughout the rest of the life of the larva. The rate of embryonic development is temperature dependent (it is faster when it is warmer). For the species and

Figure 3.1
Adult *Strongylocentrotus purpuratus* sea urchins (approximately 3 inches in diameter).

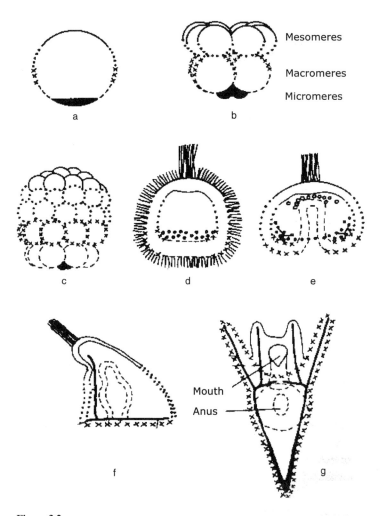

Figure 3.2
Sea urchin embryonic development. a, the uncleaved egg; b, fourth cleavage; c, sixth cleavage embryo; d, the mesenchyme blastula just before gastrulation (the mesenchyme cells have thickened to form a flat base, and the skeletogenic mesenchyme cells have ingressed); e, embryo during gastrulation; f, prism-stage embryo exhibiting distinct anatomical structures; g, the pluteus larva.

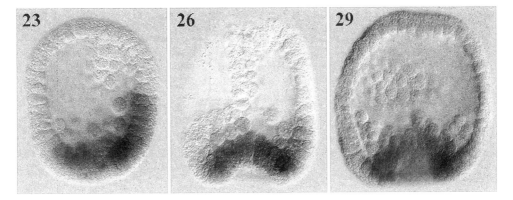

Figure 3.3
The pattern of *Endo16* expression during *S. purpuratus* embryonic growth. Numbers of hours post fertilization are shown at the top left of each figure. Side views are shown. Cells labeled brown indicate the daughters of a single macromere; cells stained blue are those expressing *Endo16*. Note that *Endo16* expression is always confined to a fraction of the macromere descendants. Adapted from Ransick and Davidson (1995). (See plate 3.)

experimental conditions reported here, the last cleavage occurs at around 20 hr postfertilization (hpf); the mesenchyme blastula stage is about 24 hpf, midgastrula about 36 hpf, prism about 60 hpf, and the full pluteus larva about 72 hpf. The experimental evaluations reported here span the period from the end of cleavage to formation of a mature larva (20–72 hpf).

Throughout all the cleavage stages, the embryo is enclosed in a "fertilization envelop" that remains roughly the same size. Thus, cleavage essentially partitions the oocyte into a large number of subvolumes. As noted by Hörstadius (1973) and indicated schematically in figure 3.2, maternal factors in the oocyte are asymmetrically distributed. Some cleavages are also asymmetric. As cleavage proceeds, cells inheriting different portions of the oocyte inherit different amounts of various maternal factors. This constitutive difference initiates differential regulation of genes in different regions of the developing embryo.

3.2.3   The *Endo16* gene

Endo16 was discovered by Ernst and colleagues in 1989 (Nocente-McGrath et al. 1989). It is a gene that encodes a polyfunctional glycoprotein found on the surface of cells in the midgut of the late embryo and larva of *S. purpuratus* (Soltysik-Espanola et al. 1994, Nocente-McGrath et al. 1989) and has proved a valuable marker for the study of cell fate in sea urchin embryos. Figure 3.3 (see also plate 3) illustrates the pattern of expression of *Endo16* in *S. purpuratus* embryos. The earliest expression of *Endo16* occurs at around the eighth to ninth cleavage (18–20 hr postfertilization) in a ring of approximately 64 cells that are descendants of the fourth-cleavage macromeres (see figure 3.2, b). In fact, the cells that express *Endo16,* and the neighboring cells (above and below) that do not, all descend from the macromeres. So the differential early expression of *Endo16* must have its roots in events that take place in the intervening four or five cleavage stages. Indeed, transplantation experiments by Ransick and Davidson (1995) have shown that *Endo16* is induced by signals from the neighboring micromeres during the fourth to sixth cleavage. The fact that *Endo16* is transcribed very early on during embryogenesis suggests that there are probably not many layers of regulatory interaction between early *Endo16* expression and the asymmetrically distributed maternal factors. As shown in figure 3.3 and Ransick and Davidson (1995), throughout this period *Endo16* is symmetrically distributed in a ring around the animal-vegetal axis. So the regulatory circuit controlling the early pattern of expression of *Endo16* may be expected to be relatively simple and amenable to analysis.

Toward the end of embryogenesis, *Endo16* expression becomes confined to the differentiating cells of the midgut (Ransick et al. 1993). Transcription is extinguished in the foregut and the delaminating mesoderm in the late gastrula, and thereafter in the hindgut. However, there is an increase in the rate of expression in the midgut, where it can still be detected in advanced feeding larval stages. So the late expression of *Endo16* involves upregulation in certain regions, and repression in other regions. This makes *Endo16* an interesting gene to model. The nature of the transcription factors that control *Endo16* expression changes with time. Initially, at least some of these factors are probably of maternal origin (Zeller et al. 1995). Later they are synthesized from zygotic transcripts.

The *cis*-regulatory system required for the complete embryonic pattern of expression of *Endo16* is included in a 2300-base pair (bp) DNA sequence extending upstream from the transcription start site (Yuh et al. 1994). Within this region, target sites have been mapped for 15 different proteins that bind with high specificity, that is, $\sim 10^4$ times their affinity for a synthetic double-stranded copolymer of deoxyinosine and deoxycytidine [poly(dI-dC)/poly(dI-dC)] (Yuh et al. 1994). Though some have been identified and cloned, most of these proteins are known only by their molecular mass, their DNA binding properties, and their site specificity. The Endo16 protein is known to be differentially spliced into three different but closely related forms. However, control of the level of expression of all Endo16 activity is primarily transcriptional (Ransick et al. 1993, Yuh and Davidson 1996, Yuh et al. 1994).

In the next section, we present an experimental analysis of this *cis*-regulatory system. The result is a model in which the logical functions mediated through *Endo16* DNA target site sequences are explicitly represented. The regulatory DNA sequences of the genome may specify thousands of such information-processing devices.

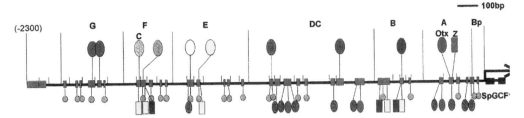

Figure 3.4
Summary of *Endo16 cis*-regulatory organization. The 2300-base pair *cis*-domain of *Endo16* comprises 55 protein binding sites (red boxes on horizontal black line denoting DNA) organized into six modular subregions labeled A – G at the top of the figure. The boundaries of the modules are delineated by the thin vertical lines incident on the DNA. Bp represents the basal promoter and the region to its right, marked by the bold arrow, marks the coding region of the gene. The large colored symbols above the DNA line indicate the unique binding sites within each module. The smaller colored symbols below the DNA line mark nonunique binding sites as labeled. (See plate 4.)

3.3 Modeling Framework

Like many other eukaryotic genes (see Kirchhamer et al. 1996 for a review), the *Endo16 cis*-regulatory system is composed of subelements (modules) that function positively or negatively in different embryonic territories and during different temporal phases of development. A regulatory module is defined as a fragment of DNA containing multiple *trans*-regulatory factor target sites that execute a particular regulatory subfunction of the overall *cis*-regulatory system. In a series of earlier studies, Davidson and colleagues (Yuh and Davidson 1996, Yuh et al. 1996) mapped out the physical and logical organization of the *Endo16 cis*-regulatory domain. In particular, it was shown that the region is divided into six distinct modules and that the module nearest to the basal transcription apparatus mediates the activities of the other five modules. Figure 3.4 summarizes the physical *cis*-regulatory organization of *Endo16* (see also plate 4). Figure 3.5 summarizes our modeling framework.

3.3.1 Experimental Procedure and Data Processing

A detailed description of the experimental procedure is beyond the scope of this chapter. This information is available elsewhere (Yuh et al. 1996). Here we provide only a brief overview of the most pertinent issues and pointers to the literature. To develop a more detailed understanding of Endo16 regulation, we first need to formulate an experimentally quantifiable modeling framework that relates the gene expression being measured to the concentrations of regulatory transcription factors.

Expression Constructs   The expression constructs used in the experiments are shown schematically in figure 3.6 (see also plate 5). They were assembled in various ways from

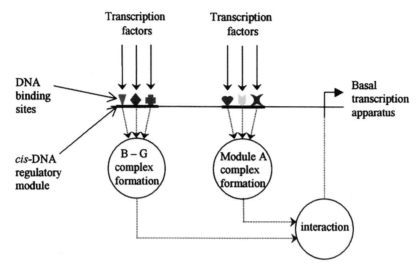

Figure 3.5
Schematic of the modeling framework used. The *cis*-regulatory DNA of the gene is divided into a number of physical domains or modules. Factors binding to specific sites within these domains form regulatory complexes that can activate the basal transcription apparatus through interactions with factors bound on module A. The dotted lines indicate interactions between molecules bound to DNA.

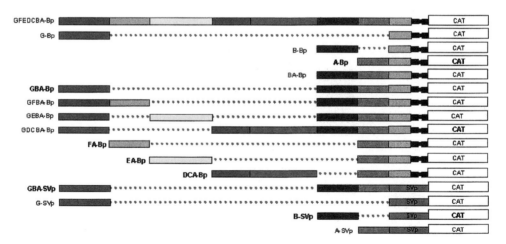

Figure 3.6
Schematic representation of some of the expression constructs used. (See plate 5.)

restriction fragments, bounded either by the restriction sites indicated in figure 3.4, by polylinker sites present in subclones, or by restriction sites present in an earlier generation of vectors that were derived from the *Endo16* CAT (chloramphenical acetyltransferase) expression construct described by Yuh et al. (1994). Constructs GBA(OTX$_m$) − Bp·CAT and BA(OTX$_m$) − Bp·CAT were assembled from a cloned insert that linked G, B, and A to Bp, into which the mutated OTX target site had been introduced by a polymerase chain reaction (PCR). The outside primers for this purpose were the Bluescript vector forward and reverse primers. The complementary inside primers included the normal *Endo16* sequence immediately upstream of the target OTX site (Yuh et al. 1994), with an appended sequence that replaces this site with an XbaI site. The XbaI served later to check the success of the construction. GBA(OTX$_m$) − Bp·CAT was assembled from GBA(OTX$_m$Xba) + (XbaOTX$_m$)A − Bp + Bluescript CAT, where OTX$_m$Xba and XbaOTX$_m$ represent the two parts of module A to be joined by ligation at the XbaI site. Other constructs were similarly generated.

Earlier studies (Yuh et al. 1994) were carried out with Endo 16 fusion constructs that included not only the upstream 2300-bp DNA fragments represented in figure 3.4, but also a 1.4-kb intron that interrupts the N-terminal portion of the protein coding sequence. This construct (Endo16·CAT) included the *Endo16* transcription start site, the whole of exon I, in which *Endo16* transcription begins, intron I and exon II, down to residue 40. At this point, a fusion with the CAT reporter gene just upstream of its translation initiation site was effected using a naturally occurring SalI site (see Yuh et al. 1994). For the experiments reported here, all but 106 bp from the 5′ end and 11 bp from the 3′ end of the intron were removed. A new fusion gene including the same upstream sequence as Endo16·CAT but lacking all but these 117 bp of intron I was constructed (GFEDCBA-Bp·CAT, the top construct in figure 3.6; "Bp" represents the basal promoter). The pattern of expression and amount of these two constructs was experimentally shown to be identical (Yuh and Davidson 1996).

CAT Enzyme Measurements   CAT enzyme activity was determined in lysates of 100 embryos. Samples were collected at various stages in development and CAT enzyme concentration was measured by chromatography. The final measure of enzyme concentration reported here includes fairly large error bars, owing to biological variation between replicate samples. The modeling implications of this are discussed later.

Data Preparation   CAT enzyme activity was typically measured at 10, 20, 30, 48, and 72 hr after fertilization. The smooth curves in the figures are generated by regression. Figure 3.7 shows two examples of data recorded from different experiments ranging over a period of 1 year. To emphasize time-course characteristics, measured CAT enzyme molecules per embryo are normalized with respect to their peak value. Each data point is

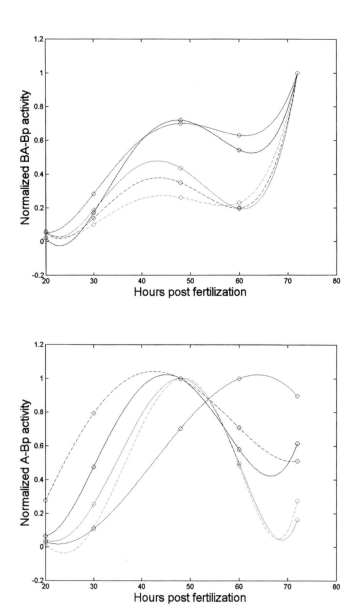

Figure 3.7
(A and B) Two examples of variations in patterns of construct expression in different experiments.

averaged over 100 embryos. As we can see, the pattern of expression over time is very similar from experiment to experiment. Nonetheless, there is considerable variation in the normalized value at any given time point. The absolute values of the data vary more widely. Averaging data from different experiments would therefore result in very large standard errors. On the other hand, the relationships among expression patterns are well preserved within experimental batches, as illustrated in figure 3.8. We therefore analyzed the relationships between regulatory elements per experimental batch, seeking models that minimized the least mean-squared error across all experiments.[1]

As noted earlier, the number of cells expressing *Endo16* changes during *S. purpuratus* embryonic development [see figure 3.9(A)]. For the analysis here, we are only interested in the relationships between individual plot points, so there is no need to normalize by cell count except in the specific case of figure 3.9(B), where we are interested in the absolute value of the expression levels.

3.3.2   Model of *Endo16* Regulatory Functions

Figure 3.10 shows the completed *Endo16 cis*-regulatory model. The process leading to the construction of this model was necessarily one of iterative hypothesis building and refinement. Indeed, the process by which the model was built engendered many of the experiments presented in figures 3.11–3.14, in that these experiments were designed to test predictions of the model or to decide between alternatives. Describing the sequence of reasoning that led to the model of figure 3.10 would make for a very long and tedious chapter. For brevity, we will only discuss the details of one of the more interesting aspects of the model in detail: the relationship between regulatory modules A and B. For the rest, only a summary of the sequence of deductions leading to the model in figure 3.10 is presented, but we will refer to some of the more significant decisions made along the way.

In the next two subsections, we first describe and deduce the interactions between whole regulatory modules (F–A), then go on to identify the individual elements in module A that mediate each interaction. The numbered statements summarize the conclusions of each discussion. Together, these observations amount to the model previewed in figure 3.10.

Modular Interactions

1. Module A by itself is sufficient to cause expression of *Endo16*.
Module A is probably responsible for initiating expression in the vegetal plate in the early embryo. In a construct that includes no other *cis*-regulatory subelements, the transcription-enhancing activity of module A increases early in development, but it then declines when expression is becoming confined to the midgut. The set of curves in figure 3.7(B) shows the activity profile of several module A constructs. However, as shown in figure 3.9(B), *Endo16* expression (as measured by the GFEDCBA-Bp construct) continues to rise after module A activity subsides. So although module A is sufficient for the early activation of

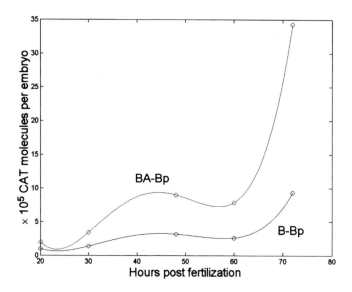

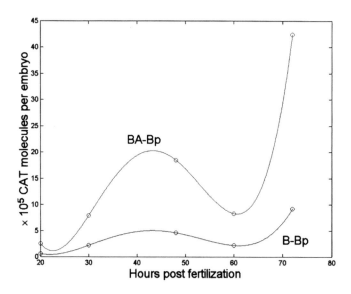

Figure 3.8
(A and B) Two examples that relationships among *Endo16* expression patterns are preserved within experimental batches.

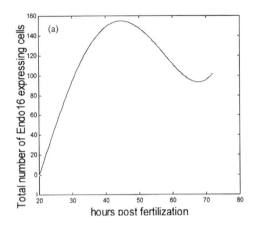

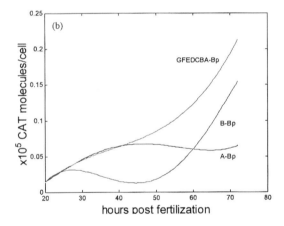

Figure 3.9
(A) The number of cells expressing *Endo16* during embryonic development. (B) Regulated activity per cell. The full *Endo16* construct initially follows the pattern of expression due to module A alone. At around the time of onset of gastrulation (30 hr post fertilization), its activity level switches to follow that of the B-Bp construct. Adapted from Yuh et al. (1996).

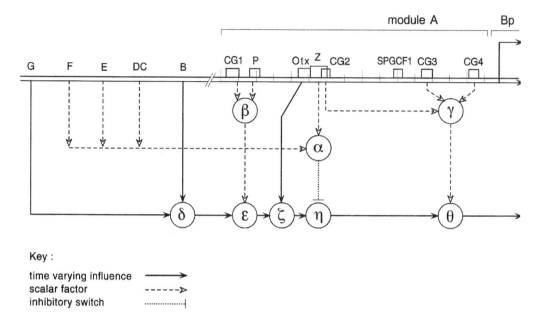

Figure 3.10
Computational model for module A regulatory functions. Schematic diagram of interrelations and functions. Interrelations between upstream modules (G to B; figure 3.4) and specific module A target, and among the module A target sites, are indicated beneath the double line at top, representing the DNA. The region from module G to module B is not to scale. Each circle or node represents the locus in the system of a specific quantitative operation, conditional on the state of the system. Operations at each node are carried out on inputs designated by the arrows incident on each circle, and produce outputs designated by arrows emergent from each circle. In the logic statements, *"inactive"* indicates a site or module site that has been mutationally destroyed or is inactive because its factor (or factors) is missing or inactive; *"active"* indicates that the site or module is present and productively occupied by its cognate transcription factor. For the case of modules F, E, and DC, a Boolean representation is chosen because ectopic expression is essentially zero (beyond technical background) in ectoderm and mesenchyme when these modules (together with module A) are present in the construct (Yuh and Davidson 1996); otherwise, ectopic expression occurs. Similarly, when they are present, LiCl response occurs; otherwise, it does not (Yuh and Davidson 1996) (figure 3.13). The logic sequence specifies the values attained at each operation locus, either as constants determined experimentally and conditional on the state of the relevant portions of the system, or in terms of time-varying, continuous inputs designated by the symbol (t). The values $\gamma=2$ or $\gamma=1$ derive from the experiments of figure 3.14 (see text). The kinetic output of modules B and G and of the Otx site are represented as B(t), G(t), and Otx(t), respectively. The input B(t) can be observed as the CAT activity profile generated by module B over time in figure 3.11A [see also figure 2 in Yuh et al. (1998) and figure 3 in Yuh and Davidson (1996)]. G(t) is shown in the same figures in Yuh et al. (1996) and Yuh and Davidson (1996). Otx(t) is the time course generated by the construct OtxZ in figure 12B (unpublished observations). The final output, Θ(t), can be thought of as the factor by which, at any point in time, the endogenous transcriptional activity of the BTA is multiplied as a result of the interactions mediated by the *cis*-regulatory control system.

α: if (F or E or DC are *active*) and (Z is *active*) then
           α = *active*
     else   α = *inactive*

β: if (P is *active* and CG_1 is *active*) the
           β = 2
     else   β = 0

γ: if (CG_2 is *active* and CG_3 is *active* and CG_4 is *active*) then
           γ = 2
     else   γ = 1

δ: $\delta(t) = B(t) + G(t)$

ε: $\varepsilon(t) = \beta * \delta(t)$

ζ: if ($\varepsilon(t)$ is *inactive* (i.e. =0)) then
           $\zeta(t) = Otx(t)$
     else $\zeta(t) = \varepsilon(t)$

η: if (α is *active*) then
           $\eta(t) = 0$ (*repressed*)
     else $\eta(t) = \zeta(t)$

Θ: $\Theta = \gamma * \eta(t)$

α: the repressive action of modules F, E, and DC mediated by site Z.

β: both P and CG1 are needed for amplification of module B influence.

γ: final, non-specific boosting of transcription level.

δ: positive input from modules B and G.

ε: amplification of module B effect by CG_1-P subsystem.

ζ: switch determining whether (a) Otx in module A or (b) modules B & G control level of activity of Endo16 (triggered from (a) to (b) at gastrulation).

η: complete inhibition of Endo16 transcriptional activity due to repressive effects of modules DC (in the primary/skeletogenic mesenchyme cells) and F and E (in the vegetal ectoderm) before gastrulation (mediated by site Z as in a above).

Θ: total regulatory effect communicated to the basal transcription apparatus

Endo16 expression, it cannot be the only activating regulatory element of *Endo16*. As reported in Yuh et al. (1994), earlier experiments identified modules B and G as positive regulators of *Endo16*. We therefore ask: How do modules A, B, and G coordinately enhance *Endo16* transcription?

2. All regulatory activity is mediated through module A.

In Yuh et al. (1996) it was shown that constructs with only regulatory modules B or G express only relatively low levels. However, when module B was combined with a module A regulatory element, the pattern of expression of the construct closely resembled the *Endo16* expression profile (see BA-Bp curves in figures 3.7 and 3.8). Thus, module A synergistically steps up the activity of modules B (and also G), boosting their output severalfold.

We will quantify this relationship shortly, but first let us consider figure 3.9(B) again. Note that the expression levels of the constructs A-Bp and GFEDCBA-Bp are identical early on, whereas later in development the slope of the expression curve for the full Endo16 construct is approximately equal to that of the B-Bp construct. Module G activity is constant throughout this period, and the repressive factors that bind modules FEDC are present in significant amounts only in cells that do not express Endo16. We therefore concluded that Endo16 upregulation is initially performed by elements within module A alone, but at around 45 hr postfertilization (when module B activity begins to rise) Endo16 upregulation becomes due primarily to module B activity (mediated through module A).

As we will see later, module A is also required for the repressive function of modules F, E, and DC (Yuh et al. 1996). Module A therefore mediates all enhancing and repressive regulatory actions on Endo16. It clearly performs a number of different regulatory functions. So the next question of interest is: How does module A interact with each of the other Endo16 regulatory modules?

3. G and B are amplified by a factor of 4 by module A.

When tested individually, the most distal element, module G, has the capacity to cause expression in the endoderm, as do modules B and A (Yuh and Davidson 1996). However, their functions differ. Module G is relatively weak and appears to act throughout as an ancillary element; module B functions mainly in later development (Yuh and Davidson 1996, Yuh et al. 1996), and after gastrulation it alone suffices to produce accurate midgut expression (Yuh and Davidson 1996).

When modules B and A are physically linked and joined to the Bp (BA construct), the transcriptional output is enhanced relative to the output of either module alone (figures 3.11A and 3.12C; see also plate 6). This provides a classic example of synergism, in this case clearly mediated by interactions between module B and some elements of module A (Yuh et al. 1996). Surprisingly, the output of the BA construct turns out to be exactly modeled by a simple linear amplification of the output of module B over the whole time course

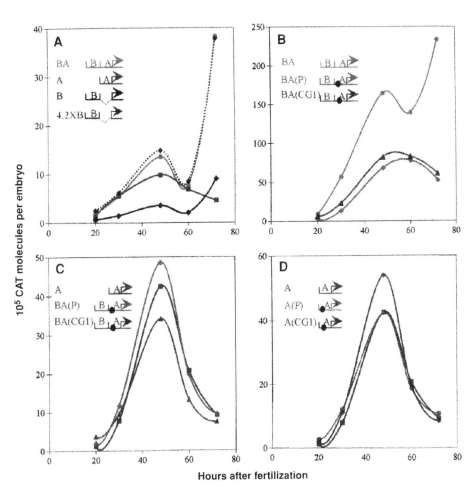

Figure 3.11
Roles of the CG1 and P sites of module A. (A) Time courses of modules A and B and of the BA construct, and demonstration that the time course of BA is a scalar amplification of that of B. The green dotted line shows the time function generated by multiplying the data for the B construct by a factor of 4.2 at each measured time point. (B–D) Time-course data generated by constructs in which CG1 and P sites were mutated. In (B), the effects of CG1 and P mutations on the output of BA are shown. In (C), BA(CG1) and BA(P) are compared with wild-type module A. In (D), A(CG1) and A(P) are compared with wild-type module A. (A) adapted from Yuh et al. (1996). (See plate 6.)

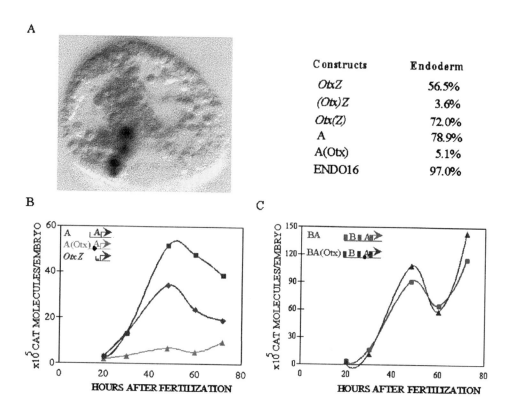

Constructs	Endoderm
OtxZ	56.5%
(Otx)Z	3.6%
Otx(Z)	72.0%
A	78.9%
A(Otx)	5.1%
ENDO16	97.0%

Figure 3.12
Role of the OTX site in generating spatial and temporal expression of module A. (A) Spatial expression of the OTX-Bp construct in the gut endoderm of a gastrula stage embryo. the ability of relevant constructs to specify expression in endoderm (80 to 200 embryos per construct) is shown at the right. Italics denote synthetic oligonucleotides included in the construct, together with the Bp and CAT reporter; parentheses denote that the site named was mutated, either in the normal module A sequence [A(OTX)] or in an oligonucleotide [(OTX)Z]. for simplicity, the additional ectopic expression generated by all constructs lacking negative modules (Yuh and Davidson 1996) is not shown. (B) Temporal and quantitative expression. (C) Lack of effect of OTX mutation on expression of the BA construct.

of module B activity. Thus, as also shown in figure 3.11A (Yuh et al. 1996), the function executed by module A is to "multiply" the output of module B by a constant factor of about 4.

4. F, E, and DC disable A action and thereby shut off *Endo16* transcription.

Under normal conditions, the central regions of the *cis*-regulatory system—that is, modules F, E, and DC [figure 3.4]—have no inherent transcription-enhancing activity. Their role is to prevent ectopic *Endo16* expression in ectodermal cells descendant from blastomeres overlying the vegetal plate (modules F and E) and in skeletogenic cells (module DC). Thus, the positive regulators that bind in modules G, B, and A are initially active in all of these domains as well as in the vegetal plate—that is, roughly in the whole bottom half of the embryo.

Yuh and Davidson (1996) previously discussed an interesting and useful effect of LiCl (lithium cloride) on the three repressor modules. This teratogen expands the domain of endoderm specification at the expense of the adjacent ectoderm (Horstadius 1973, Cameron and Davidson 1997), and concomitantly, it expands the domain of *Endo16* expression (Ransick et al. 1993). LiCl treatment abolishes the negative effect of the three repressor modules, thus increasing the total amount of *Endo16* expressed per embryo. We have used this response, which is easy to assay quantitatively, in some of the following experiments.

Enhanced expression resulting from LiCl treatment requires that both module A and one or more of modules F, E, and DC be present in the construct. Evidence with respect to module F is abstracted from Yuh and Davidson (1996) in the bottom portion of figure 4.4 of Yuh and Davidson (1996). When linked to module F, module A generates a clear LiCl response, whereas modules B and G are blind to LiCl treatment, as is module A in the absence of module F. LiCl treatment causes an enhancement of expression by a factor of 2 to 3. This is attributable to the expansion of the spatial domain of expression at the expense of the ectoderm, as well as to the intensification of expression (Yuh and Davidson, 1996), which was unequivocally observed [figure 3.13, lines indicate standard deviations (SD)]. The key sequence element of module F is a target site that binds a factor of the cyclic adenosine $3,'5'$-monophosphate response element-binding protein (CREB) family. When oligonucleotide C, which includes this target site, is linked to module A(construct CA), it confers LiCl sensitivity almost as well as does the whole of module F (figure 3.13).

Interactions with Module A Elements   Module A functions are mediated through interactions at eight different target sites for DNA-binding proteins (figure 3.4). At least four different factors interact at these sites, only two of which have been cloned: an orthodenticle transcription factor family called SpOtx1 (12, 13) and a protein termed SpGCF1 (Zeller et al. 1995). Because SpGCF1 multimerizes on binding to DNA, it may serve to mediate regionally specific DNA looping (Zeller et al. 1995). There is one

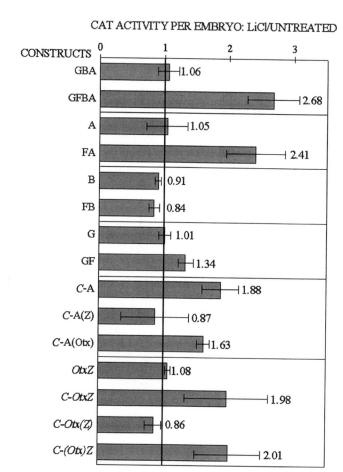

Figure 3.13
Response of expression constructs to LiCl treatment of embryos. The bars and numerical values give the mean ratios of CAT activity measured at 48 hr in samples of 100 LiCl-treated embryos to CAT activity in untreated embryos of the same batch; SDs as indicated. If a construct lacks elements required to respond to LiCl, the ratio will be about 1. Constructs are indicated at the left; as before, capital letters indicate the subelements shown in figure 3.6, parentheses denote mutations of the indicated target sites, and italics indicate oligonucleotides. The C oligonucleotide includes the target site for a CREB factor binding in module F (Michale et al. 1996).

SpGCF1 site in module A, but two more occur downstream in the basal promoter (Bp) region [figure 3.4]. SpGCF1 sites occur commonly in sea urchin genes (Yuh et al. 1994, Kirchhamer and Davidson 1966, Akasaka et al. 1994), and their function is usually manifested in gene transfer experiments as a weak stimulation of transcription (Akasaka et al. 1994, Coffman et al. 1997). Because the properties of this factor are known, the SpGCF1 sites of module A were not studied further. The module A sites labeled CG1, CG2, CG3, and CG4 in figure 3.10 bind the same protein (Yuh et al. 1994). However, the functional role of the CG1 site differs from that of sites CG2 to CG4. The other two sites in A (P and Z in figure 3.10) occur only within module A, as does the OTX site.

1. G and B are additive, and require P and CG1 for the amplification effect.

Two questions arise: What element or elements of module A are specifically responsible for this function? Why does the combined BA output display the characteristics of the time course of module B, rather than some combination of the time courses generated by modules A and B when these are tested in isolation? The answer to both questions arises from a study of the effects of mutating the target sites designated CG1 and P (figure 3.4)

Although not shown here, the rather low and almost constant output of module G is apparently added to that of module B in the complete *cis*-regulatory system, before the synergistic amplification performed by module A (Yuh et al. 1996, Yuh and Davidson 1996). The experiments shown in figure 3.11B–D, demonstrate that these sites provide the obligatory link between module B and module A and thence to the BTA, and that they mediate part of the synergistic enhancement of module B output. If either the CG1 or P site in the BA construct is mutated, the output drops by about half during the early to middle period of development. However, the late rise in activity that is characteristic of both module B and the BA construct is completely absent (figure 3.11B). The BA(CG1) and BA(P) constructs (we adopt the convention that a site in parentheses is mutated) now produce an output over time that is indistinguishable from that of module A alone (figure 3.11C). However, the same mutations have no discernible effect on the output of module A in itself (figure 3.11D). The following conclusions can be drawn:

(a) Both the CG1 and P sites are needed for function, in that a mutation of either produces the same quantitative effect (figure 3.11B and C).

(b) The CG1 and P sites constitute essential sites of module B interaction with module A because in the absence of either, the BA construct behaves exactly as if module B were not present (figure 3.11C).

(c) The CG1 and P sites constitute the exclusive sites of interaction with module B because no other mutations of module A sites have the effects shown in figure 3.11(C).

(d) The CG1 and P sites are dedicated to function (b), because the mutation of these sites has no effect whatsoever on the output of module A.

(e) When linked to module A, module B does not communicate directly with the BTA except through the CG1 and P sites of module A, confirming Yuh et al. (1996) on this point.

2. If G and/or B as well as P and CG1 are active, then they control the activity pattern of Endo16.

Module A functions as a switch. When there is input from module B through the CG1 and P sites, this input is amplified and transmitted to the BTA, and module A no longer has input to the amount of expression. Thus, the input is switched from that of module A to that of module B, even though module A displays activity over much of the same time period. When there is no (or little) module B input, the output has the form of that generated by module A, completely lacking the late rise in the rate of expression. There is no module B input in the physical absence of module B (figure 3.11A and D), when the CG1 or P site of module A is mutated (figure 3.11B), or early in development (Yuh et al. 1996). Module A becomes active first and is probably responsible for installing expression in the vegetal plate shortly after endoderm specification. Thus, as shown in figure 3.9(B), the early output of module A alone equals the output of either the BA (Yuh et al. 1996) or GBA (Yuh and Davidson 1996) construct [because the activity is relatively low at this point, it does not much affect the overall comparison of BA output and the calculated 4.2 times B output; see figure 3.11(A)].

3. In the absence of activity in the B-G pathway, element OTX in module A controls Endo16 activity.

The rise and fall in the time course of expression generated by module A when it is linked either to an SV40 (Yuh and Davidson 1996, Yuh et al. 1996) or its own Bp (Yuh and Davidson 1996, Yuh et al. 1996) (figure 3.14; see also plate 7) depends for its form exclusively on interactions mediated by the OTX site (figure 3.12). This site is also necessary and sufficient to perform the early spatial regulatory function of module A, namely, to direct expression to the primordial endoderm lineages (as well as to the surrounding cell tiers). Double-stranded oligonucleotides that included the Otx and Z sites were linked to a fragment bearing the BTA plus the CG3 or CG4 site (figure 3.4). The CG3 CG4 Bp fragment itself has very low transcriptional activity and virtually no endoderm activity (Yuh and Davidson 1996). In additional constructs, the Z and OTX sites were alternatively altered, and spatial and temporal activity were assessed. In the following discussion, mutations are indicated by parentheses around the affected sites.

In a typical embryo bearing the OTXZ construct and expressing CAT mRNA in endoderm cells (figure 3.12A), the Otx site alone suffices to generate endoderm expression, and the Z site is irrelevant. Furthermore, mutation of four base pairs in the core of the OTX site in an otherwise wild-type module A sequence [A(OTX) construct] abolishes its ability to promote expression in the endoderm (figure 3.12b). This mutation also destroys most of the transcriptional activity of module A. On the other hand, the OTXZ construct is

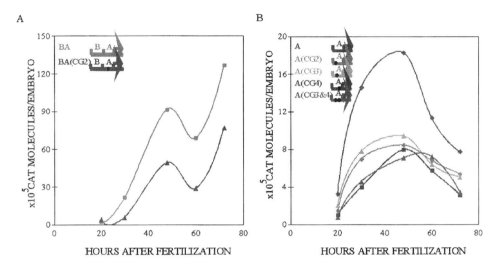

Figure 3.14
Role of CG2, CG3, and CG 4 sites. (A) Effect of mutations of the CG2 site on expression of the BA construct. Although this mutation also destroys part of the Z target site, the latter does not affect expression unless one or more of modules F, E, or DC is present (see text). This mutation prevents binding of the CG factor to the CG2 site. (B) Effect of mutations of CG2, CG3, and CG4 sites on module A activity. (See plate 7.)

able to produce a typical module A expression time course, although of lower amplitude (figure 3.12B). This result is also dependent only on the OTX site, and the Z site of the OTXZ oligonucleotide is again irrelevant. These experiments demonstrate discrete functions of module A, which are mediated exclusively by the Otx site.

As we saw earlier (see figure 3.9), early *Endo16* expression is due to module A activity. Figure 3.12B shows that the OTX and Z sites in A (construct OTXZ) are sufficient to reproduce the early expression pattern of *Endo16*, but a mutation at the Otx site [construct A(OTX)] demolishes this capability. Although the Otx mutation abolishes the activity of module A when it is tested by itself [A(OTX$_m$) construct, figure 3.12(B)], we may predict that this same mutation should not affect the expression of the BA construct over the later period of measurement. This quantitative prediction is confirmed in the experiments summarized in figure 3.12C.

4. The repressive role of F, E, and DC is mediated by element Z in module A.

The interaction mediated by the C oligonucleotide requires the Z site of module A, whereas the Otx site is irrelevant. This is shown both by mutations of the Z and Otx sites in an otherwise intact module A that has been linked to the C oligonucleotide [CA(OTX) and CA(Z) constructs, figure 3.13] and by experiments in which all three sites are represented only as oligonucleotides [C-OTXZ, C-OTX(Z), and C-(OTX)Z constructs, figure 3.13]. These experiments identified the Z site of module A as the element specifically required for functional interactions with module F. Because the other repressor modules, E and DC, have identically to F (Yuh and Davidson 1996), we presume the Z site is used for all of these interactions. It thus appears that the obligatory and exclusive role of the Z site is to transduce the input of these upstream modules. In the absence of module A, these elements have no effect on the output of the *cis*-regulatory system (Yuh and Davidson 1996) (figure 3.13).

The (F, E, DC)-Z repressive pathway disables *Endo16* transcription completely. Therefore, it also succeeds all other regulatory interactions in *Endo16*. Since the effects of the CG amplification pathway and the (F, E, DC)-Z repressive pathway are mutually exclusive, their order in the model is immaterial. However, the CG elements are physically closer to the basal transcription apparatus. Furthermore, the DNA sequence for the CG2 binding site physically overlaps with the binding site for factor Z, thus raising the possibility of competitive binding at this location. We therefore model the CG amplification effect as the last stage in the regulatory interactions. Repression through Z then has two complementary effects. First, all transcription due to B-G or OTX pathways is disabled. Second, any residual transcription due to the CG elements is disrupted by the Z binding factor.

5. CG2, CG3, and CG4 cooperate to boost the transcription rate of *Endo16* by a factor of about 2.

An indication that the CG3 and CG4 sites are directly involved with interactions between module A and the adjacent basal transcription apparatus (BTA) came from a comparison of the activities of the SV40 and endogenous promoters linked to *Endo16 cis*-regulatory elements (Yuh and Davidson 1996, Yuh et al. 1996). When combined with a truncated version of module A lacking the CG3 and CG4 sites, the SV40 promoter was less active by a factor of ~2, but if an oligonucleotide bearing only these sites was inserted, its activity became indistinguishable from that of the endogenous promoter. However, this enhancement was not seen with module B, implying that module A contains an additional element that mediates interaction with CG3 and CG4 sites and is important for communication with the BTA. An obvious guess was that this site is the nearby CG2 site (figure 3.4). Mutation of the CG2 site of the BA construct caused its activity to decrease by half without affecting the characteristic shape of the time course (figure 3.14A). The overall fourfold amplification of the output of module B by module A is attributable to the combined effects of the CG 1 and P sites ($2\times$ amplification) and the CG2 through CG4 sites ($2\times$ amplification). However, unlike the CG1 and P sites, CG2 is not dedicated to synergism with module B, because in contrast (figure 3.11D) the CG2 site affects module A output by the same factor (figure 3.14B). Thus, the CG2 site appears to process both positive upstream inputs. However, CG2 function requires CG3 and CG4 sites as well. Mutation of either or both of these sites has exactly the same effect as does the CG2 mutation on module A output (figure 3.14B). All three sites are evidently used for interaction with the adjacent BTA, in the process of which the level of expression is approximately doubled. The amplification of CG2-CG3-CG4 affects every regulatory path in *Endo16*. Therefore, it must come last in the order of interactions.

3.3.3   Implications of the *Endo16 cis*-Regulatory Model

There is an important difference between the modular organization of the *cis*-regulatory DNA segment of *Endo16,* and the functional modularity exhibited by the *cis*-regulatory logic of figure 3.10, which exhibits additional hierarchy. This can be illustrated with the aid of a few simple manipulations of the logical statements in figure 3.10. The if-then-else relationships described in figure 3.10 can be rewritten in formal logic:

$$\Theta(t) = \gamma \cdot \eta t$$

$$\gamma = (cg_2 \wedge cg_3 \wedge cg_4 + 1)$$

$$\eta_t = \overline{\alpha} \cdot \xi t$$

$$\gamma_\alpha = Z \wedge (F \mathbf{V} E \mathbf{V} DC)$$

$$\overline{\alpha} = \overline{Z \wedge (F \mathbf{V} E \mathbf{V} DC)}$$

$$= \overline{Z} \mathbf{V} \overline{(F \mathbf{V} E \mathbf{V} DC)}$$

$$= \overline{Z} \mathbf{V} (\overline{F} \wedge \overline{E} \wedge \overline{DC})$$

$$\xi_t = (OTX_t \wedge \overline{\varepsilon}_\tau) \varepsilon_t$$

$$\varepsilon_t = \beta \wedge \delta_t$$

$$\beta = 2(pc \wedge g_1)$$

$$\delta_t = B_t + G_t$$

Here, A–F, cg_i, p, and OTX all refer to normalized activity levels and have values between 0 and 1. The relationship $X \wedge Y$ requires that the outcome be a function of the product of the (normalized) activity levels of X and Y. On the other hand, $X \mathbf{V} Y$ indicates that the effects of X and Y are independent, so they may be superimposed (summed). Replacing $\wedge$ with a center dot (multiply), $\mathbf{V}$ with a plus sign (add) and keeping in mind that for our normalized variables $\overline{X} = 1 - X$, we have:

$$\gamma = (cg_2 \cdot cg_3 \cdot cg_4 + 1)$$

$$\eta_t = \overline{\alpha} \cdot \xi_t$$

$$\alpha = Z \cdot (F + E + DC)$$

$$\overline{\alpha} = \overline{Z} + (\overline{F} \cdot \overline{E} \cdot \overline{DC})$$

$$\xi_t = (OTX_t \cdot \overline{\varepsilon}_t) + \varepsilon_t$$

$$\varepsilon_t = \beta \cdot \delta_t$$

$$\beta = 2 \cdot p \cdot cg_1$$

$$\delta_t = B_t + G_t$$

Substituting variables into the expression for Θ gives:

$$\Theta(t) = (cg_2 \cdot cg_3 \cdot cg_4 + 1) \cdot (\overline{Z + \overline{F} \cdot \overline{E} \cdot \overline{DC}}) \cdot \{[\overline{2 \cdot p \cdot cg_1})(B+G)]$$

$$\cdot OTX + 2 \cdot p \cdot cg_1(B+G)\} \tag{3.1}$$

The nested brackets in equation (3.1) reveal the functional modularity and hierarchy of the *Endo.16 cis*-regulatory domain. Figure 3.15 is an illustration of equation (3.1) as a computational graph and illustrates the functional hierarchy more clearly. Activity levels on DNA segments B–G form some of the inputs to the graph. The other inputs are activity levels of the individual elements of module A, each representing a separate functional module. Three distinct regulatory controls are highlighted in color (magenta, red/orange, and green; see plate 8).

3.4 Conclusions

In this chapter we demonstrated the application of simple logic and regression to the characterization of the multiple regulatory functions of *Endo16*. The various functions mediated by module A are precisely encoded in the DNA sequence. Each target site sequence has a specific, dedicated function: Take the site away and the function is abolished; put it back in the form of a synthetic oligonucleotide and it reappears. The properties of the module A regulatory apparatus enable it to process complex informational inputs and to support the modular, polyfunctional organization of the *Endo16 cis*-regulatory system.

Endo16 is a peripheral terminus of a genetic regulatory network that includes all the genes encoding the transcription factors that direct its activity as well as the genes controlling them. But this network is to be considered not only a collection of genes, but also a network of linked regulatory devices that specify the operational logic of embryonic development. We have illustrated how qualitative and quantitative regulatory interactions can be captured with a logical model of gene regulation. We hope the framework and notation developed here will provide a basis for future large-scale models of regulatory gene networks.

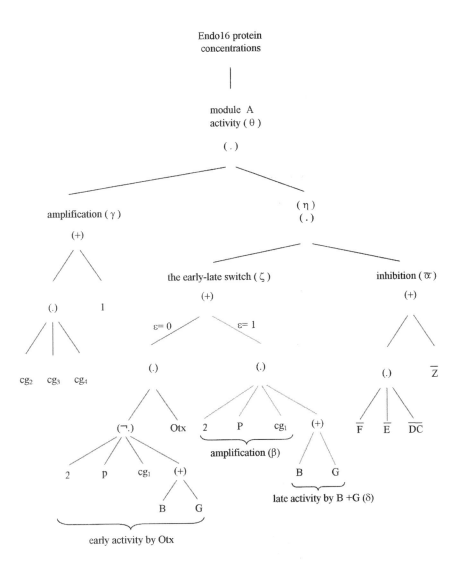

Figure 3.15
The *Endo16 cis*-regulatory computational graph. Each pathway indicates an independent *cis*-regulatory effect.
(See plate 8).

Note

1. This comment also applies to time shifts, as exhibited by one of the graphs of A−Bp·CAT expression in figure 3.7. We avoided the need to register the time axes by comparing only curves from the same experimental batches.

References

Akasaka, K., Frudakis, T. N., Killian, C. E., George, N. C., Yamasu, K., Khaner, O., and Wilt, F. H. (1994). Organization of a gene encoding the spicule matrix protein SM30 in the sea urchin *Strongylocentrotus purpuratus*. *J. Biol. Chem.* 269(13): 20592–20598.

Arnone, M., and Davidson E. H. (1997). The hardwiring of development: organization and function of genomic regulatory systems. *Development* 124: 1851–1864.

Cameron, R. A., and Davidson, E. H. (1997). LiCl perturbs endodermal Veg1 lineage allocation in *Strongylocentrotus purpuratus* embryos. *Dev. Biol.* 187: 236–236.

Coffman, J. A., Kirchhamer, C. V., Harrington, M. G., and Davidson, E. H. (1997). SpMyb functions as an intramodular repressor to regulate spatial expression of CYIIIa in sea urchin embryos. *Development* 124(23): 4717–4727.

Hörstadius, S. (1973). *Experimental Embryology of Echinoderms,* chapter 6. Clarendon Press, Oxford.

Kirchhamer, C. V., and Davidson, E. H. (1996). Spatial and temporal information processing in the sea urchin embryo: modular and intramodular organization of the CYIIIa gene *cis*-regulatory system. *Development* 122: 333–348.

Kirchhamer, C. V., Yuh, C. H., and Davidson, E. H. (1996) Modular cis-regulatory organization of developmentally expressed genes: two genes transcribed territorially in the sea urchin embryo, and additional examples. *Proc. Natl. Acad. Sci. U.S.A.* 93: 9322–9328.

Michael, L. F., Alcorn, J. L., Gao, E., and Mendelson, C. R. (1996). Characterization of the cyclic adenosine 3,′ 5′-monophosphate response element of the rabbit surfactant protein-A gene: evidence for transactivators distinct from CREB/ATF family members. *Mol. Endocrinol.* 10(2): 159–170.

Nocente-McGrath, C., Brenner, C. A., and Ernst, S. G. (1989). Endo16, a lineage-specific protein of the sea-urchin embryo is first expressed just prior to gastrulation. *Mol. Endocrinol.* 136(1): 264–272.

Ransick, A., and Davidson, E. H. (1995). Micromeres are required for normal vegetal plate specification in sea urchin embryos. *Development* 121: 3215–3222.

Ransick, A., and Davidson, E. H. (1998). Late specification of veg1 lineages to endodermal fate in the sea urchin embryo. *Dev. Biol.* 195: 38–48.

Ransick, A., Ernst, S., Britten, R. J., and Davidson, E. H. (1993). Whole-mount *in situ* hybridization shows Endo16 to be a marker for the vegetal plate territory in sea urchin embryos. *Mech. Dev.* 42: 117–124.

Soltysik-Espanola, M., Klinzing, D. C., Pfarr, K., Burke, R. D., and Ernst, S. G. (1994). Endo16, a laqrge multidomain protein found on the surface and ECM of endodermal cells during sea-urchin gastrulation, binds calcium. *Dev. Biol.* 165(1): 73–85.

Yuh, C. H., and Davidson, E. H. (1996). Modular *cis*-regulatory organization of Endo16, a gut-specific gene of the sea-urchin embryo. *Development* 122: 1069–1082.

Yuh, C. H., Ransick, A., Martinez, P., Britten, R. J., and Davidson, E. H. (1994). Complexity and organization of DNA–protein interactions in the 5′ regulatory region of an endoderm-specific marker gene in the sea-urchin embryo. *Mech. Dev.* 47: 165–186.

Yuh, C. H., Moore, J. G., and Davidson, E. H. (1996). Quantitative functional interrelations within the *cis*-regulatory system of *S. purpuratus* Endo16 gene. *Development* 122: 4045–4056.

Zeller, R. W., Griffith, J. D., Moore, J. G., Kirchhamer, C. V., Britten, R. J., and Davidson, E. H. (1995). A multimerizing transcription factor of sea-urchin embryos capable of looping DNA. *Proc. Natl. Acad. Sci. U.S.A.* 92(&): 2989–20598.

4 Trainable Gene Regulation Networks with Applications to *Drosophila* Pattern Formation

Eric Mjolsness

4.1 Introduction

This chapter very briefly introduces and reviews some computational experiments in using trainable gene regulation network models to simulate and understand selected episodes in the development of the fruit fly, *Drosophila melanogaster*. For details, the reader is referred to the papers mentioned in this chapter. A new gene regulation network model is presented that can describe promoter-level substructure in gene regulation.

As described in chapter 1, gene regulation may be thought of as a combination of *cis*-acting regulation by the extended promoter of a gene (including all regulatory sequences) by way of the transcription complex, and of *trans*-acting regulation by the transcription factor products of other genes. If we simplify the *cis*-action by using a phenomenological model that can be tuned to data, such as a unit or other small portion of an artificial neural network, then the full *trans*-acting interaction between multiple genes during development can be modeled as a larger network that can again be tuned or trained to data. The larger network will in general need to have recurrent (feedback) connections since at least some real gene regulation networks do. This is the basic modeling approach taken in 1(199), which describes how a set of recurrent neural networks can be used as a modeling language for multiple developmental processes, including gene regulation within a single cell, cell–cell communication, and cell division. Such network models have been called *gene circuits, gene regulation networks,* or *genetic regulatory networks,* sometimes without distinguishing the models from the actual systems modeled.

In Mjolsness et al. (1991) a number of choices were made in formulating the trainable gene regulation network models; these choices affect the spatial and temporal scales at which the models are likely to be useful. The dynamics were chosen to operate deterministically and continuously in time, on continuous-valued concentrationlike variables, so that the dynamic equations for the network are coupled systems of ordinary differential equations (ODEs). One such form was

$$\tau_i \frac{dv_i}{dt} = g\left(\sum_j T_{ij} v_j + h_i\right) - \lambda_i v_i \tag{4.1}$$

in which v_i is the continuous-valued state variable for gene product i, T_{ij} is the matrix of positive (zero) or negative connections by which one transcription factor can enhance or repress another, and $g()$ is a nonlinear monotonic sigmoidal activation function. When a particular matrix entry T_{ij} is nonzero, there is a regulatory "connection" from gene product j to gene i. The regulation is enhancing if T is positive and repressing if it is negative. If T_{ij} is zero, there is no connection. Figure 4.1 sketches the model, showing a few representative

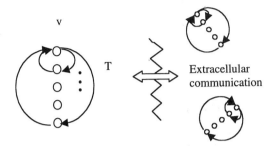

Figure 4.1
Sketch of recurrent analog neural network model for gene regulation networks. A set of analog-valued units v are
connected in a continuous-time, recurrent circuit by a connection matrix T. Communication with circuits in other
cells may require additional connections, e.g., as formulated in Mjolsness et al. (1991).

nonzero connections as arrows between genes represented by open circles. The entire network is localized to a cell, but communicates with other such networks in nearby cells.

Such equations are often stiff, owing to the nonlinear transfer function $g(u)$. Optimizing
the unknown parameters T, λ, and τ has so far proven to be computationally difficult. Special versions of simulated annealing optimization (Lam and Delosme 1988a,b)have been
required for good results, for example, to start from expression patterns derived from a
known model and recover its parameters reliably (Reinitz and Sharp 1995). As discussed
in chapter 1, this kind of training is quite different and much slower than the usual
backpropagation of error training used with feedforward (nonrecurrent) artificial neural
networks. Informatics work on improving this situation could be important.

In addition to the analog circuit model, the framework of Mjolsness et al. (1991) also
proposes a dynamic grammar by which multiple biological mechanisms can be modeled
by networks and then combined into a consistent overall dynamic model. The grammar–
circuit combination has some similarities to hybrid models incorporating discrete event
systems and ODEs. In this way one can, for example, combine intracellular and
intercellular regulation network submodels. The grammar is also suitable for implementation in object-oriented computer simulation programs.

4.2   Three Examples

In this section, some of the literature on trainable gene circuit models that have been fit to
Drosophila gene expression patterns is reviewed. Three applications to pattern formation
are discussed to demonstrate the generality of the methods. First, a model of gap gene expression patterns along the anterior–posterior (A–P) is described. Second, the extension of
this model to incorporate the important pair-rule gene *even-skipped (eve)* and a number of

other improvements is introduced. Finally, a gene circuit model of neurogenesis incorporating nonlinear signaling between nearby cells through the Notch receptor and the Delta ligand is briefly described.

4.2.1   Gap Gene Expression

Gap gene regulation network models can be tuned or "trained" with real gene expression data, and then used to make robust and at least qualitatively correct experimental predictions, as was shown in Reinitz et al. (1995). In that study, the goal was to understand the network of gap genes expressed in bands (domains) along the anterior–posterior axis of the very early embryo (the syncytial blastoderm) of *Drosophila*. This experimental system has the advantage that there are no cell membranes between adjacent cell nuclei, so elaborate cell–cell signaling mechanisms do not need to be modeled. Also, *Drosophila* is an easy species to manipulate genetically because, for example, "saturation mutagenesis"— finding all the genes affecting a particular process—is possible.

Positional information along the A–P axis of the syncytial blastoderm is encoded in a succession of different ways during development. At first the main encoding is a roughly exponential gradient of Bicoid (bcd) protein imposed by the mother fly, along with maternal Hunchback (hb) expression. These provide gene regulation network inputs to the gap genes: *Kruppel (Kr), knirps (kni), giant (gt), tailless (tll),* and *hunchback (hb)* again. These each establish one or two broad domains of expression along the A–P axis. The gap genes then serve as network inputs to the pair-rule genes, including *eve* and *fushi tarazu (ftz),* which establish narrow, precise stripes of expression and precise positional coding. These in turn provide input to segment-polarity genes such as *engrailed* and *wingless,* which are the first to retain their expression pattern into adulthood. For example, *engrailed* is expressed in bands just one cell wide that define the anterior borders of the parasegments. Introductions to the relevant *Drosophila* developmental biology may be found in Lawrence (1992) and Alberts et al. (1994).

An example of a spatial gene expression pattern along the A–P axis of a triple-stained embryo is shown in figure 4.2 (see also plate 9). Here, fluorescently labeled antibodies simultaneously label those nuclei in the syncytial blastoderm expressing *Kruppel, giant,* and *even-skipped.* The first computer experiments with fitting such analog gene regulation nets to real expression data concerned the establishment of the broad gap gene domains (excluding the extreme ends of the A–P axis) from maternally supplied initial conditions, by a gene regulation network in which all gap genes interact with all others and *bcd* provides input to, but does not receive any input from, the gap genes.

Figure 4.3 shows the experimentally observed and model-fitted curves for gap gene expression. They are in qualitative agreement, which is the most that can be expected from the expression data that were available at the time. The extra dip in *gt* expression could not

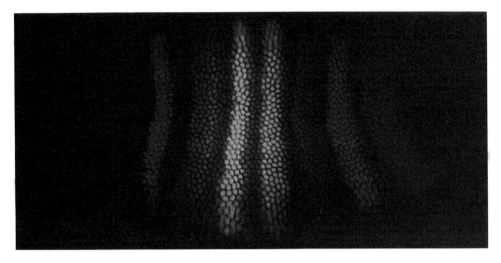

Figure 4.2
Spatial pattern of gene expression in a *Drosophila* syncytial blastoderm for two gap genes and one pair-rule gene. Immunofluorescent staining of nuclei for *Kruppel* (green), *giant* (blue), and *even-skipped* (red). Overlapping areas of *Kr* and *eve* appear yellow, and overlaps of *gt* and *eve* appear purple. Image courtesy of John Reinitz. (See plate 9.)

be predicted by the model, which can be interpreted as an indication of the role of circuit components not included in the model.

The most important predictions of the model concerned the anomalous dose response observed by Driever and Nusslein-Volhard (1988). Figure 4.4 shows the prediction in detail; it may be summarized by saying that positional information for the gap gene system is specified cooperatively by maternal *bcd* and *hb*. This qualitative behavior was observed to be robust over many runs of the simulated annealing parameter-fitting procedure, and therefore taken to be a prediction of the model. Essential features of the cooperative control of positional information by maternal *bcd* and *hb* were verified experimentally in Simpson-Brose et al. (1994). The gap gene model prediction and the experiment occurred independently of one another.

4.2.2   *eve* **Stripe Expression**

Following the gap gene computer experiments, Reinitz and Sharp (1995) went on to perform a detailed study of the gap gene circuit as extended to include the first of the pair-rule genes, *eve* (figure 4.5; see also plate 10). The additional observations that could be included in this model allowed an important milestone to be reached. Not only qualitative behaviors, but also the circuit parameter signs and rough magnitudes became reproducible from one optimization run to another, and some parameters such as connections to *eve*

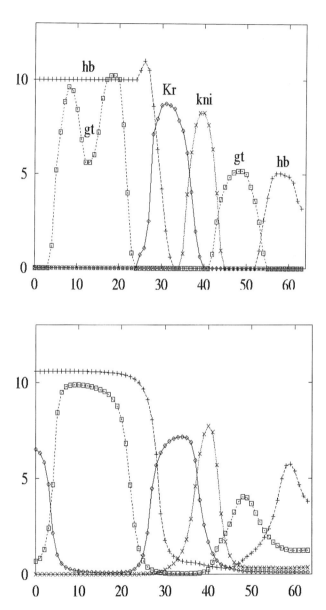

Figure 4.3
Data and model for gap gene circuit. Horizontal axes are nuclei along lateral midline from anterior to posterior. Vertical axe are relative concentrations. (a) Data estimated from immunofluorescence images similar to Figure 4.2 for pairs of gap genes. (b) Output of a circuit model fit to expression data using a nonlinear least-squares criterion and simulated annealing optimization. From Reinitz et al. (1995).

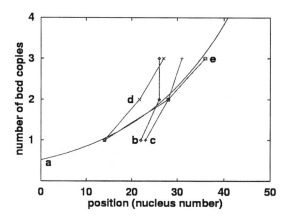

Figure 4.4
Predictions of the model as *Bicoid* dosage is increased: location of selected landmarks along A–P (horizontal) axis vs. number of bcd copies (vertical axis). a, Displacement of a landmark (anterior margin of the *Kr* domain) expected if it were determined by reading off a fixed concentration value of maternal Bicoid protein alone. b–c, Smaller displacement of the same landmark (anterior margin of the *Kr* domain) predicted by the model (a retrodiction). d, Observed anomalously small displacement of a related landmark: the first *eve* stripe, not available in the gap gene model but expected to be offset anteriorly from the *Kr* landmark. Note anomalously high slope compared with a, but as in b,c. e, Prediction: return to the behavior of a if maternal *hunchback* is set equal to zero. From Reinitz et al. (95).

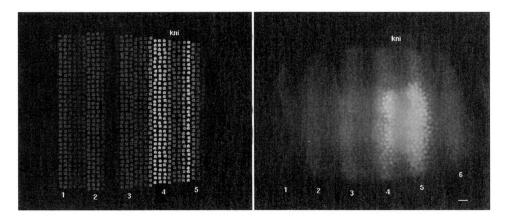

Figure 4.5
Drosophila eve stripe expression in model (right) and data (left). Green, *eve* expression; red, *kni* expression. From Reinitz and Sharp (1995). Figure supplied by courtesy of J. Reinitz and D. H. Sharp. (See plate 10.)

were still more reproducible. Hence, far more could be predicted. For example, the diffusion constant for *eve* was much lower than that for other transcription factors in successful runs. This has an experimental interpretation: *eve* mRNA is expressed in the outer part of each future cell just as the cell membranes are invaginating into the blastoderm embryo, providing an apical obstruction to diffusion.

More important, each of the eight boundaries of the four central stripes of *eve* expression could be assigned a particular gap gene as the essential controller of that boundary. This picture is in agreement with experimental results, with the possible exception of the posterior border of *eve* stripe 3, the interpretation of which is an interesting point of disagreement (Small et al. 1996, Reinitz and Sharp 1995, Frasch and Levine 1995) and a possible focal point for further laboratory and/or computer experiments.

Further experimental understanding of the gap genes' influence on *eve* expression is obtained in Reinitz et al. (1998), where it is shown that the fact that *eve* is unregulated by other pair-rule genes can be understood by the phase of its periodic spatial pattern. No other phase-offset pattern of pair-rule expression (e.g., the phase-shifted patterns of *hairy* or *fushi-tarazu*) can be produced from gap gene input alone. Related work on modeling the gap gene and *eve* system of A–P axis positional information in *Drosophila* includes that by Hamahashi and Kitano (1998).

4.2.3 Neurogenesis and Cell–Cell Signaling

The syncytial blastoderm is very favorable but also very unusual as morphogenetic systems go, because there is no cell membrane interposed between nearby cell nuclei and therefore the elaborate mechanisms of cell–cell signaling do not come into play. However, if we are to model development in its generality, it is essential to include signaling along with gene regulation networks. As a first attempt in this direction, we have modeled the selection of particular cells in an epithelial sheet (later in *Drosophila* development) to become neuroblasts. Virtually the same gene network is thought to be involved in the selection of particular cells in wing imaginal disks to be sensor organ precursors. The essential molecule to add is the Notch receptor, a membrane-bound receptor protein responsible for receiving the intercellular signals that mediate this selection process. It binds to a ligand molecule ("Delta" for this system) on neighboring cells. Recent experiments by Schroeter et al. (1998) indicate that it acts on the nucleus (following activation by a ligand on another cell) by having an intracellular domain cleaved off and transported there. Variants of the Notch receptor occur in many developmental subsystems where a subpopulation of cells must be picked out, in *Drosophila* and homologously across many species.

Marnellos (1997) and Marnellos and Mjolsness (1998a,b) reported computer experiments incorporating both intracellular and intercellular components in a gene regulation network model of neurogenesis. A minimal gene circuit model with lateral inhibition (such

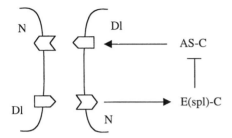

Figure 4.6
A hypothesized minimal gene regulation circuit for lateral inhibition mediated by Notch and Delta. Two neighboring cells express Notch (N) and Delta (Dl) at their surfaces. Notch positively regulated transcription of genes of the *Enhancer-of-split* complex E(spl)-C, which negatively regulate transcription of genes of the *achaete-scute*complex (AS-C), which positively regulate transcription of *Delta*. Curved boundaries are the cell membranes between two neighboring cells. Related circuit diagrams have been suggested elsewhere (e.g., Lewis 1996). Redrawn from figure 6, Heitzler et al. (1996).

as that depicted in figure 4.6) was not quite sufficient to robustly produce the observed patterns of selection. Incorporating a denser intracellular connection matrix and/or the dynamic effects of delamination on the geometry of cell–cell contact area produced better results. However, the "data" to which the fits were made were highly abstracted from real gene expression data, so it is premature to form a unique biological hypothesis from the model. Figure 4.7 shows the resulting model behavior in the case of dense interconnections (see also plate 11). Related work on Notch-mediated signaling in *Drosophila* developmental models includes the appearance of Notch and Delta in the ommatidia model of Morohashi and Kitano (1998).

4.3   Extending the Modeling Framework to Include Promoter Substructure

A very important scientific problem is understanding the influence of promoter substructure on *eve* stripe formation. The *eve* promoter has many transcription factor binding sites, some of which are grouped more or less tightly into promoter elements such as the stripe 2 "minimal stripe element" (MSE2) (Small et al. 1992), or a similar less tightly clustered element for stripes 3 and 7 (Small et al. 1996). As an example of the scientific problems that are raised, if *hb* is an enhancer for MSE2 but an inhibitor for MSE3, what is its net effect on *eve,* and can it change sign (Reinitz et al. 1998)? And how are we to understand the action of "silencer" elements such as the one apparently responsible for long-range repression of *zen* by *dorsal* (Gray et al. 1995)? Such questions point to the need for at least one additional level of complexity in the phenomenological models of gene networks whose application is described above, to describe the substructure of promoters: binding sites,

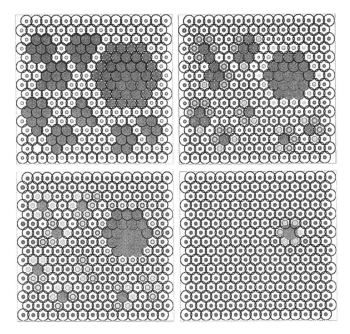

Figure 4.7
Cluster resolution. A circuit "trained" to resolve simple proneural cluster configurations into individual
neuroblasts (or sensory organ precursor cells) is tested on more complex and irregular configurations. In this case,
each cluster was successfully resolved into a single neuroblast, but the large clusters resolved more slowly.
Times: $t = 1$ (top left), $t = 76$ (top right), $t = 106$ (bottom left), and $t = 476$ (bottom right). The figure is similar to
Marnellos and Mjolsness (1998a). Supplied by courtesy of George Marnellos. (See plate 11).

their interactions, and promoter elements. Otherwise the relevant experiments cannot even
be described, let alone predicted, with network models.

In Small et al. (1992) an informal model for activation of MSE2 is suggested. It is activated by *bcd* and *hb* "in concert," and repressed by *gt* anteriorly and *Kr* posteriorly. A simple "analog logic" expression for the activation of MSE2 in terms of variables taking
values in [0,1] might then be (GRN 1998):

$$u_{MSE2} = (bcd + \gamma \times hb)(1 - gt)(1 - Kr)$$

$$v_{MSE2} = g(u_{MSE2}),$$

where γ is a weight on the relative contribution of *hb* compared with *bcd*. A similar
simplified formula for the model of Small et al. (1996) (figure 4.8) for MSE3 could be, for
example,

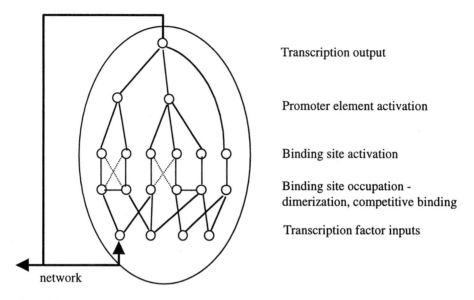

Transcription output

Promoter element activation

Binding site activation

Binding site occupation -
dimerization, competitive binding

Transcription factor inputs

network

Figure 4.8
The hierarchical cooperative activation model fr promoter substructure within a gene "node" in a gene regulation network. Different layers of subnodes have different forms of dynamics. This network could be used to selectively expand some or all of the nodes in figure 4.1, for example, only the *eve* gene in a network for the gap genes and *eve*.

$$u_{MSE3} = D\text{stat}(1 - hb)(1 - kni)$$

$$v_{MSE3} = g(u_{MSE3}).$$

[We omit direct activation of MSE3 by *tailless (tll)* since *tll* represses *kni* (Pankratz et al. 1989), which represses MSE3.] The rate of *eve* transcription would be approximated by a further analog logic formula including a weighted "or" of the MSE activations v_{MSE2} and v_{MSE3}.

The validation or invalidation of such formulas and their interpretation in terms of more detailed models will require a quantitative treatment of the relevant expression data, which are not yet available. It may also lead to fitting the parameters in quantitative network models of a promoter-level substructure within a gene regulation network.

4.3.1   Hierarchical Cooperative Activation

An example of such a gene network model incorporating a promoter-level substructure is introduced here as a hierarchical cooperative activation (HCA) model for the degree of activation of a transcription complex. It at least seems more descriptive of known mechanisms than a previous attempt to derive phenomenological recurrent neural network equations as an approximation of gene regulation dynamics (Mjolsness et al. 1991). An earlier suggestion for including a promoter-level substructure in gene regulation networks is described in Sharp et al. (1993). The present HCA model is more detailed, but has not been fit to any experimental data yet and is therefore quite speculative; perhaps a next stage of successful modeling will include some of the following ingredients.

The basic idea of the model is to use an equilibrium statistical mechanics model [complete with partition functions valid for dilute solutions (Hill 1985)] of "cooperative activation" in activating a protein complex. Such a model can be constructed from the following partition function, which is essentially the Monod–Wyman–Changeux (MWC) model for a concerted state change among subunits (Hill 1985):

$$Z = K \prod_b (1 + K_b v_{j(b)}) + \prod_b (1 + \hat{K}_b v_{j(b)}), \tag{4.2}$$

in which the probability of activation of some complex is determined by relative binding constants for each component b of the complex in the active and inactive states, but there are no other interactions. As before, v_j represents the concentration of gene product j of a gene circuit. In this formula, j is a function of b so that each binding site is specialized to receive only one particular transcription factor. To remove this assumption, one could write instead:

$$Z = K \prod_b (1 + \sum_j A_{bj} K_{bj} v_j) + \prod_b (1 + \sum_j A_{bj} \hat{K}_{bj} v_j), \tag{4.3}$$

where $A = 0$ or 1 specifies which transcription factors may bind to which sites by its sparse nonzero elements. For either expression, K is the relevant binding constant for a binding site when the complex is in its "active" state and $\hat{K}$ is the binding constant when the complex is inactive.

For this partition function, given a global active or inactive state, all binding sites are independent of one another. For example, the components could be the occupants of all the binding sites b within a particular regulatory region of a eukaryotic promoter. This conditional independence leads to the products over the binding sites in the expression for Z. There are two such products because there is one additional bit of global state, which can be "active" or "inactive."

For this model, the probability of activation of the complex under consideration can be calculated and is

$$P = g(u) = \frac{Ku}{1 + Ku} \tag{4.4}$$

$$u = \prod_b \left(\frac{1 + K_b v_{j(b)}}{1 + \hat{K}_b v_{j(b)}} \right);$$

so (if $Kv << 1$),

$$u \approx 1 + \sum_b (K_b - \hat{K}_b) v_{j(b)}.$$

[Further simplifications result if the binding constants K_b specific for a given transcription factor $j(b)$ are all roughly equal to a common value K_j. The final line above suggests a neural networklike approximation for u, although in that regime g could be linearized also.] We will use this model as a building block to construct a more detailed one.

Given the MWC-style model of "cooperative activation," we'd like to use it hierarchically to describe the activation of promoter "modules" or "elements" in terms of transcription factor concentrations, and then to describe the activation of the whole transcription complex in terms of the "concentrations" of active promoter elements, which are proportional to their activities. An additional wrinkle is to allow either monomeric or homodimeric transcription factor binding. (Heterodimers will be introduced later with appropriate notation.) The resulting bare-bones hierarchical model would replace the neural-net activation dynamics (cf. equation 4.1):

$$\tau_i \frac{dv_i}{dt} = [\text{transcribing}]_i - \lambda_i v_i$$

$$[\text{transcribing}]_i = g(u_i)$$

$$u_i = \sum_j T_{ij} v_j + h_i$$

with the two-level model

$$\tau_i \frac{dv_i}{dt} = [\text{transcribing}]_i - \lambda_i v_i$$

$$[\text{transcribing}]_i = g(u_i) = \frac{Ju_i}{1 + Ju_i} \tag{4.5}$$

$$u_i = \prod_{\alpha \in i} \left(\frac{1 + J_\alpha P_\alpha}{1 + \hat{J}_\alpha P_\alpha} \right)$$

(the product is taken over enhancer elements α which regulate gene i) and

$$P_\alpha = g_\alpha(\tilde{u}_\alpha) = \frac{\tilde{K}_\alpha \tilde{u}_\alpha}{1 + \tilde{K}_\alpha \tilde{u}_\alpha} \tag{4.6}$$

$$\tilde{u}_\alpha = \prod_{b \in \alpha} \left(\frac{1 + K_b v_{j(b)}^{n(b)}}{1 + \hat{K}_b v_{j(b)}^{n(b)}} \right).$$

Here $n(b)=1$ for monomers and 2 for homodimers. Note that for this simple feedforward version of the model, the parameters K_b and $\hat{K}_b$ are related to observables

$$f_{\alpha b} = \frac{K_b v_{j(b)}^{n(b)}}{1 + K_b v_{j(b)}^{n(b)}}; \hat{f}_{\alpha b} = \frac{\hat{K}_b v_{j(b)}^{n(b)}}{1 + \hat{K}_b v_{j(b)}^{n(b)}},$$

where $f_{\alpha b}$ is the probability that site $b \in \alpha$ is occupied if α is active, and $\hat{f}_{\alpha b}$ is the probability that site b is occupied if α is inactive. In principle, these quantities could be observed by *in vivo* footprinting. Such observations could be used to evaluate the parameters K_b used in the first expression for u_a for arbitrary inputs v_j. If we are modeling a network rather than a single gene, then some of the quantities listed above require an additional i index.

We have the opportunity to include a few more important biological mechanisms at this point. One is the possibility that, as in the *Endo16* model of Yuh et al. (1998), the hierarchy could go much deeper than two levels—especially if transcription complex formation is a sequential process. Another significant mechanism is competitive binding within a promoter element. This could arise if several transcription factors bind to a single site, as we have formulated earlier, or if binding at one site eliminates the possibility of binding at a nearby site and vice versa. In this case the four-term product of two two-term binding-site partition functions is replaced with one three-term function by excluding the configuration in which both competing sites are occupied:

$$Z_{bb'} = (1 + \sum_j A_{bj} K_{bj} v_j + \sum_k A_{b'k} K_{b'k} v_k) \tag{4.7}$$

$$\hat{Z}_{bb'} = (1 + \sum_j A_{bj} \hat{K}_{bj} v_j + \sum_k A_{b'k} \hat{K}_{b'k} v_k),$$

(where again $A = 0$ or 1 describes which transcription factors bind to which sites by its sparse nonzero elements) with corresponding modifications to the updated equations. Also, homodimeric and heterodimeric transcription factor binding is easy to accommodate with appropriate concentration products in more general one-site and two-site partition functions:

$$Z_b^{(1)} = (1 + \sum_j A_{bj} K_{bj} v_j + \sum_{jk} A_{bjk} K_{bjk} v_j v_k)$$

$$\hat{Z}_b^{(1)} = (1 + \sum_j A_{bj} \hat{K}_{bj} v_j + \sum_{jk} A_{bjk} \hat{K}_{bjk} v_j v_k) \tag{4.8}$$

$$\hat{Z}_{bb'}^{(2)} = (1 + \sum_j A_{bj} K_{bj} v_j + \sum_{jk} A_{bjk} K_{bjk} v_j v_k + \sum_{lm} A_{b'l} K_{b'l} v_l + \sum_{lm} A_{b'lm} K_{b'lm} v_l v_m)$$

$$\hat{Z}_{bb'}^{(2)} = (1 + \sum_j A_{bj} \hat{K}_{bj} v_j + \sum_{jk} A_{bjk} \hat{K}_{bjk} v_j v_k + \sum_{lm} A_{b'l} \hat{K}_{b'l} v_l + \sum_{lm} A_{b'lm} \hat{K}_{b'lm} v_l v_m).$$

Transcription factor trimers and higher-order subcomplexes at adjacent binding sites could be described by suitable generalizations of these expressions, at the cost of introducing more parameters.

Similarly, constitutive transcription factor binding with activation by phosphorylation or dephosphorylation can be described with minor modifications of the appropriate one-site or two-site partition functions. For example, one could use Michaelis–Menton kinetics in steady state for phosphorylation and dephosphorylation, and the one-site dimeric partition functions would become

$$Z_b^{(1)} = [1 + \sum_{jk} A_{bjk,lm} K_{bjk}^{\text{eff}} v_j v_k x_l / (x_l + y_m)] \tag{4.9}$$

$$\hat{Z}_b^{(1)} = [1 + \sum_{jk} A_{bjk,lm} \hat{K}_{bjk}^{\text{eff}} v_j v_k x_l / (x_l + y_m)]$$

where x_l is proportional to the concentration of a kinase for the bound j/k dimer (with proportionality constants depending on the catalytic reaction rates) and y_m is proportional to a corresponding phosphatase concentration. Also, $A_{bjk,lm} \leq A_{bjk}$, so that the extra indices l and m only specify the relevant kinase(s) and phosphatase(s) from a kinase network. For example, MAP kinase-mediated signaling could be modeled as activating a gene regulation network by this mechanism.

In this model formulation, we have omitted lateral interactions other than competitive binding between activation of nearby binding sites. Such interactions could be modeled in the manner of an Ising model. For simplicity we use a tree topology of states and partition functions here.

Given such one-site and two-site partition functions, the overall partition function for a promoter element in terms of its binding sites is

$$Z_\alpha = K_\alpha \left(\prod_{b|C=0} Z_b^{(1)} \right) \left(\prod_{bb'|C=1} Z_{bb'}^{(2)} \right) + \left(\prod_{b|C=0} \hat{Z}_b^{(1)} \right) \left(\prod_{bb'|C=1} \hat{Z}_{bb'}^{(2)} \right). \tag{4.10}$$

Here each binding site competes with at most one other site as determined by the 0/1-valued paramters C_b, $C_{bb'}$. In this picture, silencers are just particular promoter elements with sufficiently strong negative regulation of transcription to veto any other elements.

The full hierarchical cooperative activation (HCK) dynamics then become

$$\tau_i \frac{dv_i}{dt} = [\text{transcribing}]_i - \lambda_i v_i \tag{4.11}$$

$$[\text{transcribing}]_i = g(u_i) = \frac{J u_i}{1 + J u_i},$$

$$u_i = \prod_{\alpha \in i} \left(\frac{1 + J_\alpha P_\alpha}{1 + \hat{J}_\alpha P_\alpha} \right)$$

and

$$P_\alpha = g_\alpha(u_\alpha) = \frac{\tilde{K}_\alpha \tilde{u}_\alpha}{1 + \tilde{K}_\alpha \tilde{u}_\alpha} \tag{4.12}$$

$$\tilde{u}_\alpha = \prod_{b \in \alpha | C = 0} \left(\frac{Z_b^{(1)}}{\hat{Z}_b^{(1)}} \right) \prod_{b,b' \in \alpha | C = 1} \left(\frac{Z_{b,b'}^{(2)}}{\hat{Z}_{b,b'}^{(2)}} \right)'$$

with Z's as before:

$$Z_b^{(1)} = (1 + \sum_j A_{bj} K_{bj} v_j + \sum_{jk} A_{bjk} K_{bjk} v_j v_k) \tag{4.13}$$

$$Z_{bb'}^{(2)} = (1 + \sum_j A_{bj} K_{bj} v_j + \sum_{jk} A_{bjk} K_{bjk} v_j v_k + \sum_{lm} A_{b'l} K_{b'l} v_l + \sum_{lm} A_{b'lm} K_{b'lm} v_l v_m)'$$

and likewise for inactive-module (hatted) Z's and K's. These partition functions encode monomeric, homodimeric, and heterodimeric protein–DNA binding using the various A parameters.

The resulting HCA model (figure 4.8) can describe promoter elements, silencer regions, dimeric and competitive binding, and constitutive transcription factor binding, among other mechanisms. The price is that there are considerably more unknown parameters in the model than in the previous recurrent neural network models—not exponentially as many as in the general N-binding site partition function, but enough to pose a challenge to model-fitting procedures and data sets.

4.4   Conclusion

Gene regulation networks have been used to model several episodes in the development of *Drosophila,* successfully approximating experimental results. A variety of biological mechanisms, including intercellular signaling, can now be included in such models. We have proposed a new version of gene regulation network models for describing experiments that involve promoter substructure, such as transcription factor binding sites or promoter regulatory elements.

Acknowledgments

This work reflects ideas that arose in discussions with Hamid Bolouri, Michael Gibson, George Marnellos, John Reinitz, David Sharp, and Barbara Wold. It was supported by funding from the Office of Naval Research grant N0014-97-1-0422 and the National Aeronautics and Space Administration Advanced Concepts program.

References

Alberts, B., Bray, D., Lewis, J., Raff, M., Roberts, K., and Watson, J. D. (1994). *The Molecular Biology of the Cell,* 3rd ed. pp. 1077–1103. Garland Publishing, New York.

Driever, W., and Nusslein-Volhard, C. (1988). A gradient of Bicoid protein in *Drosophila* embryos. *Cell* 54: 83–93.

Frasch, M., and Levine, M. (1995). Complementary patterns of *even-skipped* and *fushi tarazu* expression involve their differential regulation by a common set of segmentation genes in *Drosophila. Genes Dev.* 2: 981–995.

Gray, S., Cai, H., Barol, S., and Levine, M. (1995). Transcriptional repression in the *Drosophila* embryo. *Phil. Trans. R. Soc. Lond.* B 349: 257–262.

Hamahashi, S., and Kitano, H. (1998). Simulation of *Drosophila* embryogenesis. *Artificial Life VI: Proceedings of the Sixth International Conference on Artificial Life,* C. Adami, R. Belew, H. Kitano, and C. Taylor, eds. pp. 151–160. MIT Press, Cambridge, Mass.

Heitzler, P., Bourouis, M., Ruel, L., Carteret, C., and Simpson, P. (1996). Genes of the *Enhancer of split* and *achaete-scute* complexes are required for a regulatory loop between *Notch* and *Delta* during lateral signaling in *Drosophila. Development* 122: 161–171.

Hill, T. L. (1985). *Cooperativity Theory in Biochemistry: Steady-State and Equilibrium Systems.* Springer Series in Molecular Biology, Springer-Verlag, New York. See especially pp. 6–8, 35–36, and 79–81.

Lam, J., and Delosme, J. M. (1988a). An Efficient Simulated Annealing Schedule: Derivation. Technical Report 8816, Yale University Electrical Engineering Department, New Haven, Conn.

Lam, J., and Delosme, J. M. (1988b). An Efficient Simulated Annealing Schedule: Implementation and Evaluation. Technical Report 8817, Yale University Electrical Engineering Department, New Haven, Conn.

Lawrence, P. (1992). *The Making of a Fly.* Blackwell Scientific, Oxford.

Lewis, J. (1996). Neurogenic genes and vertebrate neurogenesis. *Curr. Opin. Neurobiol.* 6: 3–10.

Marnellos, G. (1997). Gene Network Models Applied to Questions in Development and Evolution. Ph.D. Dissertation, Yale University, New Haven, Conn.

Marnellos, G., and Mjolsness, E. (1998a). A gene network approach to modeling early neurogenesis in *Drosophila.* In *Pacific Symposium on Biocomputing '98,* pp. 30–41. World Scientific, Singapore.

Marnellos, G., and Mjolsness, E. (1998b). Probing the dynamics of cell differentiation in a model of *Drosophila* neurogenesis. In *Artificial Life VI: Proceedings of the Sixth International Conference on Artificial Life,* C. Adami, R. Belew, H. Kitano, and C. Taylor, eds. pp. 161–170. MIT Press, Cambridge, Mass.

Mjolsness, E., Sharp, D. H., and Reinitz, J. (1991). A connectionist model of development. *J. Theor. Biol.* 152: 429–453.

Morohashi, M., and Kitano, K. (1998). A method to reconstruct genetic networks applied to the development of *Drosophila's* eye. In *Artificial Life VI: Proceedings of the Sixth International Conference on Artificial Life,* C. Adami, R. Belew, H. Kitano, and C. Taylor, eds. pp. 72–80. MIT Press, Cambridge, Mass.

Pankratz, M. J., Hoch, M., Seifert, E., and Jackle, H. (1989). Kruppel requirement for Knirps enhancement reflects overlapping gap gene activities in the *Drosophila* embryo. *Nature* 341: 337–340.

Reinitz, J., and Sharp, D. H. (1995). Mechanism of *eve* Stripe formation. *Mech. Devel.* 49: 133–158.

Reinitz, J., Mjolsness, E., and Sharp, D. H. (1995). Model for cooperative control of positional information in *Drosophila* by Bicoid and Maternal Hunchback. *J. Exp. Zool.* 271: 47–56. (First available in 1992 as Los Alamos National Laboratory Technical Report LAUR-92-2942, Los Alamos, N.M.)

Reinitz, J., Kosman, D., Vanario-Alonso, C. E., and Sharp, D. H. (1998). Stripe Forming Architecture of the Gap Gene System, Technical Report LAUR-98-1762, Los Alamos National Laboratory, Los Alamos, N.M.

Schroeter, E. H., Kisslinger, J. A., and Kopan, R. (1998). Notch-1 signaling requires ligand-induced proteolytic release of intracellular domain. *Nature* 393: 382–386.

Sharp, D. H., Reinitz, J., and Mjolsness, E. (1993). Multiscale models of developmental processes. *Open Systems and Information Dynamics* 2(1): 1–10.

Simpson-Brose, M., Triesman, J., and Desplan, C. (1994). Synergy between two morphogens, Bicoid and Hunch-back, is required for anterior patterning in *Drosophila*. *Cell* 78: 855–865.

Small, S., Blair, A., and Levine, M. (1992). Regulation *of even-skipped* stripe 2 in the *Drosophila* embryo. *EMBO J.* 11(11): 4047–4057.

Small, S., Blair, A., and Levine, M. (1996). Regulation of two pair-rule stripes by a single enhancer in the *Drosophila* embryo. *Dev. Biol.* 175: 314–324.

Yuh, C. H., Bolouri, H., and Davidson, E. H. (1998). Genomic *cis*-regulatory logic: experimental and computa-tional analysis of a sea urchin gene. *Science* 279: 1896–1902.

5 Genetic Network Inference in Computational Models and Applications to Large-Scale Gene Expression Data

Roland Somogyi, Stefanie Fuhrman, and Xiling Wen

5.1 Introduction

Analysis of large-scale gene expression is motivated by the premise that the information on the functional state of an organism is largely determined by the information on gene expression (based on the central dogma). This process may be conceptualized as a genetic feedback network in which information flows from gene activity patterns through a cascade of inter- and intracellular signaling functions back to the regulation of gene expression. The genetic network perspective provides a particularly useful conceptual framework for development. Gene sequence information in *cis*-regions (regulatory inputs) and protein coding regions (regulatory outputs; determines biomolecular dynamics) is expanded into spatiotemporal structures defining the organism. Long-term responses of the organism to stimuli, chronic illness (including cancer), and degenerative disorders may also be suitably viewed in terms of genetic network behavior. Owing to the complexity of biological genetic networks, one may pursue a path of crystallizing essential features and principles of network behavior in relatively simple theoretical models, such as Boolean networks.

In order to draw meaningful inferences from gene expression data, it is important that each gene be surveyed under a variety of conditions, preferably in the form of expression time series in response to perturbations. Such data sets may be analyzed using a range of methods with increasing depth of inference. Abstract computational models may serve as a test-bed for the development of these inference techniques. Only in such models can the dynamic behavior of many elements (trajectories, attractors = transient responses, final outcomes) be unequivocally linked to a selected network architecture (wiring and rules = molecular links and interactions). Beginning with cluster analysis and determination of mutual information content, one can capture control processes shared among genes.

The ultimate goal of analysis of expression data is the complete reverse engineering of a genetic network (functional inference of direct causal gene interactions). The challenge here lies in the amount of available data. If the number of genes surveyed is much larger than the number of conditions under which their activity is assessed, it is difficult to meaningfully separate out individual gene functions. Fortunately, large amounts of response data can be easily generated in model networks. Within Boolean networks, we are now able to completely reverse engineer the functional connections using the reverse engineering algorithm (REVEAL).

We have applied genetic network inference techniques to an extensive survey of gene expression in central nervous system (CNS) development. Detailed cluster analysis has uncovered waves of expression that characterize distinct phases of the development of spinal cord and hippocampus. Indeed, recognized functional classes and gene families clearly map to particular expression profiles, suggesting that the definition of pathways may be recast in terms of gene expression clusters. These pathways may be likened to modules in genetic programs. Analysis of gene activity patterns following injury-induced responses in the hippocampus [kainic acid (KA)-induced seizures and excitotoxic cell death] indicates a general recapitulation of developmental programs.

Finally, we have pursued a comprehensive reconstruction of genetic network connections by applying a linear modeling framework to the expression time series data from spinal cord and hippocampus. This inference procedure produces a diagram of putative gene interactions akin to a biochemical pathways chart. While the results from this analysis essentially appear plausible, in-depth experimental testing of model predictions will tell us more about the strengths and weaknesses of this approach. As we establish an appropriate modeling framework, it will be a question of generating experimental data that systematically test theoretical predictions; these data will in turn be used to update and improve the model. Once a significant level of predictive accuracy is achieved, we may hope to effectively identify and target key biological processes relevant to therapeutics and bioengineering.

5.2   Learn to Walk with the Model before You Run with the Data

Multigenic and pleiotropic regulation are the basis of genetic networks. From a strictly reductionistic viewpoint, we may begin by asking which gene underlies a disease or with which molecule does a protein interact. The resulting investigations have shown us that more often than not, several genes contribute to a disease, and that molecules interact with more than one partner (figure 5.1). The challenge now lies in developing methods that allow us to systematically identify the significant connections in these molecular networks. As our understanding of biomolecular networks increases, we may gain deeper insights into which abstract principles allow them to function in a reliable yet flexible and evolvable manner (Kauffman 1993, Savageau 1998).

5.2.1   From Gene Expression to Functional Networks

Gene expression is regulated through a circuit of signaling functions. The expression patterns represent the variables, while the signaling functions are determined by the gene structure (sequence). In this feedback network, the expression of mRNA can be viewed as both the origin and the target of information flow (figure 5.2).

We now have at our disposal technologies that allow us to precisely monitor the activity of many genes in parallel. We hope to use this information to infer major regulatory connections between genes, resulting in a gene interaction diagram reminiscent of traditional

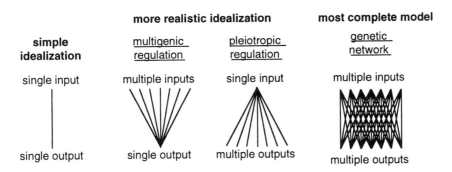

Figure 5.1
From pairwise interactions to networks. (From Somogyi et al. 1997).

biochemical pathways. Data analysis methods are now being generated that allow us to accomplish just that, as shown in the example of proposed functional connections in the GABAergic (γ-aminobutryic acid) signaling gene family (figure 5.3). Using a linear, additive model, D'haeseleer computed a gene interaction diagram based on gene expression time series data from spinal cord development, hippocampal development, and hippocampal injury (D'haeseleer et al. 1999). Coupled with targeted perturbation experiments, such inference procedures may be validated and improved until we can routinely make reliable predictions of important molecular interactions. In the following sections we discuss several approaches for functional inference from expression data. We begin by explaining important principles using simplified, abstract models.

Conceptualizing a Distributed Biomolecular Network   Perhaps a model based on idealized, elemental mechanisms can illustrate the nature of complex behavior. Boolean networks constitute such a model: Each gene may receive one or several inputs from other genes or itself (figure 5.1). Assuming a highly cooperative, sigmoid input/output relationship, a gene can be modeled as an idealized binary element (figure 5.4). The output (time = $t + 1$) is computed from the input (time = t) according to logical or Boolean rules. This is reminiscent of our view of promoter regulation (Arnone and Davidson 1997), in which multiple stimulatory and inhibitory elements act in combination through AND, OR, and NOT logical functions and their nested logical combinations. Of course, this idealization cannot capture features such as "weak" and "strong" promoters, since information on intermediate expression levels is lost in the binary idealization. For calculating Boolean network behavior, one uses discrete and simultaneous (as opposed to sequential) updating of all genes.

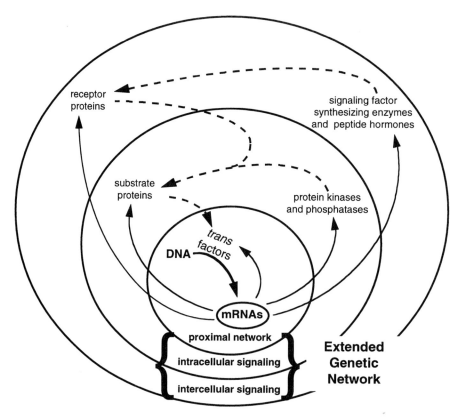

Figure 5.2
The genetic signaling network. The solid lines refer to information flow from primary sources (DNA, mRNA).
The broken lines correspond to information flow from secondary sources back to the primary source (Somogyi
and Sniegoski 1996).

Wiring and Rules Determine Network Dynamics   The dynamics of Boolean networks
of any complexity are determined by the gene connections (wiring) and functional interactions (rules); wiring and rules determine the state-transition tables, just as the laws of molecular interactions determine the behavior of a biological system in time. A simple wiring
diagram representing functional connections between genes is shown in the top panel of
figure 5.5. The lines connect the upper row of output genes to the lower row of input genes.
The number of inputs and the pertaining decimal representation of the logical rule are
shown underneath each wiring diagram (see Somogyi and Sniegoski 1996). As a simple
exercise, the temporal behavior of this simple system (lower panels) can be determined by
hand from the wiring and logical rules, resulting in a short time space pattern or trajectory.

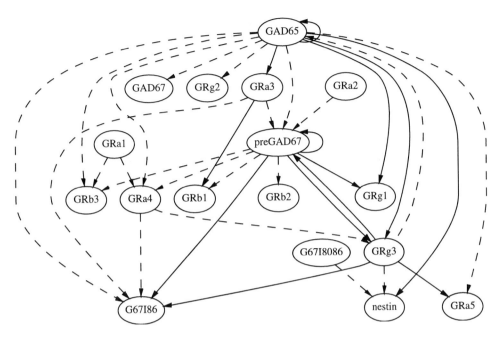

Figure 5.3
Gene interaction diagram for the GABA signaling family inferred from developmental gene expression data (spinal cord and hippocampus data). Solid lines correspond to positive interactions, broken lines suggest inhibitory relationships (D'haeseleer et al. 1999).

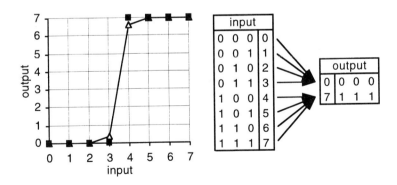

Figure 5.4
From continuous to discrete kinetics. Sigmoid interactions represent a form of data reduction, a "many to one mapping," which has profound implications for systems stability (Somogyi 1998).

Roland Somogyi et al.

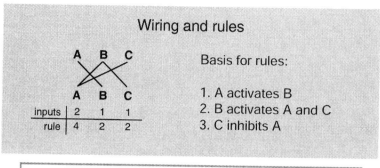

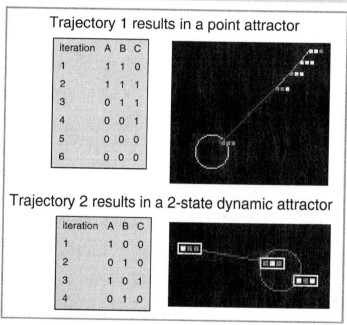

Figure 5.5
Wiring diagram and logical rules for a three-gene network. The temporal pattern of binary gene activity (bottom half) can be easily calculated from the logical rules governing the state transitions (top half). This simple network produces two attractors, or final, repeating state cycles of gene activity (Somogyi and Sniegoski 1996).

Since a system of three binary genes can occupy only a finite number of states (on/off combinations of gene activities), precisely $2^3 = 8$, and there is no boundary on time, states must be cyclically revisited after a certain time period. This repeating pattern is referred to as an *attractor*. The middle panel shows a point attractor, its basin of attraction including five states. The lower panel illustrates a two-state dynamic attractor (repeating pattern), with a basin of three states.

Many States Converge on One Attractor   All Boolean network time series, independent of the size of the network, terminate in specific, repeating state cycle or attractor patterns (consider cell-cycle or circadian rhythms as biological examples of attractors). These can be visualized as basins of attraction graphs (figure 5.6; see also plate 12). The network trajectory (upper right panel) inexorably leads to a final state or state cycle—an attractor, as evident in the repeating pattern of activities. The center graph also shows an alternative way to plot this trajectory; each state of the trajectory is shown as a point (labeled by its timestep number). The labeled series of state transitions is one of many trajectories converging on the repeating, six-state attractor pattern. All of the centripetal trajectories leading to the attractor form the basin of attraction.

This is an important realization from a biological perspective, and applies not only to Boolean networks, but also generally to nonlinear network models, discrete or continuous (Glass 1975, Glass and Kauffman 1973, Thomas 1991, Stucki and Somogyi 1994). We begin to see how it is possible for a complex network of many parts to show predictive and stable behavior; all trajectories are strictly determined, and many states converge on one attractor. This suggests that for living systems small perturbations altering a particular state of the system may have no effect on general system outcome. In addition, a network may have more than one attractor, each with a basin of attraction covering many states. These may be likened to the different cell types in an organism (Kauffman 1993), or diseased cells such as in cancer or in degenerative diseases.

As another example, wound healing corresponds to a trajectory leading from a perturbed state (injury; e.g., the state at the edge of the basin of attraction in figure 5.6) back to the original attractor (repaired tissue; the attractor pattern in figure 5.6). In addition, the cyclical nature of attractors explains why rhythms and cycles are such prevalent features in complex, biomolecular networks. We can now apply the perspectives, terminology, and insights introduced through the study of these simple models to the data sets that are emerging from large-scale gene expression studies (table 5.1). Just as new measurement technology is providing us with a new kind of data on the biological systems level, we need appropriate conceptual frameworks of systems for meaningful interpretation (Somogyi and Sniegoski 1996).

Figure 5.6
Basin of attraction of 12-gene Boolean genetic network model. The trajectory in the top right panel is depicted as a series of connected points in the center, the basin of attraction graph. The selected trajectory is one of many that converge to the central, six-state cycle attractor. While this basin of attraction covers many different states of the system, they all converge on a simple attractor, essentially corresponding to a form of data reduction (Somogyi and Sniegoski 1996, Somogyi 1998). (See plate 12.)

Table 5.1
Terminology for dynamic systems

Network model	Biological system
Wiring	Biomolecular connections
Rules (functions, codes)	Biomolecular interactions
State	Set of molecular activity values; e.g., gene expression, signaling molecule concentration
State transition	Response to previous state or perturbation
Trajectory	Series of state transitions; e.g., differentiation, perturbation response
Attractor	Final outcome; e.g., phenotype, cell type, chronic illness

5.2.2 Inference of Shared Control Processes

In addition to helping us identify organizational and dynamic principles, models provide an exploratory framework for the development of analytical tools. One of the major challenges in molecular signaling biology today lies in extracting functional relationships from gene expression time series. We will show how this can be facilitated using model networks.

Cluster Analysis of a Model Network   We need to address the classification of gene activity patterns as a first step in functional analysis. It seems plausible that similarities in gene expression patterns suggest shared control or common wiring. Therefore, clustering gene expression patterns according to a heuristic distance measure is the first step toward constructing a wiring diagram.

Clustering requires that we define a similarity measure between patterns, such as the Euclidean distance. A gene expression pattern over N time points is a point in N-dimensional parameter space. The Euclidean distance between two points in N dimensions is simply the square root of the sum of the squared distances in each dimension, that is, $D = \sqrt{\Sigma(a_i - b_i)^2}$, for $i = 1$ to N.

In the example in figure 5.7, we used hierarchical clustering based on a Euclidean distance metric to generate dendrograms of model genes grouped according to shared inputs and shared dynamics in a model, logical network of binary genes (Somogyi et al. 1997). Note that the clustering pattern in the dendrogram of functional time series (lower panel) closely resembles the gene groupings according to wiring (upper panel). This analysis suggests that such clustering may be applied to biological activity data for the inference of shared control processes.

Biological Information Flow: It's not about mechanism, it's about information flow
As our experimental technologies become more sophisticated in sensitivity and throughput, we are generating vast amounts of information at all levels of biological organization.

Wiring (Molecular Interaction) Clusters

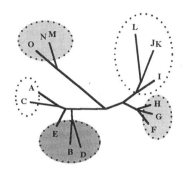

gene	Boolean rule
A	F and H and J
B	G and H and J
C	F and H and I
D	G and H and I
E	H and I and J
F	I and J and K and L and (not G)
G	I and J and K and L and (not O)
H	I and J and K and L
I	J and K and L
J	K and L
K	K or L
L	L or M
M	N or O
N	N and O
O	N and O and (not E)

Trajectory (Gene Expression) Clusters

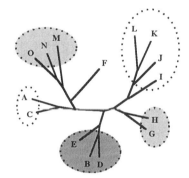

trajectory	I											II				III				IV		
time	1	2	3	4	5	6	7	8	9	10	1	2	3	4	1	2	3	4	1	2	3	4
A	0	0	0	0	0	0	1	0	0	1	1	0	0	1	0	0	0	1	0	0	0	
B	0	0	0	0	0	0	1	1	1	1	1	0	0	1	0	0	0	1	0	0	0	
C	0	0	0	0	0	0	1	0	0	1	1	0	0	1	1	0	0	1	0	0	0	
D	0	0	0	0	0	0	1	1	1	1	1	0	0	1	1	0	0	1	1	0	0	
E	0	0	0	0	0	0	1	1	1	0	1	0	0	0	0	0	0	0	0	0	0	
F	0	0	0	0	0	1	0	0	0	1	0	0	0	1	0	0	0	1	0	0	0	
G	0	0	0	0	0	1	1	1	1	1	0	0	0	1	0	0	0	1	0	0	0	
H	0	0	0	0	0	1	1	1	1	1	0	0	0	1	0	0	0	1	0	0	0	
I	0	0	0	0	1	1	1	1	1	1	0	0	0	1	0	0	0	1	0	0	0	
J	0	0	0	1	1	1	1	1	1	1	0	0	0	0	0	0	0	0	0	0	0	
K	0	0	0	1	1	1	1	1	1	0	0	0	0	0	0	0	0	0	0	0	0	
L	0	0	1	1	1	1	1	1	1	0	0	0	0	0	0	0	0	0	0	0	0	
M	0	1	0	0	0	0	0	0	0	0	0	0	0	0	0	0	0	0	0	0	0	
N	0	0	0	0	0	0	0	0	0	0	0	0	0	0	0	0	0	0	0	0	0	
O	1	0	0	0	0	0	0	0	0	0	0	0	0	0	0	0	0	0	0	0	0	

Figure 5.7
Inference of shared control through Euclidean cluster analysis. (Top) Clustering according to shared inputs. (Bottom) Clustering according to similarity in temporal behavior (Somogyi et al., 1997).

The challenge lies in inferring important functional relationships from these data. This problem is becoming acute because we are preparing to generate molecular activity data (e.g., mRNA and protein expression) for organisms in health and disease. This new perspective suggests that we should treat the organism as an information processing system. But how can we conceptualize information flow in living systems and how can we quantify this question? The next sections deal with this issue.

Information Can be Quantified: Shannon Entropy (H)   From a mathematical-statistical standpoint, information can be quantified as the Shannon entropy (named after Claude Shannon, the founder of information theory; Shannon and Weaver 1963). The Shannon entropy (H) can be calculated from the probabilities of occurrences of individual or combined events as shown below:

a

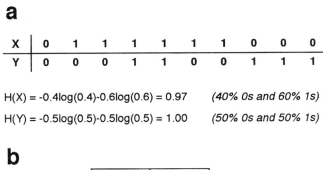

X	0	1	1	1	1	1	1	0	0	0
Y	0	0	0	1	1	0	0	1	1	1

H(X) = -0.4log(0.4)-0.6log(0.6) = 0.97   *(40% 0s and 60% 1s)*

H(Y) = -0.5log(0.5)-0.5log(0.5) = 1.00   *(50% 0s and 50% 1s)*

b

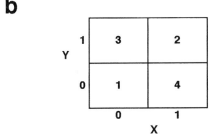

H(X,Y) = -0.1log(0.1)-0.4log(0.4)-0.3log(0.3)-0.2log(0.2) = 1.85

Figure 5.8
Calculation of Shannon entropy from a series of observations. (a) Calculation of *H* from the probabilities of occurrences for parameters *X* and *Y*. (b) Calculation of *H* for particular co-occurrences of *X* and *Y* from measurement series in (a) (Liang et al. 1998).

$$H(X) = -\Sigma p_x \log p_x$$

$$H(Y) = -\Sigma p_y \log p_y$$

$$H(X,Y) = -\Sigma p_{x,y} \log p_{x,y}.$$

The determination of *H* is illustrated in a numerical example (figure 5.8). A single element is examined in figure 5.8(a). Probabilities (*p*) are calculated from the frequency of on/off values of *X* and *Y*. The distribution of value pairs is shown in figure 5.8(b). *H(X,Y)* is calculated from the probability of occurrence of combinations of *x, y* values over all measurements.

The Shannon entropy is maximal if all states are equiprobable. The Shannon entropies for a two-state information source (0 or 1) are graphed in figure 5.9. Since the sum of the state probabilities must be unity, $p(1) = 1-p(0)$ for two states. Note that the maximal Shannon entropy, or information content, occurs when the occurrence of 0 and 1 are equiprobable, that is, $p = 0.5$. Maximal information content is essentially equivocal to maximal diversity.

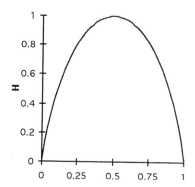

Figure 5.9
Shannon entropy vs. probability of event. The Shannon entropy is maximal when all events are equiprobable, corresponding to maximal diversity (Liang et al. 1998).

Mutual Information: The Information (Shannon Entropy) Shared by Nonindependent Elements   Beyond looking at an absolute information measure, we would like to find a way to quantify relationships between different parameters in terms of overlapping diversity, or "mutual information." Figure 5.10 illustrates the relationships between the information content of individual and combined, nonindependent information sources using Venn diagrams. Mutual information, *M,* is defined as the sum of the individual entropies minus the entropy of the co-occurrences:

$$M(X,Y) = H(X) + H(Y) - H(X,Y).$$

In each case, add the shaded portions of both squares in the figure to determine one of the following: $[H(X) + H(Y)]$, $H(X,Y)$, and $M(X,Y)$. The small corner rectangles represent information that X and Y have in common, that is, $M(X,Y)$. $H(Y)$ is shown smaller than $H(X)$ and with the corner rectangle on the left instead of the right to indicate that X and Y are different, although they have some mutual information.

5.2.3   A Candidate Boolean Network for Reverse Engineering

Can measures of individual and mutual information be used to quantify causal relationships between elements fluctuating in a dynamic network? Using the example of figure 5.11, we will attempt to reconstruct the wiring diagram (figure 5.11 a), and the Boolean rules (figure 5.11b) from the state transition table (figure 5.11c). The state transition table contains the complete dynamic output of this network (the input column shows that all states at time $= t$, outputs (prime) correspond to the matching states at time $= t + 1$), which is analogous to perturbation–response measurements in an experimental biological system.

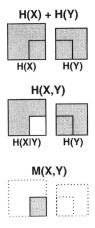

Figure 5.10
Venn diagrams of information relationships. The areas shaded in gray correspond to : top panel, H(X) + H(Y); middle panel, H(X, Y); lower panel, M(X, Y) (Liang et al. 1998).

a

A B C

A' B' C'

b

A' = B
B' = A or C
C' = (A and B) or (B and C) or (A and C)

c

input			output		
A	**B**	**C**	**A'**	**B'**	**C'**
0	0	0	0	0	0
0	0	1	0	1	0
0	1	0	1	0	0
0	1	1	1	1	1
1	0	0	0	1	0
1	0	1	0	1	1
1	1	0	1	1	1
1	1	1	1	1	1

Figure 5.11
Target Boolean network for reverse engineering. (a) The network wiring and (b) logical rules determine (c) the dynamic output. The challenge lies in inferring (a) and (b) from (c) (Liang et al. 1998).

The Principle Behind REVEAL   Figure 5.12 details the steps taken in REVEAL (Liang et al. 1998) for the inference of functional connections and rules from the dynamics (i.e., state transition tables) of the Boolean network shown in figure 5.11. H's and M's are calculated from the time series or lookup tables according to the definitions (shaded in Figure 5.10). The wiring of the example Boolean network can be inferred from the state transition table using progressive M-analysis (left, odd steps). If the mutual information of an input or a combination of inputs with a single output equals the entropy of that output, then that particular output is determined by this input combination. Once the inputs (wiring) to a gene are known, one can construct the rule table by simply matching the states of the inputs to those of the output from the state transition table (right, even steps).

Inference from Incomplete Time Series or State Transition Tables.   REVEAL will quickly find a minimal solution for a Boolean network given any set of time series (figure 5.13). For $n = 50$ (genes) and $k = 3$ or less (number of inputs per gene), the correct or full solution can be unequivocally inferred from 100 state transition pairs. Note that for $n = 50$ and $k = 3$, only a vanishingly small fraction (~ 100) of all possible state transitions ($2^{50} \sim 10^{15}$ states!) is required for reliable inference of the network wiring and rules. This demonstrates that only a relatively small sampling of the dynamic space is required for meaningful inference. In principle, analogous state-transition data sets can be acquired from biological experiments, and therefore the principle of REVEAL may be applicable for inferring functional connections in biomolecular networks.

5.3   You're on Your Own: Start Running with the Large-Scale Expression Data

5.3.1   Functional Inference from Large-Scale Gene Expression Data

Gene expression patterns contain much of the state information of the genetic signaling network and can be measured experimentally. We are facing the challenge of inferring or reverse engineering the internal structure of this genetic network from measurements of its output. This will require high precision in data acquisition and sufficient coherence among the data sets, as found in measurements of time series and perturbation responses.

5.3.2   High-Precision, High-Sensitivity Assay

There are several methods for measuring gene expression patterns, each with its advantages and disadvantages (table 5.2). In our exploration of reverse transcription polymerase chain reaction (RT-PCR) technology, we have found that it goes beyond simple hybridization-based expression assays in that it enables (1) absolute, quantitative analysis (which is necessary for reference expression), (2) distinction of closely related genes within a gene family that differ in only a few base pairs, (3) distinction of and search for alternative splicing variants, (4) detection in the 10 molecule per sample range (which is required for small

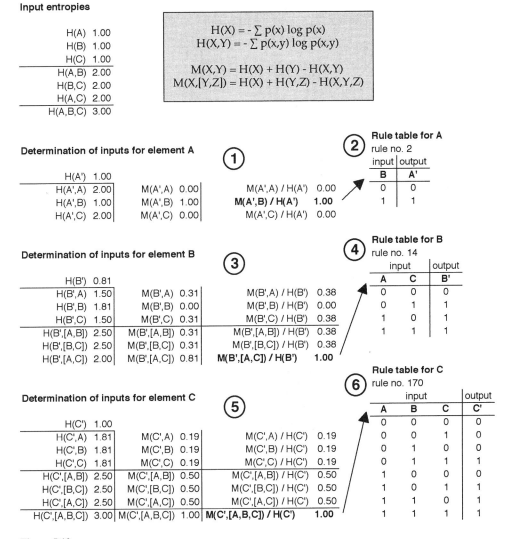

Input entropies

H(A)	1.00
H(B)	1.00
H(C)	1.00
H(A,B)	2.00
H(B,C)	2.00
H(A,C)	2.00
H(A,B,C)	3.00

$$H(X) = - \sum p(x) \log p(x)$$
$$H(X,Y) = - \sum p(x,y) \log p(x,y)$$

$$M(X,Y) = H(X) + H(Y) - H(X,Y)$$
$$M(X,[Y,Z]) = H(X) + H(Y,Z) - H(X,Y,Z)$$

Determination of inputs for element A   ①   ②

Rule table for A
rule no. 2

input	output
B	A'
0	0
1	1

H(A')	1.00				
H(A',A)	2.00	M(A',A)	0.00	M(A',A) / H(A')	0.00
H(A',B)	1.00	M(A',B)	1.00	**M(A',B) / H(A')**	**1.00**
H(A',C)	2.00	M(A',C)	0.00	M(A',C) / H(A')	0.00

Determination of inputs for element B   ③   ④

Rule table for B
rule no. 14

input		output
A	C	B'
0	0	0
0	1	1
1	0	1
1	1	1

H(B')	0.81				
H(B',A)	1.50	M(B',A)	0.31	M(B',A) / H(B')	0.38
H(B',B)	1.81	M(B',B)	0.00	M(B',B) / H(B')	0.00
H(B',C)	1.50	M(B',C)	0.31	M(B',C) / H(B')	0.38
H(B',[A,B])	2.50	M(B',[A,B])	0.31	M(B',[A,B]) / H(B')	0.38
H(B',[B,C])	2.50	M(B',[B,C])	0.31	M(B',[B,C]) / H(B')	0.38
H(B',[A,C])	2.00	M(B',[A,C])	0.81	**M(B',[A,C]) / H(B')**	**1.00**

Determination of inputs for element C   ⑤   ⑥

Rule table for C
rule no. 170

input			output
A	B	C	C'
0	0	0	0
0	0	1	0
0	1	0	0
0	1	1	1
1	0	0	0
1	0	1	1
1	1	0	1
1	1	1	1

H(C')	1.00				
H(C',A)	1.81	M(C',A)	0.19	M(C',A) / H(C')	0.19
H(C',B)	1.81	M(C',B)	0.19	M(C',B) / H(C')	0.19
H(C',C)	1.81	M(C',C)	0.19	M(C',C) / H(C')	0.19
H(C',[A,B])	2.50	M(C',[A,B])	0.50	M(C',[A,B]) / H(C')	0.50
H(C',[B,C])	2.50	M(C',[B,C])	0.50	M(C',[B,C]) / H(C')	0.50
H(C',[A,C])	2.50	M(C',[A,C])	0.50	M(C',[A,C]) / H(C')	0.50
H(C',[A,B,C])	3.00	M(C',[A,B,C])	1.00	**M(C',[A,B,C]) / H(C')**	**1.00**

Figure 5.12
REVEAL principle: progressive *M* analysis. By determining the mutual information between input/output pairs (odd-numbered steps), we can unequivocally find the combinations of elements that share sufficient information to predict every output. The rule tables can be read directly from the state transition table once the input/output wiring has been determined (Liang et al. 1998).

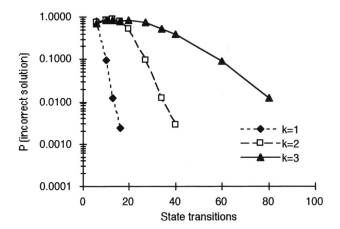

Figure 5.13
Performance of REVEAL. For random networks of $n = 50$ binary genes and $k = 1, 2,$ and 3 inputs per gene, we calculated the probability of finding the incorrect solution (100 trial runs) as a function of the number of state transitions used as input data. Note that the likelihood of finding an incorrect solution vanishes quickly with increasing training data (Liang et al. 1998).

specimens), and (4) an unsurpassed dynamic range covering 6–8 orders of magnitude (figure 5.14) (Somogyi et al. 1995).

RT-PCR is an optimal technology for the sensitive and precise measurement of temporal gene expression patterns for multiple genes expressed in parallel. In this technique, a plasmid-derived RNA standard serves as an internal control. Measurement scales linearly with RNA copy number on log scales (Somogyi et al. 1995). RT-PCR is described in more detail in the next subsection. This assay combines high sensitivity (down to 10 template molecules) with a superb dynamic range (covering up to 8 orders of magnitude). The challenge today lies in scaling up to cover thousands of genes; efforts in robotics and miniaturization are currently under way to combine maximal sensitivity and dynamic range with throughput.

RT-PCR Analysis of Gene Expression in Developing Rat Central Nervous System
RT-PCR, RNA is isolated and purified from the tissue and is reverse transcribed along with the control RNA sample. The resulting sample and control cDNAs are then amplified using PCR, with sample and control in the same reaction. The amplified cDNAs are subjected to polyacrylamide gel electrophoresis (PAGE). Using densitometry, the ratio of the sample band to the control band is determined and used as a measure of relative gene expression within the set of time points. Figure 5.15 demonstrates the results of this method for five genes expressed at nine stages during the development of the rat spinal cord (Wen et al. 1998). Each gel contains bands for both sample and control (150 bp). The triplicate ratios

Table 5.2
Comparison of methods for measuring gene expression

QUALITY	TECHNIQUE				
	Northern	RNAse protection	Monitored PCR ("TaqMan")	Membrane arrays and microarrays[a]	Quantitative RTpcr (QRTPCR)
Sensitivity	Good	Good	Excellent	Poor[b]	Excellent
Absolute quantification	Good	Excellent	Good	Poor	Excellent
Dynamic range	2 orders of magnitude	3 orders of magnitude	5 orders of magnitude	2 orders of magnitude	6–8 orders of magnitude
Distinction of gene family members	Good	Good	Good	Poor	Excellent
Distinction of splice variants	Good	Good	Poor	Poor	Good
Simultaneous detection of multiple genes (high throughput)	Poor (<100)	Poor (<100)	Good (about 1000)	Arrays: good (about 1000) Microarrays: excellent (about 10,000)	Good (about 1000)
Adaptability to automation	Poor	Poor	Good	Good	Good
Cross-platform utility	Poor	Poor	Excellent	Good	Excellent
Reference	Yamaguchi et al. (1994)	Gautreau and Kerdelhue (1998)	Heid et al. (1996)	Bowtell (1999)	Somogyi et al. (1995)

[a]DNA hybridization microarrays fall into two general categories: those that use cDNA fragments (Shalon et al. 1996) and those that use oligonucleotides (Lipshutz et al. 1999). Both are useful as survey tools.
[b]Linear amplification of transcripts by T7 polymerase methods (Phillips and Eberwine 1996) is likely to signicantly improve the sensitivity of microarrays.

of sample to control bands are averaged for each time point and used to construct the corresponding graphs on the right in the figure. The graphs show relative expression levels over time for each gene.

The data can then be used to construct a graph (a gene expression matrix) of the type shown in figure 5.16; see also plate 13) by normalizing the data to the maximal expression value for each gene over the set of time points (Wen et al. 1998). This permits us to compare temporal gene expression patterns for different genes, both within and between recognized functional categories. It must be stressed that while this technique is useful for comparing the temporal expression pattern of one gene with another, absolute expression values at individual time points may not be compared between genes. The reason for this is that PCR does not operate on all cDNA sequences with equal efficiency. Only by using quantitative PCR can we obtain a measure of absolute quantities of cDNA, and compare genes at individual time points (Somogyi et al. 1995).

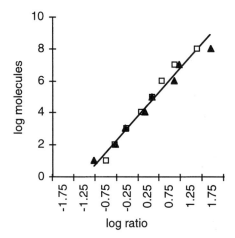

Figure 5.14
RT-PCR sensitivity and dynamic range. The ratio of internal control over sample PCR product is plotted against
the number of sample starting molecules. The assay can accurately measure within a range of 10 to 1 million target molecules (Somogyi et al. 1995).

5.3.3   Gene Expression Matrix of Rat Spinal Cord Development

By measuring relative gene expression levels for multiple genes at multiple time points, we
obtain a measure of the dynamic output of a genetic network. The time points for figure
5.16 were selected because of their appropriateness to the time scale of rat central nervous
system development. Time points range from embryonic day 11 (E11, when spinal cord
development begins) to adult or postnatal day 90 (P90). The darkest color in the figure indicates the highest expression level detected; white indicates undetectable levels. For each
gene, expression levels are normalized to maximal expression for that gene. Each data
point (colored square) is the average of results from three animals. Although the genes we
selected are only a tiny fraction of the total number of genes expressed during spinal cord
development, they are representative of the different types of proteins that are expressed
during the differentiation of tissue. The neurotransmitter- and peptide-related genes encode the proteins and peptides responsible for intercellular communication during CNS
development. We have focused on intercellular signaling genes since these are directly responsible for differentiation in a multicellular organism.

Inference of Shared Control Processes   Genes with similar temporal expression patterns may share common genetic control processes and may therefore be related functionally (see figure 5.7). Clustering gene expression patterns according to a similarity or
distance measure is the first step toward constructing a wiring diagram for a genetic

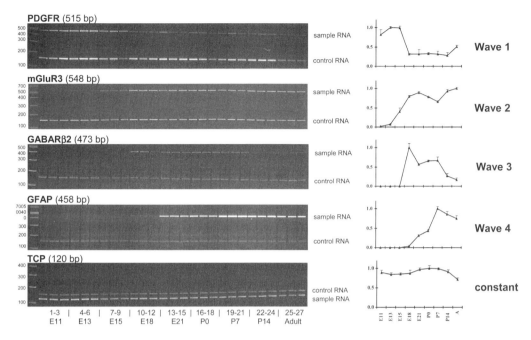

Figure 5.15
RT-PCR/PAGE analysis of five genes over nine developmental stages. Target gene products are products corresponding to the top layer of bands, while the internal standard control RNA product is shown in the lower bands of the PAGE chromatograms. The normalized ratio of sample to control product over the developmental time points in shown in the graphs on the right (Wen et al. 1998).

network. Such diagrams should permit the development of new hypotheses concerning gene interactions during recovery from injury and disease, or during therapeutic drug treatments.

Euclidean Cluster Analysis of Gene Expression Dynamics in Development Analogous to the previous theoretical examples (figure 5.7), we have used hierarchical clustering based on a Euclidean distance metric to group genes according to similarities in their temporal expression patterns. This method is similar to clustering according to positive linear correlations. Figure 5.17 shows a tree generated by the Fitch-Margoliash clustering algorithm using the Euclidean distance matrix for 112 genes expressed in the rat spinal cord. Similarities in temporal expression patterns are indicated by common branch points. According to visual inspection of the tree, the genes appear to cluster into groups, or "waves." Each wave (shown as an inset) corresponds to an average pattern for all the genes of the corresponding cluster: wave 1 genes are expressed at a high level early in development and then decrease in expression toward adult; wave 2 genes are expressed at low levels early

Neuro-glial Markers

Markers

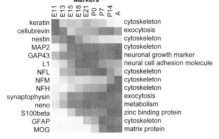

	E11	E13	E15	E18	E21	P0	P7	P14	A	
keratin										cytoskeleton
cellubrevin										exocytosis
nestin										cytoskeleton
MAP2										cytoskeleton
GAP43										neuronal growth marker
L1										neural cell adhesion molecule
NFL										cytoskeleton
NFM										cytoskeleton
NFH										cytoskeleton
synaptophysin										exocytosis
neno										metabolism
S100beta										zinc binding protein
GFAP										cytoskeleton
MOG										matrix protein

Neurotransmitter Metabolizing Enzymes

GAD65		neurotransmitter synthesis
pre-GAD67		neurotransmitter synthesis
GAD67		neurotransmitter synthesis
G67I80 / 86		neurotransmitter synthesis
G67I86		neurotransmitter synthesis
GAT1		neurotransmitter transport
ChAT		neurotransmitter synthesis
AChE		neurotransmitter degradation
ODC		neurotransmitter synthesis
TH		neurotransmitter synthesis
NOS		neurotransmitter synthesis

Neurotransmitter Receptors

GABA-A Receptors

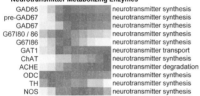

GRa1		receptor / ion channel
GRa2		receptor / ion channel
GRa3		receptor / ion channel
GRa4		receptor / ion channel
GRa5		receptor / ion channel
GRb1		receptor / ion channel
GRb2		receptor / ion channel
GRb3		receptor / ion channel
GRg1		receptor / ion channel
GRg2		receptor / ion channel
GRg3		receptor / ion channel

Glutamate Receptors

mGluR1		G-protein coupled receptor
mGluR2		G-protein coupled receptor
mGluR3		G-protein coupled receptor
mGluR4		G-protein coupled receptor
mGluR5		G-protein coupled receptor
mGluR6		G-protein coupled receptor
mGluR7		G-protein coupled receptor
mGluR8		G-protein coupled receptor
NMDA1		receptor / ion channel
NMDA2A		receptor / ion channel
NMDA2B		receptor / ion channel
NMDA2C		receptor / ion channel
NMDA2D		receptor / ion channel

Acetylcholine Receptors

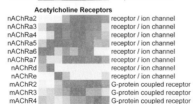

nAChRa2		receptor / ion channel
nAChRa3		receptor / ion channel
nAChRa4		receptor / ion channel
nAChRa5		receptor / ion channel
nAChRa6		receptor / ion channel
nAChRa7		receptor / ion channel
nAChRd		receptor / ion channel
nAChRe		receptor / ion channel
mAChR2		G-protein coupled receptor
mAChR3		G-protein coupled receptor
mAChR4		G-protein coupled receptor

Serotonin Receptors

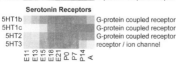

5HT1b		G-protein coupled receptor
5HT1c		G-protein coupled receptor
5HT2		G-protein coupled receptor
5HT3		receptor / ion channel

E11	E13	E15	E18	E21	P0	P7	P14	A

Peptide Signaling

Neurotrophins

	E11	E13	E15	E18	E21	P0	P7	P14	A	
NGF										signaling peptide
NT3										signaling peptide
BDNF										signaling peptide
CNTF										signaling peptide
trk										receptor tyrosine kinase
trkB										receptor tyrosine kinase
trkC										receptor tyrosine kinase
CNTFR										receptor tyrosine kinase

Heparin-binding Growth Factors

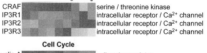

MK2		signaling peptide
PTN		signaling peptide
GDNF		signaling peptide
EGF		signaling peptide
bFGF		signaling peptide
aFGF		signaling peptide
PDGFa		signaling peptide
PDGFb		signaling peptide
EGFR		receptor tyrosine kinase
FGFR		receptor tyrosine kinase
PDGFR		receptor tyrosine kinase
TGFR		receptor tyrosine kinase

Insulin / IGF

Ins1		signaling peptide
Ins2		signaling peptide
IGF I		signaling peptide
IGF II		signaling peptide
InsR		receptor tyrosine kinase
IGFR1		receptor tyrosine kinase
IGFR2		receptor tyrosine kinase

Diverse

Intracellular Signaling

CRAF		serine / threonine kinase
IP3R1		intracellular receptor / Ca^{2+} channel
IP3R2		intracellular receptor / Ca^{2+} channel
IP3R3		intracellular receptor / Ca^{2+} channel

Cell Cycle

cyclin A		cell cycle regulator
cyclin B		cell cycle regulator
H2AZ		DNA binding protein
statin		cell cycle regulator

Transcription Factor

cjun		immediate early gene
cfos		immediate early gene
Brm		transcriptional facilitator
TCP		transcriptional facilitator

Novel / EST

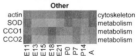

SC1		unknown
SC2		unknown
SC6		unknown
SC7		unknown
DD63.2		unknown

Other

actin		cytoskeleton
SOD		metabolism
CCO1		metabolism
CCO2		metabolism

E11	E13	E15	E18	E21	P0	P7	P14	A

and then plateau, etc. One cluster, "constant," contains the genes whose expression levels are relatively invariant over the time course. Within each wave, the genes may be said to share the same expression kinetics over the time course E11 to P90. The generation of Euclidean distance trees may provide clues as to which genes share a common genetic control process. For example, the members of wave 3 may all be regulated by a particular regulatory process involving, for example, shared transcriptional regulatory elements.

Mutual Information Cluster Analysis   As an alternative to clustering based on the Euclidean distance that captures only positive correlations, one can use a more general distance measure based on mutual information (M; see earlier discussion), which also captures nonlinear relationships (Michaels et al. 1998).

M captures the information that is shared among temporal gene expression patterns (e.g., see figures 5.8 and 5.10). H refers to the Shannon entropy, which for our purposes is a measure of the variety of expression levels exhibited by a gene. H is high if a gene shows a large number of expression levels over a time course; it is low if the expression pattern is relatively invariant over time. The higher the value of H, the more information the pattern contains. A completely flat or constant expression pattern carries no information and has an H of zero. Note that H reveals nothing about specific expression levels at individual time points because it is based only on the relative frequencies, that is, the frequency of occurrence of expression levels within a time course. We introduce "coherence" (C) as a normalized mutual information measure for use as a distance measure for gene pairs in clustering: $C = M(A, B)/H_{max}(A, B)$. Coherence captures similarities in patterns independent of individual information entropies, to answer the question: How far is pattern A able to predict pattern B? Coherence is an important modification because mutual information increases with entropy. We can correct for this bias by dividing the mutual information by the maximum entropy of the pair. For example, even though two genes with relatively "constant" expression patterns have low H's (and therefore, low M), we may wish to use C to acknowledge that they nevertheless have highly similar patterns.

Mutual information (coherence) cluster analysis implies shared wiring, with no constraints on rules (figure 5.18). Unlike the Euclidean distance measure, mutual information determines negative and nonlinear, as well as positive and linear, correlations. Mutual information therefore clusters genes that may share inputs, but that respond to those inputs with different kinetics. For example, genes A and B may receive an input from C (shared wiring), but A would respond to C by increasing, while B would respond by decreasing

Figure 5.16
Temporal expression patterns for 112 genes expressed in rat spinal cord, as determined by RT-PCR. Genes are color coded according to functional classes. The dark color corresponds to the most intense expression; white corresponds to no detectable expression (Wen et al. 1998). (See plate 13.)

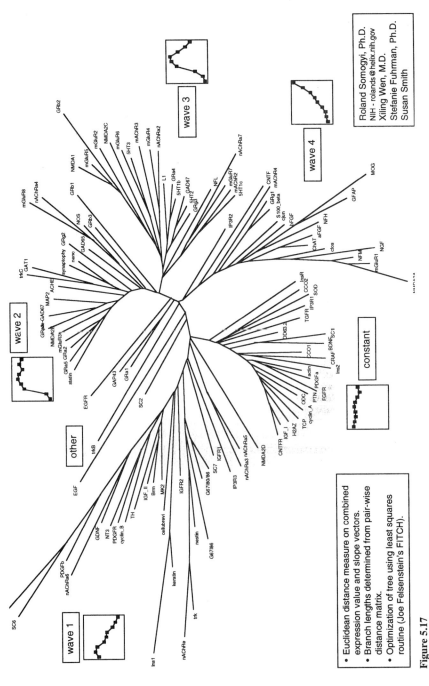

Figure 5.17
Euclidean distance tree for genes expressed in rat spinal cord. Averge gene expression patterns of the major branches of the tree are shown in the insets, describing gene expression "waves" (Carr et al. 1997).

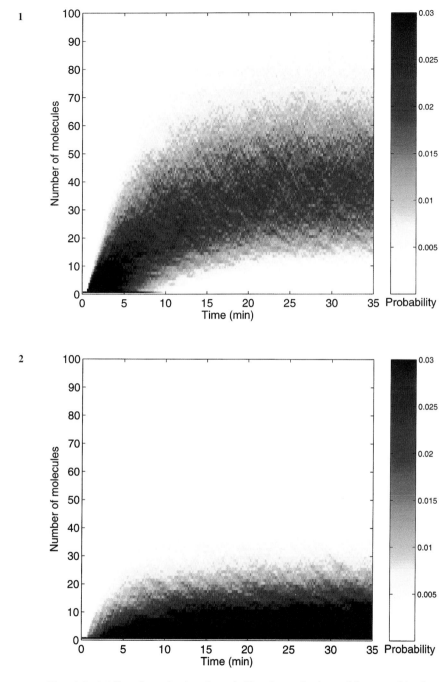

Plate 1 Probability of *n* molecules of protein N at time *t*, for the model presented in chapter 2 with [Cro2] = [Repressor2] = 0. Color represents probability. (See chapter 2.)

Plate 2 Probability of *n* molecules of protein N at time *t*, for the model presented in chapter 2 with [Cro2] = 50 nM, [Repressor2] = 10 nM. Color represents probability. (See chapter 2.)

3

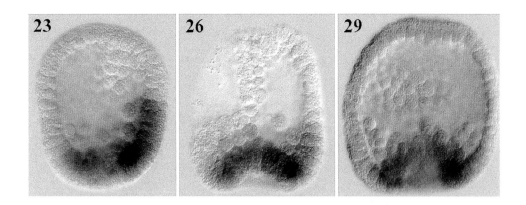

4

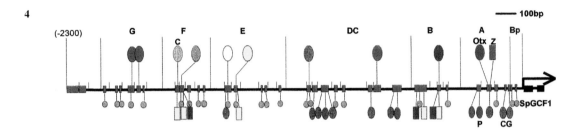

5

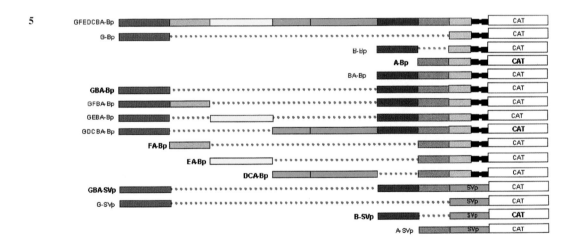

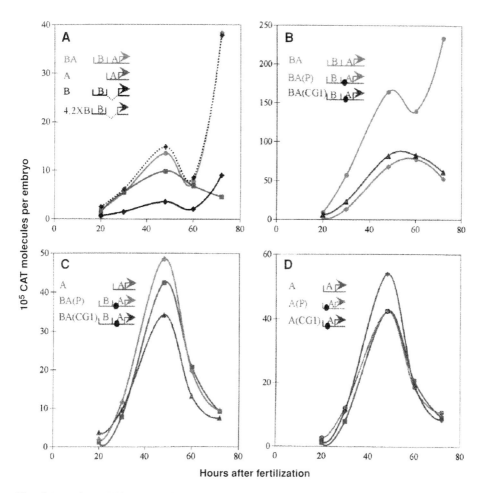

Plate 3 (opposite, top) The pattern of *Endo16* expression during *S. purpuratus* embryonic growth. Numbers of hours post fertilization are shown at the top left of each figure. Side views are shown. Cells labeled brown indicate the daughters of a single macromere; cells stained blue are those expressing *Endo16*. Note *Endo16* expression is always confined to a fraction of the macromere descendants. Adapted from Ransick and Davidson (1995). (See chapter 3.)

Plate 4 (opposite, middle) Summary of *Endo16 cis*-regulatory organization. The 2300-base pair *cis*-domain of *Endo16* comprises 31 protein binding sites (red boxes on horizontal black line denoting DNA) organized into six modular subregions labeled A – G at the top of the figure. The boundaries of the modules are delineated by the thin vertical lines incident on the DNA. Bp represents the basal promoter and the region to its right, marked by the bold arrow, marks the coding region of the gene. The large colored symbols above the DNA line indicate the unique binding sites within each module. The smaller colored symbols below the DNA line mark nonunique binding sites as labeled. (See chapter 3.)

Plate 5 (opposite, bottom) Schematic representation of some of the expression constructs used. (See chapter 3.)

Plate 6 Roles of the CG1 and P sites of module A. (A) Time courses of modules A and B and of the BA construct, and demonstration that the time course of BA is a scalar amplification of that of B. The green dotted line shows the time function generated by multiplying the data for the B construct by a factor of 4.2 at each measured time point. (B–D) Time-course data generated by constructs in which CG1 and P sites were mutated. In (B), the effects of CG1 and P mutations on output of BA are shown. In (C), BA(CG1) and BA(P) are compared with wild-type module A. In (D), A(CG1) and A(P) are compared with wild-type module A. (A) adapted from Yuh et al. (1996). (See chapter 3.)

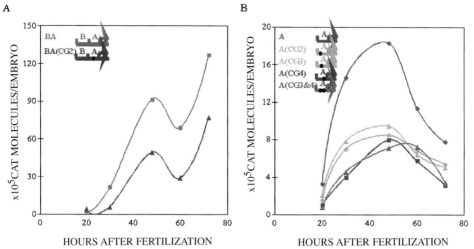

Plate 7 Role of CG2, CG3, and CG4 sites. (A) Effect of mutations of the CG2 site on expression of the BA construct. Although this mutation also destroys part of the Z target site, the latter does not affect expression unless one or more of modules F, E, or DC is present (see text). This mutation prevents binding of the CG factor to the CG2 site. (B) Effect of mutations of CG2, CG3, and CG4 sites on module A activity. (See chapter 3.)

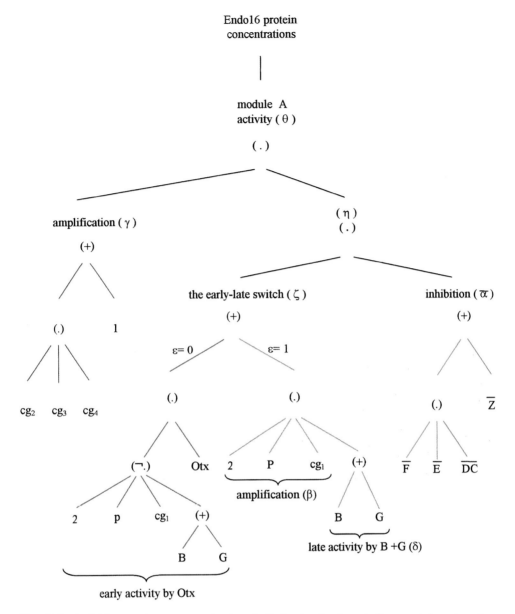

Plate 8 The *Endo16 cis*-regulatory computational graph. Each colored pathway indicates an independent *cis*-regulatory effect. (See chapter 3.)

9

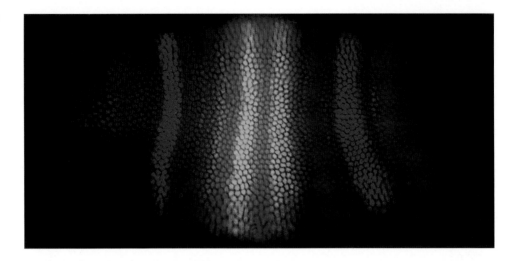

10

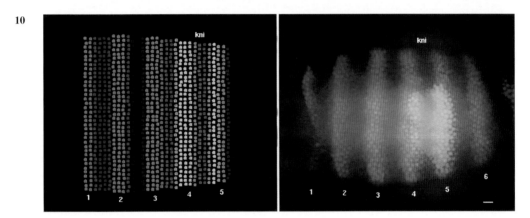

Plate 9 Spatial pattern of gene expression in a *Drosophila* syncytial blastoderm for two gap genes and one pair-rule gene. Immunofluorescent staining of nuclei for *Kruppel* (green), *giant* (blue), and *even-skipped* (red). Overlapping areas of *Kr* and *eve* appear yellow, and overlaps of *gt* and *eve* appear purple. Image courtesy of John Reinitz. (See chapter 4.)

Plate 10 *Drosophila eve* stripe expression in model (right) and data (left). Green, *eve* expression; red, *kni* expression. From Reinitz and Sharp (1995); courtesy J. Reinitz and D. H. Sharp. (See chapter 4.)

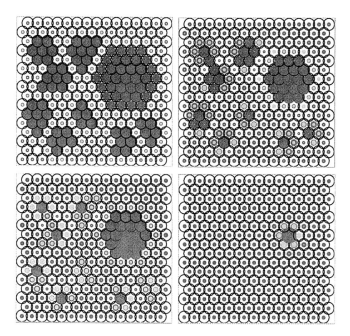

Plate 11 Cluster resolution. A circuit "trained" to resolve simple proneural cluster configurations into individual neuroblasts (or sensory organ precursor cells) is tested on more complex and irregular configurations. In this case, each cluster was successfully resolved into a single neuroblast, but the large clusters resolved more slowly. Times: $t = 1$(top left), $t = 76$ (top right), $t = 106$ (bottom left), $t = 476$ (bottom right). Similar to Marnellos and Mjolsness (1998a); courtesy of George Marnellos. (See chapter 4.)

Plate 12 Basin of attraction of 12-gene Boolean genetic network model. The trajectory in the top right panel is depicted as a series of connected points in the center, the basin of attraction graph. The selected trajectory is one of many that converge to the central, six-state cycle attractor. While this basin of attraction covers many different states of the system, they all converge on a simple attractor, essentially corresponding to a form of data reduction (Somogyi and Sniegoski 1996; Somogyi 1998). (See chapter 5.)

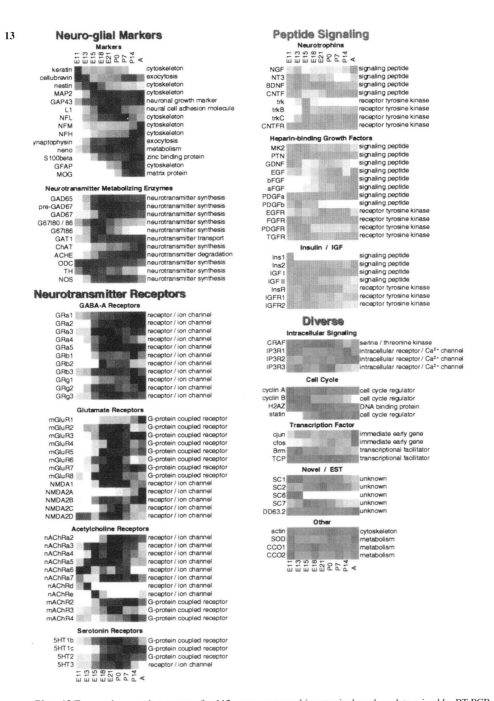

Plate 13 Temporal expression patterns for 112 genes expressed in rat spinal cord, as determined by RT-PCR. Genes are color coded according to functional classes. The dark color corresponds to the most intense expression, white corresponds to no detectable expression (Wen et al. 1998). (See chapter 5.)

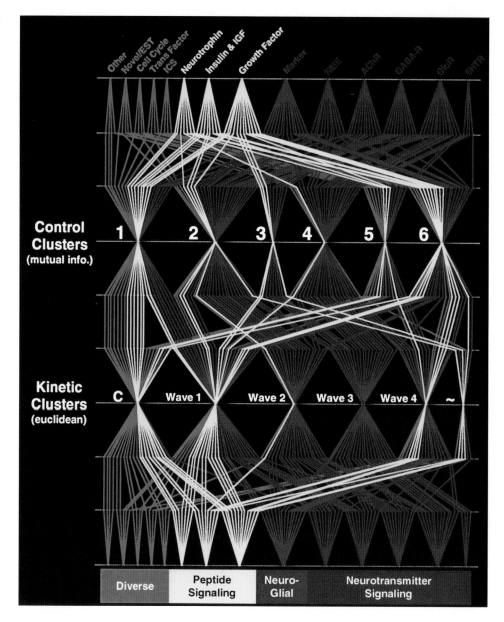

Plate 14 Functional gene families map to distinct control processes. Genes are color coded according to major functional families. The graph maps each of the 112 genes surveyed to the major clusters determined from Euclidean clustering (kinetic clusters), and mutual information-based clustering (control clusters), back all the way to functional subcategories (Carr et al. 1997). (See chapter 5.)

15

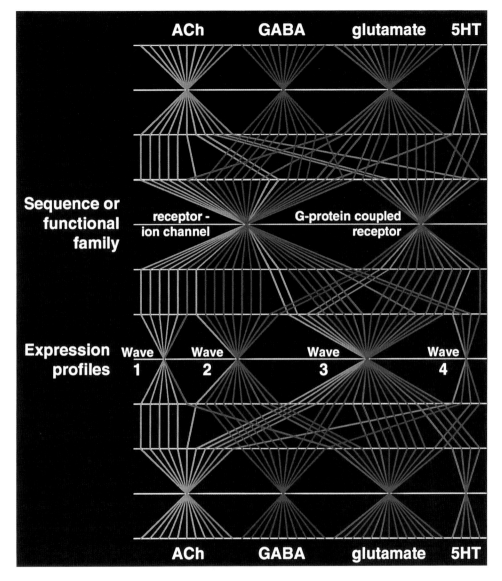

Plate 15 Mapping of expression clusters to neurotransmitter receptor gene families. Neurotransmitter receptors follow particular expression wave forms according to ligand and functional class. See Euclidean distance tree (figure 5.17) for pictograms showing typical expression profile for each wave. Note that the early expression waves 1 and 2 are dominated by ACh and GABA receptors, and by receptor ion channels in general (Agnew, 1998). (See chapter 5.)

16

development   KA-injury

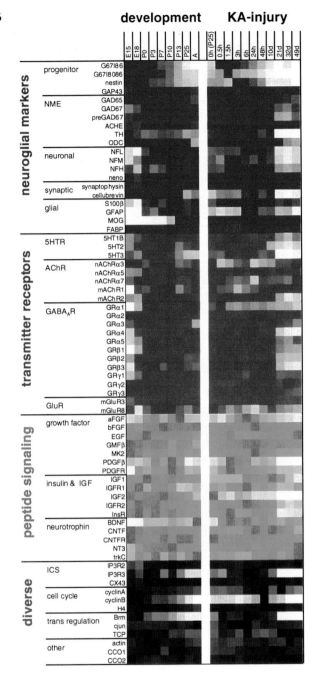

Plate 16 Temporal expression patterns for 70 genes expressed in hippocampus during development (left) and after a seizure injury (right), as determined by RT-PCR. Expression values are normalized to maximum expression for each gene (injury data were normalized separately from development data). Each data point is the average value for three animals. In this case, adult corresponds to P60. (See chapter 5.)

17

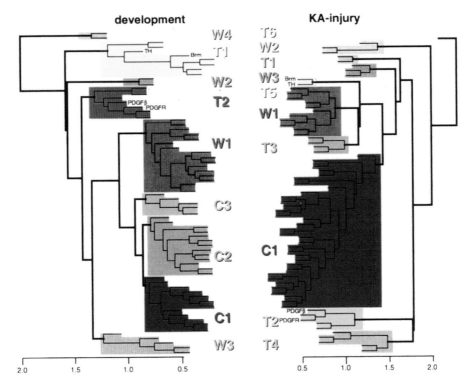

Plate 17 Cluster analysis of hippocampal gene expression time series. Gene expression clusters for developing and injured rat hippocampus. Clusters were generated using the Euclidean distance measure and S-plus software for agglomerative clustering. (See chapter 5.)

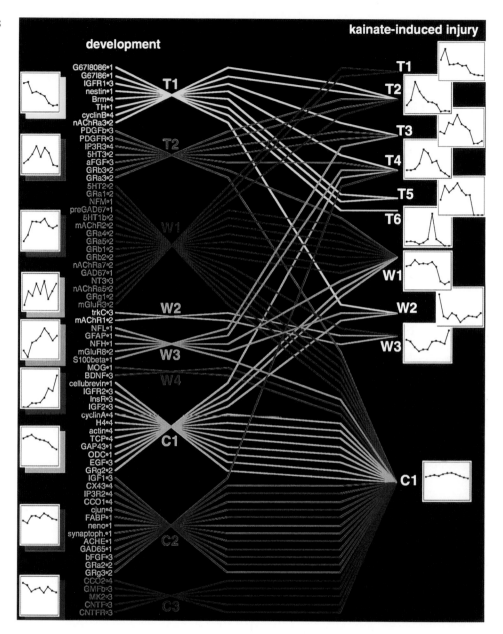

Plate 18 Mapping of hippocampal developmental gene expression clusters to KA-injury clusters. The lines connect genes from the developmental clusters on the left to their corresponding injury clusters on the right. The pictograms show the average expression pattern for each coexpression group. (See chapter 5.)

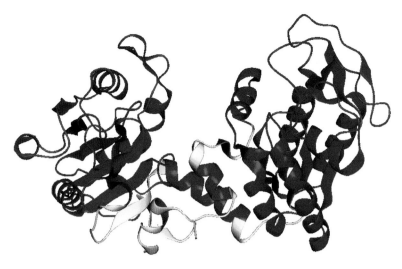

Plate 19 The H-NEIMO model for phosphoglycerate kinase. Here the parts of two domains that bind the substrates (shown in yellow) are kept rigid while the loops (shown in green) are allowed to have full torsional freedom. H-NEIMO MD simulations exhibit large-scale domain motions that bring the substrates close together and then apart. (See chapter 6.)

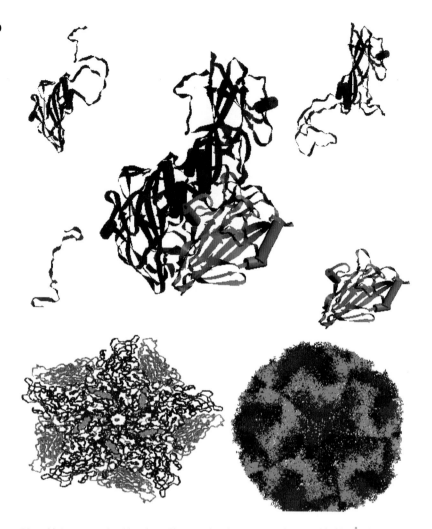

Plate 20 Structure of a rhinovirus. The complete icosahedral viral capsid (300 Å diameter) is shown at the bottom right. The four viral proteins making up the asymmetric unit of the virus are shown in the top half. The five asymmetric units forming the pentamer are shown in the bottom left. Twelve such pentamers form the complete viral capsid. (See chapter 6.)

Plate 21 (opposite, top) A pictorial description of the effect of buffer binding on calcium diffusion for slowly diffusing buffers. The yellow dots represent calcium ions, the green blobs buffer molecules. The purple blobs are calcium channels and the lightning symbol represents chemical interaction. The reacting dot is gradient colored to distinguish it from the other intracellular dot. Note that the figure only shows the net effect for the respective populations. In reality, individual molecules follow erratic paths that do not directly show the buffering effect. (See chapter 7.)

Plate 22 (opposite, bottom) Effect of immobile buffer binding rates on calcium diffusion. The changes in calcium concentration at several distances from the plasma membrane due to a short calcium current into a spherical cell are shown. (A) Calcium concentration for simulation with no buffers present. (B) Same with buffer with slow forward rate f (similar to EGTA) present. (C) Same with buffer with fast f (similar to BAPTA) present. Fainter colors: approximation using Dapp [eq. (7.24)]. (D) Bound buffer concentration for simulation shown in B. (E) Same for simulation shown in (C). (See chapter 7.)

21

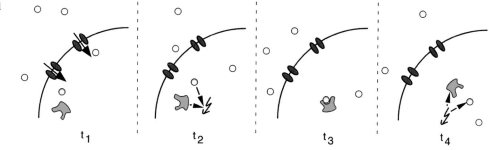

22

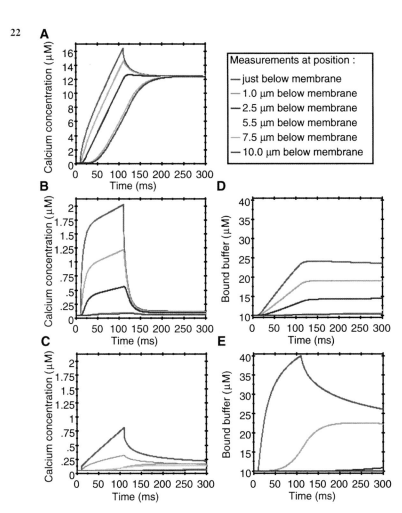

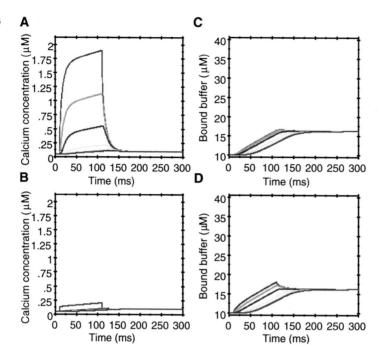

A Calcium concentration (μM) — Time (ms)

B Calcium concentration (μM) — Time (ms)

C Bound buffer (μM) — Time (ms)

D Bound buffer (μM) — Time (ms)

Plate 23 Effect of diffusible buffer binding rates on calcium diffusion. Simulation results similar to those in figure 7.3 (B–E), but the buffer is now diffusible. (A) Calcium concentration for simulation with diffusible buffer with slow forward binding rate f. (B) Same with diffusible buffer with fast f. (C) Bound buffer concentration for simulation shown in (B). (D) Same for simulation shown in (C). (See chapter 7.)

Plate 24 (opposite, top) Effect of competition between two diffusible buffers. Simulations similar to those in figure 7.3 (B–E), with both diffusible buffers present. (A) Calcium concentrations. (B) Bound buffer concentrations: upper traces, slow buffer; lower traces, fast buffer. (C) Superposition of bound buffer traces for fast buffer and slow buffer (fainter colors; traces have been superimposed by subtracting the steady-state bound buffer concentration). Concentrations just below the plasma membrane (red curves) and at a 30-mm depth (blue curves) are shown. (See chapter 7.)

Plate 25 (opposite, bottom) Concentration profiles along length axis of the dendritic shaft. (Above) Smooth dendrite case (see text). (Below) Spiny dendrite case (see text). (See chapter 7.)

A

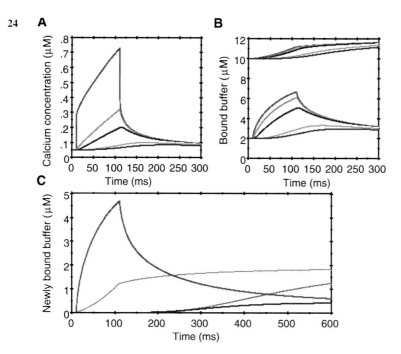

C

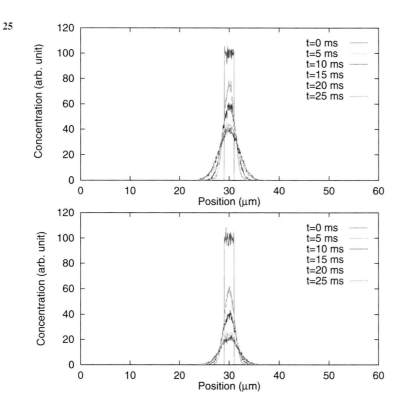

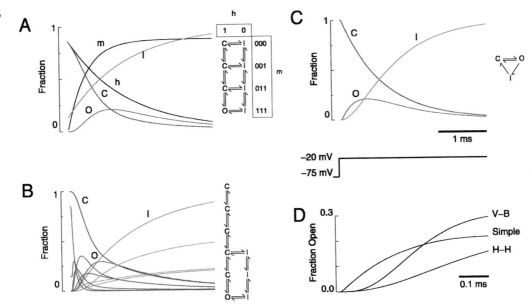

Plate 26 Three kinetic models of a squid axon sodium channel produce qualitatively similar conductance time courses. A voltage-clamp step from rest, V = −75 mV, to V = −20 mV was simulated. The fraction of channels in the open state (O, red), closed states (C, blue), and inactivated states (I, green) are shown for the Hodgkin-Huxley model, a detailed Markov model, and a simple Markov model. (A) An equivalent Markov scheme for the Hodgkin-Huxley model is shown [right insert, eq. (8.20)]. Three identical and independent activation gates (m) give a form with three closed states (corresponding to 0, 1, and 2 activated gates) and one open state (three activated gates). The independent inactivation gate (h) adds four corresponding inactivated states. Voltage-dependent transitions were calculated using the original equations and constants of Hodgkin and Huxley (1952). (B) The Markov model of Vandenberg and Bezanilla (1991) [eq. (8.23)]. Individual closed (violet) and inactivated (yellow) states are shown, as well as the sum of all five closed states (C, blue), the sum of all three inactivated states (I, green) and the open state (red). (C) A simple three-state Markov model fit to approximate the detailed model [eq. (8.23)]. (D) Comparison of the time course of open channels for the three models on a faster time scale shows differences immediately following a voltage step. The Hodgkin-Huxley (H-H) and Vandenberg-Bezanilla (V-B) models give smooth, multiexponential rising phases, while the three-state Markov model (Simple) gives a single exponential rise with a discontinuity in the slope at the beginning of the pulse. Modified from Destexhe et al. (1994c), where all parameters are given. (See chapter 8.)

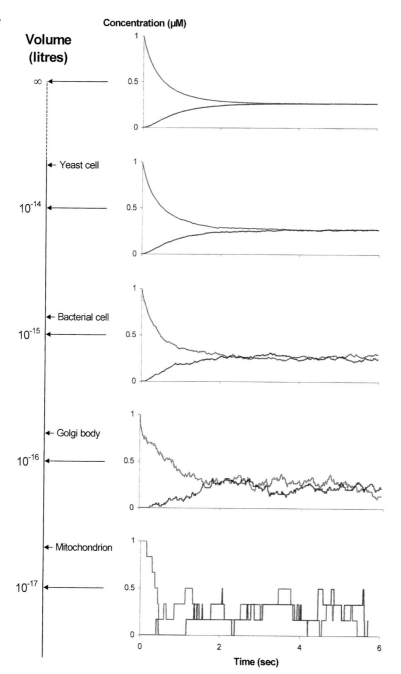

Plate 27 Predicted time course of an enzyme-catalyzed cyclic reaction showing the progressive emergence of random behavior as the reaction volume becomes smaller. The reaction is the simple interconversion of substrate (green), S, and product (red), P, by two enzymes. The top graph shows the time course of the reaction simulated by MIST, a conventional continuous, deterministic simulator (Ehlde and Zacchi 1995). The next four graphs show changes in S and P predicted by StochSim for different volumes of the reaction system. Note that at the smallest volume considered, each step in concentration corresponds to a change of one molecule. (See chapter 9.)

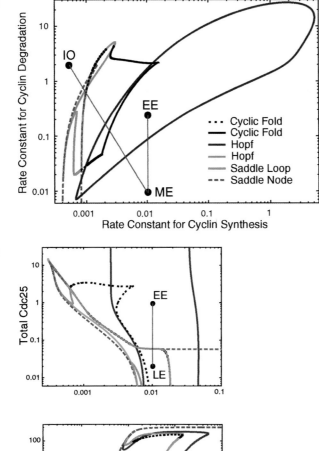

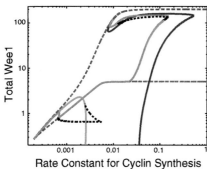

Plate 28 Two-parameter bifurcation diagram for the full model in figure 10.1. See text for explanation. From Borisuk and Tyson (1998). (See chapter 10.)

Plate 29 Alternative slices through the "avocado." See plate 28 for a key to the colors. See text for explanation. From Borisuk and Tyson (1998). (See chapter 10.)

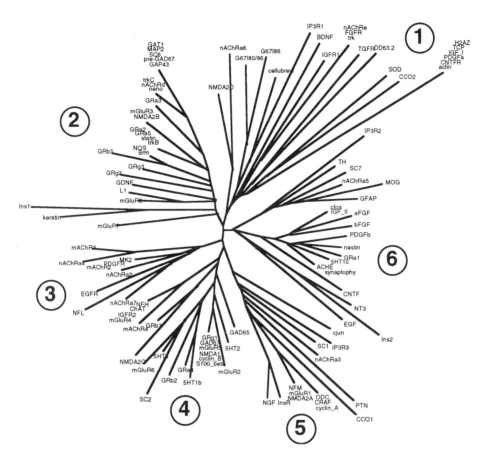

Figure 5.18
Mutual information tree for genes expressed in rat spinal cord. Hierarchical clustering was carried out using "coherence" as a distance measure (Michaels et al. 1998).

(different rules). This relationship between A and B would be recognized by mutual information, but not by the Euclidean distance measure.

Cluster Analysis Suggests Five to Six Primary Pathways   Developmental gene expression exhibits apparent redundancy, that is, it is far from maximally diverse. The fact that we have been able to cluster more than a hundred genes into a small number of temporal expression patterns suggests that the number of genetic control processes is much smaller than the number of regulated genes. Of course, the present sample of genes is quite small compared with the whole genome (estimated at 60,000 genes in the rat). Without further studies, it is impossible to know whether any of the remaining genes exhibit as-yet unobserved expression patterns, such as a U-shaped pattern. However, analyses of large-scale microarray surveys of gene expression suggest that only a fraction of genes, on the order of ~10%, generally fluctuate in a particular system (unpublished observations, S. Furhman). For extended studies, only a subset of genes may be subject to detailed analysis following large-scale surveys designed to identify the active population of genes.

It is interesting to note that two Euclidean distance clusters (waves) consist almost entirely of neurotransmitter signaling genes from both the ionotropic and metabotropic classes (figure 5.19; see also plate 14). This particular category of genes therefore appears to be confined to specific genetic control processes and may share a functional role in development despite differences in DNA sequences. It is also interesting that while another cluster, Euclidean wave 1, contains neurotransmitter signaling genes, these are exclusively ionotropic, suggesting that some ionotropic receptors have a function distinct from that of other neurotransmitter receptors (figure 5.20; see also plate 15). Moreover, the order in which the different neurotransmitter receptor genes appear in spinal cord development is also linked to ligand class, beginning with acetylcholinesterase (Ach) (wave 1), transitioning to GABA (wave 2), then glutamate, and 5-hydroxytryptamine (5HT).

Some proportion of the 60,000 genes of the rat will fall into the constant category, having relatively invariant expression levels over time. For our purposes, constant genes convey no information about phenotypic change, although they may fluctuate under other conditions. In higher organisms, it is a priority to focus on intercellular signaling genes because these are essential for development in multicellular organisms. Some genes (involved in both intra- and intercellular signaling) can be expected to fall into the constant cluster, even after a perturbation, and may subsequently be ignored. This will allow researchers to concentrate their efforts on the genes most relevant to development or other phenotypic change, such as the response to a disease, injury, or therapeutic drug treatment.

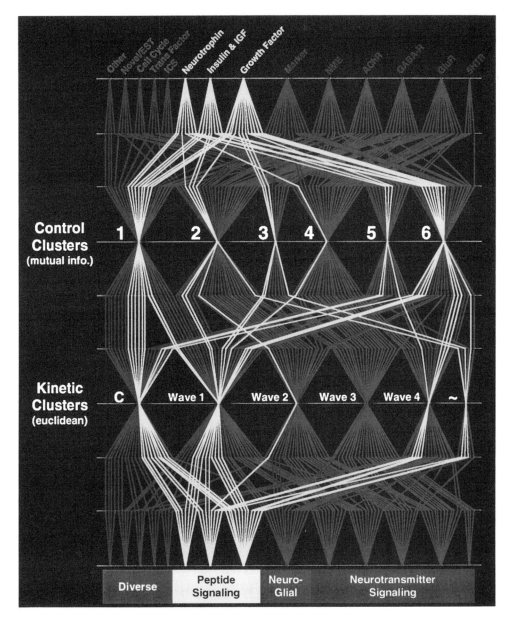

Figure 5.19
Functional gene families map to distinct control processes. Genes are color coded according to major functional families. The graph maps each of the 112 genes surveyed to the major clusters determined from Euclidean clustering (kinetic clusters), and mutual information-based clustering (control clusters), back all the way to functional subcategories (Carr et al. 1997). (See plate 14.)

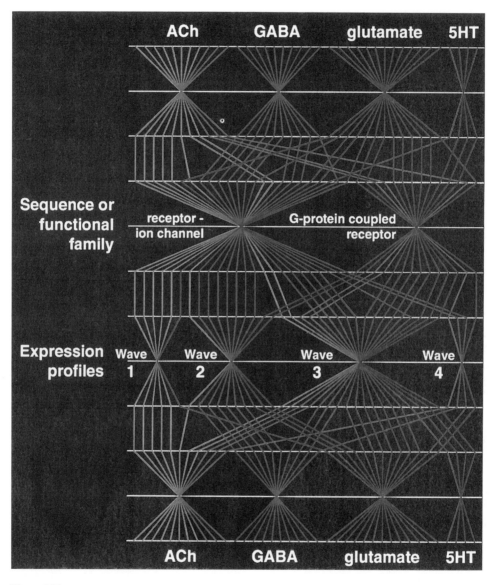

Figure 5.20
Mapping of expression clusters to neurotransmitter receptor gene families. Neurotransmitter receptors follow particular expression wave forms according to ligand and functional class. See Euclidean distance tree (figure 5.17) for pictograms showing typical expression profile for each wave. Note that the early expression waves 1 and 2 are dominated by ACh and GABA receptors, and by receptor ion channels in general (Agnew 1998). (See plate 15.)

5.3.4 Gene Expression Matrix of Rat Hippocampus

Determining an interaction diagram for a genetic network requires that the system be analyzed under a variety of conditions. The complexity of the system's response data must be of the same order as the complexity of the underlying network—otherwise reliable inference is not possible. We can generate more diversity in data by subjecting our biological model to perturbations. In that context, we carried out an RT-PCR-based expression analysis of hippocampal development and hippocampal response to a drug perturbation. Figure 5.21 displays the hippocampal development and the response to kainate-induced seizures at P25 (see also plate 16). Since the injury data were acquired at a higher resolution time scale than the development data (in accordance with the known effects of kainate), we do not have a control for the developmental period from P25 to adult. However, it is reasonable to assume that most genes have a relatively invariant pattern of expression level during this stage, given the data for P13 to adult; in addition, the responses to seizure shown in the figure are too rapid to be attributed to normal development.

We carried out an agglomerative, Euclidean cluster analysis of the developmental and injury time series (figure 5.22; see also plate 17) and mapped the corresponding expression profiles to their functional families (figure 5.23). Genes from the same functional group clearly tend to map to a limited set of expression clusters (see, e.g., enrichment of transmitter receptor transcripts in W1 in development and W1 after injury). It is interesting that a minority of genes remained unperturbed by the seizure. This suggests that portions of the genetic network operate relatively independently of one another.

Recapitulation of Developmental Programs   In figure 5.24, average expression patterns for all clusters are shown as pictograms. The colors (shown in plate 18) correspond to developmental clusters (they match shadowing of developmental pictograms). Lines connect genes in developmental clusters to their respective KA-injury clusters. Each gene can be followed from its label (left column) along a line connecting it to the first focus (developmental cluster) and then, according to the mirror image of this line, to the focus of the KA-injury cluster. Clusters are labeled T, W, and C, corresponding to transient, wave form, and constant patterns. In development, T marks genes that are expressed at significantly higher levels during early to mid-development in relation to the adult; W indicates genes that show other fluctuating patterns; and C marks clusters that exhibit relatively invariant expression over the time course. Note that T, W, and C cluster members in development generally map to the corresponding T, W, and C patterns following KA injury. In particular, while looking at the transient expression profiles of the T clusters in development and injury, it becomes apparent that developmental genes are selectively, transiently reactivated after injury. We could describe this as a recapitulation of developmental programs in response to a perturbation (seizure). This result suggests that genes may operate within ex-

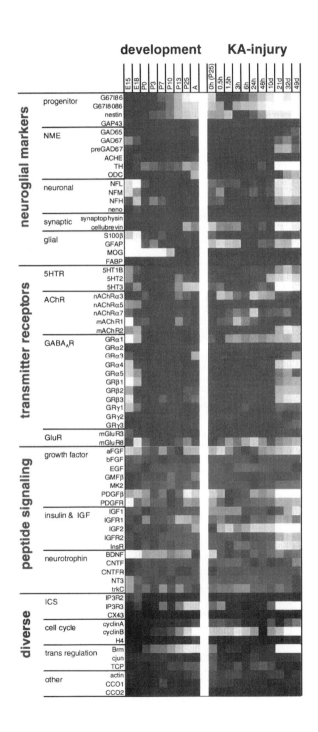

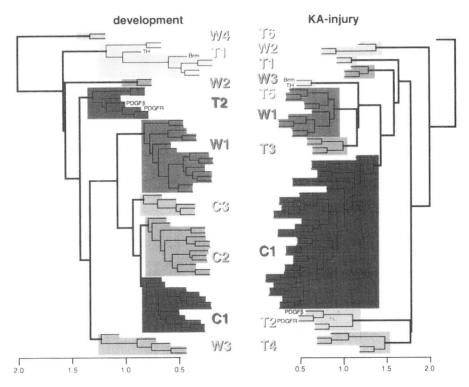

Figure 5.22
Cluster analysis of hippocampal gene expression time series. Gene expression clusters for developing and injured rat hippocampus. Clusters were generated using the Euclidean distance measure and S-plus software for agglomerative clustering. (See plate 17.)

pression modules to control differentiation and repair, and provides a clue about the organization of the genetic network.

Overlapping Control of Gene Expression in Spinal Cord and Hippocampus.   Comparisons of gene expression time series for different anatomical structures may provide clues concerning which genes are involved in basic developmental patterns. Figure 5.25 contains time series for nine genes expressed in rat spinal cord and hippocampus. Each of these genes has a very similar pattern in both CNS regions. The similarity of gene expres-

Figure 5.21
Temporal expression patterns for 70 genes expressed in hippocampus during development (left) and after a seizure injury (right), as determined by RT-PCR. Expression values are normalized to maximum expression for each gene (injury data were normalized separately from development data). Each data point is the average value for three animals. In this case, adult corresponds to P60. (See plate 16.)

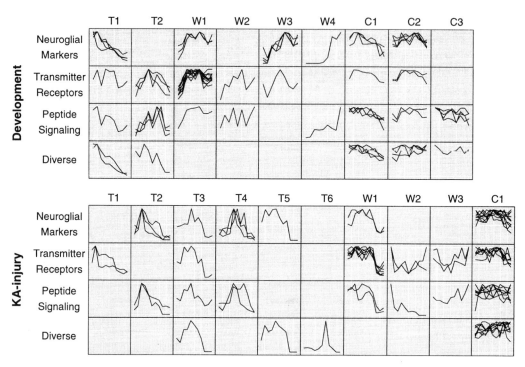

Figure 5.23
Distribution of gene expression profiles—analysis of cluster tightness. All gene expression profiles are plotted according to their representation in expression clusters (columns) and four basic functional categories (rows).

sion patterns between hippocampus and spinal cord suggests the existence of a generalized genetic program of neural development, common to all CNS regions. The extrapolation of this interpretation to other CNS structures is a reasonable one, given the evolutionary distance and anatomical differences between hippocampus and spinal cord.

5.3.5 Analysis of CNS Development and Injury Identifies Tightly Coregulated Genes: Evidence for Genetic Programs

Gene expression time series may be useful in determining putative functional connections between genes. As shown in figure 5.26, we found that some pairs of genes exhibit parallel expression patterns under three different conditions: spinal cord development (top), hippocampal development (middle), and hippocampal injury (bottom). Such parallelism may not be surprising in the case of, for example, PDGFb and its receptor. However, to our knowledge, there is no known functional relationship between the transcription factor Brm (Brahma) and the neurotransmitter metabolizing enzyme TH (tyrosine hydroxylase). It is

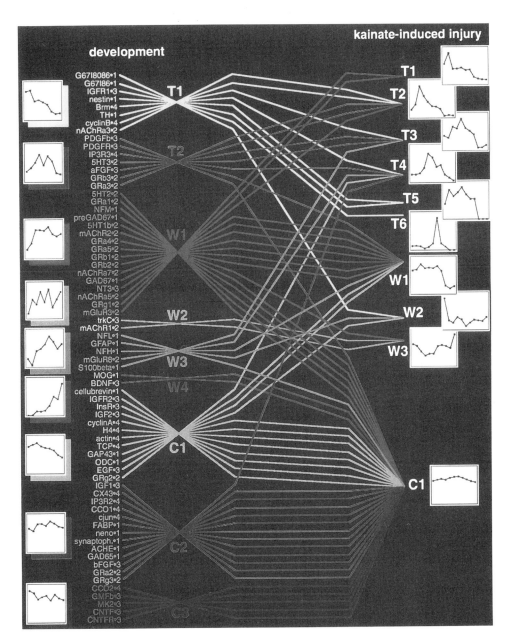

Figure 5.24
Mapping of hippocampal developmental gene expression clusters to KA-injury clusters. The lines connect genes from the developmental clusters on the left to their corresponding injury clusters on the right. The pictograms show the average expression pattern for each coexpression group. (See plate 18.)

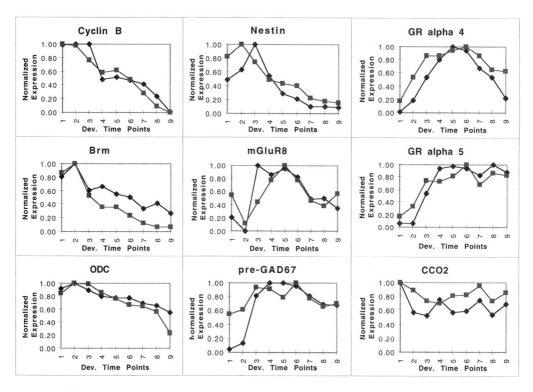

Figure 5.25
Gene coexpression pairs in spinal cord and hippocampus (dark lines = spinal cord expression; light lines = hippocampal expression) (Fuhrman et al. 1998).

particularly interesting that Brm and TH continue to fluctuate in parallel even after an injury perturbation (chemically induced seizure). Further studies will be necessary to confirm a functional connection between these two genes.

Gene expression time series may also be useful in establishing possible functions for newly discovered genes. This is particularly relevant now that whole genomes are being sequenced. A large proportion of yeast (*Saccharomyces cerevisiae*) genes, for example, have no known homologs in other organisms, leaving molecular biologists clueless as to their functions. Large-scale temporal gene expression studies in different tissues and under different conditions can provide a starting point for investigations of these novel genes by comparing their expression time series with those of known genes.

5.3.6   Causal Inference of Gene Interactions from Expression Data

Reverse Engineering of Genetic Networks   The ultimate goal of expression measurements and their analysis is to discover how precisely the genes regulate one another. The

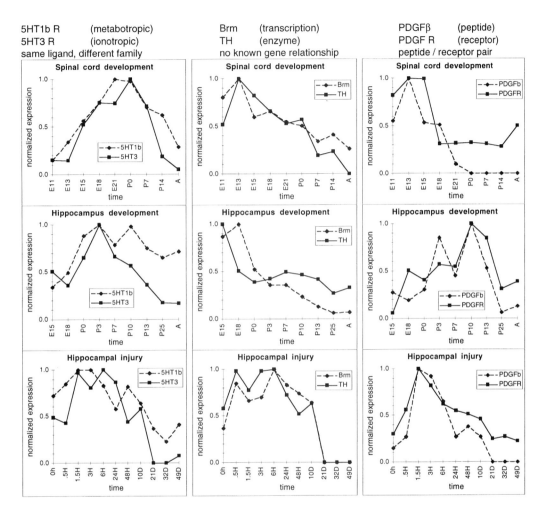

Figure 5.26
Gene coexpression pairs in CNS development and injury. Note the surprising, highly coordinate expression profiles of *Brahma* and *Tryrosine Hydroxylase*, genes that would not otherwise have been functionally associated.

end product is a model of gene interaction from which we can predict outcomes and select perturbation strategies for therapeutic and bioengineering purposes. We have described an algorithm, REVEAL, that works for genetic network models, for which we can generate large amounts of data liberally. However, given current data limitations, one may pursue a more constrained inference procedure as a first step toward reverse engineering of a genetic network.

D'haeseleer (D'haeseleer et al. 1999) has used a linear modeling approach to build a simple gene interaction model. Essentially, the expression level of each gene at a particular time point corresponds to the weighted sum of the expression levels of all other genes at a previous time point. He used this model to fit the data sets from spinal cord and hippocampal development and hippocampal injury. Indeed, the model is able to reconstruct the original data with great precision (figure 5.27). Moreover, the weight matrix capturing putative gene interactions from this analysis is sparsely populated, which is what we would expect from a biological network in which each gene receives only a limited number of critical inputs.

A subset of inferred gene interactions is shown in figure 5.28 in the form of a wiring diagram; essentially, this is simply another graphical representation of the same information shown in figure 5.3. Strong interactions, positive (solid lines) and negative (broken lines), are proposed with the family of genes covering GABA receptors and the GABA-synthesizing enzyme glutamic acid decarboxylase (GAD), which is biologically plausible—genes from a signaling group are generally expected to interact with one another. It is interesting that the model proposes a positive feedback interaction for GAD65 (arrow pointing from GAD65 to GAD65) and preGAD67, a prediction made independently in a previous publication (Somogyi et al. 1995). This type of analysis provides the predictions that the laboratory scientist requires for targeted experimental design. As more data accumulate from additional, targeted perturbation experiments, we can expect progressive improvement in the model's predictive power.

Analogous approaches have been applied to other biological systems. For example, the anatomical distribution of a handful of transcriptional regulatory genes has been studied extensively in fruit fly development. These data were sufficient to train a mathematical model incorporating nonlinear regulation with feedback and diffusion, which could accurately reproduce the observed molecular distributions (Mjolsness et al. 1991, Reinitz and Sharp 1995, Reinitz et al. 1995). For biochemical networks, statistical inference procedures were developed by Arkin and Ross for determining the major functional connections and stimulatory or inhibitory effects. These procedures were successfully used to reverse engineer biochemical pathways from experimental (Arkin et al. 1997) and simulated (Arkin and Ross 1997) data sources.

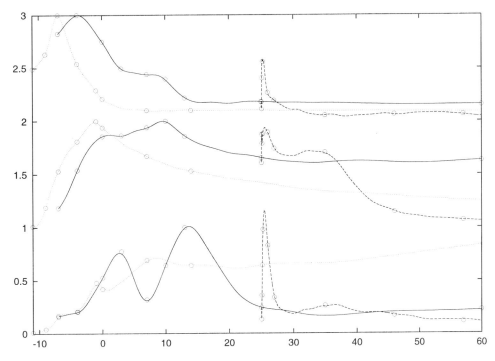

Figure 5.27
Experimental gene expression data (development and injury), and simulation using a linear model. Original (circles) and reconstructed time series (lines) for nestin (top), GABA receptor__4 (middle) and acidic FGF (bottom). *Nestin* and *GABA* receptor__4 are offset by 2.0 and 1.0 on the ordinate, respectively. Time is in days from birth (day 0). Dotted line: spinal cord, starting at day −11 (embryonic day 11). Solid line, hippocampus development, starting at day −7 (embryonic day 15). Dashed line, hippocampus kainate injury, starting at day 25 (D'haeseleer et al. 1999).

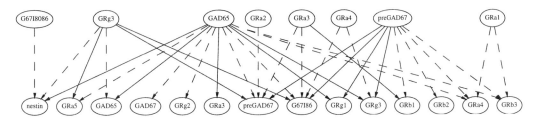

Figure 5.28
Gene interaction diagram predicted by linear model fr GABA signaling family. Solid lines correspond to positive interactions; broken lines suggest inhibitory relationships. The diagram provides a graphic alternative (akin to a wiring diagram) to the same relationships shown in figure 5.3 (D'haeseleer et al. 1999).

5.4   Conclusion

5.4.1   General Strategies for Constructing Network Models

Molecular biological research strategies have traditional been guided by a "bottom up" approach that focuses on the determination of individual biomolecular interactions, the knowledge of which is expected to result in the building of predictive models. With the exception of relatively confined systems such as viruses (McAdams and Shapiro 1995), the modeling aspect has lagged far behind because of the complexity of these networks and the lack of reliable data.

Given new large-scale measurement technologies, we are now exploring the option of a "top down" approach to predictive model building. Essentially, by determining extensive input/output patterns through time series and perturbations of the network on a genomic scale, predictive models may be constructed using statistical inference techniques and various computational reverse engineering strategies. As predictions of these models are subjected to testing with a targeted perturbation strategy, erroneous predictions may be overcome by updating the model with new corrective data. Predictive reliability is expected to steadily increase with accumulating data. The rate of progress will most likely follow a nonlinear, sigmoid learning curve.

Level-by-level Inference from Large-Scale Gene Expression Data   Any serious analysis effort requires that the experimental data be of sufficient breadth, sensitivity, and precision to accurately reflect the state of the biological system (e.g., automated, quantitative RT-PCR meets these criteria on the level of mRNA expression). Causal inference depends on the measurement of a diverse set of state transitions, such as time series of responses to a variety of perturbations. Ideally, measurements should include the activity of all genes of the organism on the RNA and protein level, extending to protein modification and accounting for sufficient anatomical resolution in multicellular organisms. However, the evidence so far indicates that not all genes are active in every major cellular process. Also, considering that there appears to be appreciable redundancy within these networks, one may expect to be able to learn much from the measurement of extensive but limited subsets of molecular activities.

The first level of functional inference deals with a systematic classification of genes into coexpression pathways. Vectors of gene activity measurements can be compared using a variety of similarity measures, each providing information on shared control processes (D'haeseleer et al. 1998). Most fundamentally, the Euclidean distance measure is used to group genes according to a simple positive correlation, suggesting that genes within a tight coexpression cluster are subject to the same molecular control functions. Moreover, coregulation clusters over a diversity of measurements that include genes from several protein function families are highly suggestive of integrated functional pathways; it would seem improbable that genes are highly coregulated without a form of evolutionary pressure enforcing such coherence. The evidence so far suggests that gene expression trajectories fall within a limited set of classes, the members of each class often originating from related functional families, which is highly suggestive of pathway definitions. The Pearson correlation coefficient provides an alternative distance measure for clustering. Linear correlation clusters should identify genes that share common regulatory inputs, independent of whether the control function is stimulatory or inhibitory. Finally, the most general measure for expression clustering, mutual information, only implies shared wiring, but allows for any conceivable, nonlinear input function.

The final goal of any functional inference strategy is the complete reverse engineering of a network's architecture from activity measurements. This is potentially a formidable task, requiring theoretical investigations to establish that it is possible in principle. Indeed, within a general discrete network model, an algorithm based on information analysis, REVEAL, was developed that enables the complete reconstruction of a network within reasonable computational bounds and input data requirements. The principle of REVEAL can be applied to experimental data, but requires a rich input set of state transition measurements. The depth of inference is dependent on the volume and resolution of the data. For the analysis of a currently available, limited data set, we have examined an alternative linear reverse engineering approach. While it is less general than REVEAL, the constraints within the linear model place less stringent requirements on data volume. The model provides plausible predictions that will provide useful clues for the design of targeted perturbation experiments.

An Integrated Network Analysis Strategy   It is now reasonable to say that an integrated networks approach to biomolecular research is the inevitable next step. Progress in molecular biology, physiology, and biochemistry clearly points to highly cross-wired networks. Moreover, the trend toward large-scale experimental assays is intimately coupled to computational technologies for data management, analysis, and predictive model building.

Vital practical applications depend on integration of systematic experimental and computational strategies. At a fundamental level, these technologies can be immediately applied to the diagnosis of complex diseases, for which correlative evidence suffices. Deeper analysis of causality within signaling networks will be required to effectively elucidate complex medical challenges such as cancer, degenerative disorders, and regeneration after injury. From a general perspective, this research direction simply points toward bioengineering. In a sense, therapeutic strategies are a specialized aspect of bioengineering. Further applications await in agriculture (growth, resistance) and microorganisms (waste treatment, chemical and metabolic engineering).

References

Agnew, B. (1998). NIH Plans bioengineering initiative. *Science* 280 (5374): 1516–1518.

Arkin, A., and Ross, J. (1997). Statistical construction of chemical reaction mechanisms from measured time-series. *J. Phys. Chem.* 99: 970–979.

Arkin, A., Shen, P., and Ross, J. (1997). A test case of correlation metric construction of a reaction pathway from measurements. *Science* 277: 1275–1279.

Arnone, M. I., and Davidson, E. (1997). The hardwiring of development: organization and function of genomic regulatory systems. *Development* 124: 1851–1864.

Bowtell, D. D. (1999). Options available—from start to finish—for obtaining expression data by microarray. *Nat. Genet.* 21: 25–32.

Carr, D. B., Somogyi, R., and Michaels, G. (1997). Templates for looking at gene expression clustering. *Stat. Comp. Graphics Newslett.* 8(1): 20–29.

D'haeseleer, P., Wen, X., Fuhrman, S., and Somogyi, R. (1998). Mining the gene expression matrix: inferring gene relationships from large-scale gene expression Data. In *Proceedings of the International Workshop on Information Processing in Cells and Tissues 1997*, R. Patow, ed., pp. 203–212. Plenum Press, New York.

D'haeseleer, P., Wen, X., Fuhrman, S., and Somogyi, R. (1999). Linear modeling of mRNA expression levels during CNS development and injury. In *Pacific Symposium on Biocomputing*, vol. 4, pp. 41–52. World Scientific, Singapore.

Fuhrman, S., Wen, X., Michaels, G., and Somogyi, R. (1998). Genetic network inference. *Interjournal.* See www.interjournal.org.

Glass, L. (1975). Classification of biological networks by their qualitative dynamics. *J. Theor. Biol.* 54: 85–107.

Glass, L., and Kauffman, S. A. (1973). The logical analysis of continuous, non-linear biochemical control networks. *J. Theor. Biol.* 39: 103–129.

Kauffman, S. A. (1993). *The Origins of Order, Self-Organization and Selection in Evolution.* Oxford University Press, New York.

Liang, S., Fuhrman, S., and Somogyi, R. (1998). REVEAL, a general reverse engineering algorithm for inference of genetic network architectures. In *Pacific Symposium on Biocomputing*, vol. 3, pp. 18–29.

Lipshutz, R. J., Fodor, S. P., Gingeras, T. R., and Lockhart, D. J. (1999). High-density synthetic oligonucleotide arrays. *Nat. Genet.* 21 (Supp. 1): 20–24.

McAdams, H. H., and Shapiro, S. (1995). Circuit simulation of genetic networks. *Science* 269: 650–656.

Michaels, G., Carr, D. B., Wen, X., Fuhrman, S., Askenazi, M., and Somogyi, R. (1998). Cluster analysis and data visualization of large-scale gene expression data. In *Pacific Symposium on Biocomputing* Vol. 3, pp. 42–53. World Scientific, Singapore.

Mjolsness, E., Sharp, D. H., and Reinitz, J. (1991). A connectionist model of development. *J. Theor. Biol.* 152: 429–453.

Phillips, J., and Eberwine, J. H. (1996). Antisense RNA amplification: a linear amplification method for analyzing the mRNA population from single living cells. *Methods* 10(3): 283–288.

Reinitz, J., and Sharp, D. H. (1995). Mechanism of *eve* stripe formation. *Mech. Dev.* 49: 133–158.

Reinitz, J., Mjolsness, E., and Sharp, D. H. (1995). Model for cooperative control of positional information in *Drosophila* by *bicoid* and *maternal hunchback*. *J. Exp. Zool.* 271: 47–56.

Savageau, M. A. (1998). Rules for the evolution of gene circuitry. In *Pacific Symposium on Biocomputing*, Vol. 3, pp. 54–65. World Scientific, Singapore.

Shalon, D., Smith, S. J., and Brown, P. O. (1996). A DNA microarray system for analyzing complex DNA samples using two-color fluorescent probe hybridization. *Genome Res.* 6: 639–645.

Shannon, C. E., and Weaver, W. (1963). *The Mathematical Theory of Communication.* University of Illinois Press, Urbana, Ill.

Somogyi, R. (1998). Many to one mappings as a basis for life. *Interjournal,* (http://rsb.info.nih.gov/mol-physiol/ICCS/ms/mappings.html).

Somogyi, R., and Sniegoski, C. (1996). Modeling the complexity of genetic networks. *Complexity* 1(6): 45–63.

Somogyi, R., Wen, X., Ma, W., and Barker, J. L. (1995). Developmental kinetics of GAD family mRNAs parallel neurogenesis in the rat spinal cord. *J. Neurosci.* 15(4): 2575–2591.

Somogyi, R., Fuhrman, S., Askenazi, M., and Wuensche, A. (1997). The gene expression matrix: towards the extraction of genetic network architectures. Nonlinear analysis. *Proc. Second World Cong. Nonlinear Analysts* (WCNA96), 30(3): 1815–1824.

Stucki, J. W., and Somogyi, R. (1994). A dialogue on $Ca^{2/+/}$ oscillations. *Biochim. Biophys. Acta* 1183: 453–472

Thomas, R. (1991). Regulatory networks seen as asynchronous automata: a logical description. *J. Theor. Biol.* 153: 1–23.

Wen, X., Fuhrman, S., Michaels, G. S., Carr, D. B., Smith, S., Barker, J. L., and Somogyi, R. (1998). Large-scale temporal gene expression mapping of CNS development. *Proc. Natl. Acad. Sci. U.S.A.* 95(1): 334–339.

Yamaguchi, K., Zhang, D., and Byrn, R. A. (1994). A modified nonradioactive method for northern blot analysis. *Anal. Biochem.* 218(2): 343–346.

II Modeling Biochemical Networks

6 Atomic-Level Simulation and Modeling of Biomacromolecules

Nagarajan Vaidehi and William A. Goddard III

6.1 Introduction

In principle, all the problems in biology could be solved by solving the time-dependent Schroedinger equation (quantum mechanics, QM). This would lead to a detailed understanding of the role that molecular-level interactions play in determining the fundamental biochemistry at the heart of biology and pharmacology. The difficulty is the vast range of length and time scales, from a nitrous oxide molecule to an organ (heart, lung), which makes a QM solution both impractical and useless. It is impractical because there are too many degrees of freedom describing the motions of the electrons and atoms, whereas in the functioning of an organ it may be only the rate of transfer across some membrane. The solution to both problems is the hierarchical strategy outlined in figure 6.1. We average over the scale of electrons (from QM) to describe the forces on atoms (the force field, FF), then average over the dynamics of atoms (molecular dynamics, MD) to describe the motions of large biomoleculcs, then average over the molecular scale to obtain the properties of membranes, then average over the components in a cell, then average over the cells to describe a part of an organ. The strategy is to develop a methodology for going between these various levels so that first principle's theory can be used to predict the properties of new systems.

In this chapter, we describe the first two levels of simulations and the attempts to coarsen the simulations to the mesoscale level in the hierarchy. The complexity of QM limits its applications to systems with only 10 to 200 atoms (depending on the accuracy), leading to distance scales of less than 20 Å and time scales of femtoseconds. The quantum mechanical forces are then translated into a set of parameters describing the bonded and nonbonded forces in a molecule. This set of parameters is known as the *force field*. Given the forces calculated with the FF, we solve Newton's equations to describe the motions of atoms with time, a process referred to as *molecular dynamics*. With MD, one can now consider systems with up to 1 million atoms, allowing practical simulations of systems as large as small viruses (say, 300 Å and 10 ns). The fundamental unit of MD is atoms (not the electrons of QM), allowing us to interpret the chemistry of the systems. A set of parameters that describe the forces between assemblies of atoms (collections of atoms or domains of a protein) in a molecule can be derived from the output of MD simulations. These parameters are further used in mesoscale dynamics.

These atomistic computational methods of chemistry and physics have evolved into sophisticated but practical tools that (using modern supercomputers) now allow many systems to be modeled and simulated. Thus, it is becoming possible to quantitatively predict the three-dimensional structure and dynamics of important biomacromolecules and to ana-

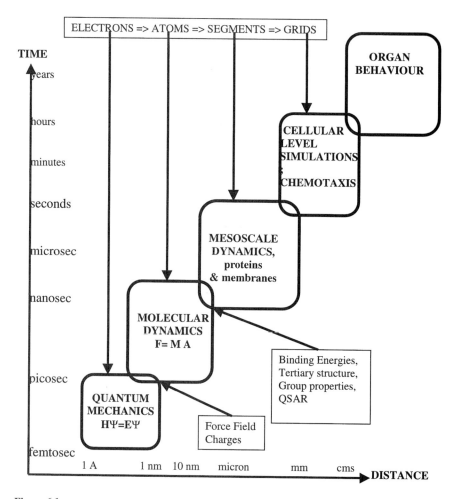

Figure 6.1
The hierarchy of biomolecular simulations. It is necesary to predict reliable properties before starting an experiment. The foundation is quantum mechanics. This allows prediction in advance, but is not practical for time and distance scales of molecular engineering. Thus one must extend from QM to large-scale, cellular-level dynamics by a succession of scales, where at each scale the parameters are determined by averaging over the finer scale.

lyze their interactions with other molecules at appropriate levels of computational resolution (ranging from electronic, to molecular, to the mesoscale cellular level). This makes it practical to begin addressing many complex issues of biological systems in terms of atomistic descriptions that provide quantitative information about the fundamental processes. Such atomistic computer simulations allow one to obtain static and dynamic molecular models for complex biosystems that describe the properties of the macroscale system or process in terms of concepts emphasizing the atomic origins of the phenomena (e.g., how the precise shape and exposed surface of a protein determines its function, which is critical to a drug design). This atomic-level description of the dynamic structure of proteins should be valuable for understanding:

- reaction chemistry at active sites in enzymes
- binding energetics and rates of small molecules or ligands to DNA, enzymes, and receptors (e.g., retinal to the signal receptor to protein rhodopsin)
- binding of antibodies to antigens
- conformational changes in proteins and how these changes modify function (relevant to neurological diseases)
- binding of proteins and other ligands to specific sites on DNA (relevant to expression)

6.2   Molecular Dynamics

Depending on the size of the system to be modeled and the accuracy required, there has evolved a hierarchy of QM methods in which approximations are often used to obtain greater speeds. The most accurate methods use no experimental data (first principles or *ab initio* QM) and are known by such names as Hartree–Fock (HF), density functional theory (DFT), and configuration interaction (CI) (Schaefer 1984, Parr and Yang 1989). However more approximate but much faster semiempirical methods [known by such names as modified intermediate neglect of differential overlap (MINDO), extended Hückel, and AM1) based partly on comparisons with experiments are quite valuable for many problems (Pople and Beveridge 1970). Such QM methods are essential for describing systems in which the nature of the bonds changes; for example, chemical reactions, excited states of molecules, and electron transfer.

In describing the structure and dynamics of large molecules such as proteins and DNA, the nature of the bonds is relatively insensitive to the environment. Instead, the focus of interest usually involves packing and conformation. For such problems, the electrons of QM can be accurately replaced with springs and the dynamics described with Newton's equations rather Schroedinger's. Here the choice of parameters in the FF is critical. They should give a description close to QM (and experiment). This has served well for many problems; however, it is usually difficult to include polarization, charge transfer, and changes in bond order in the FF. Consequently, direct theoretical descriptions of chemical reactions have been the domain of QM. Recent progress in developing FFs that allow for bond breaking and reactions has been reported (Che et al. 1999, Qi et al. 1999, Demiralp et al. 1998).

6.2.1 The Force Field

The choice of the FF is critical for accurate predictions of the properties of a system. For biomolecules, the FF is described in terms of two types of interaction energies:

$$E_{\text{tot}} = E_{\text{nonbond}} + E_{\text{valence}}, \tag{6.1}$$

where E_{valence} describes interactions involving changes in the covalent bonds and E_{nonbond} describes the nonbonded interactions.

The nonbond energy is separated into electrostatic (Coulomb), van der Waals (VDW), and sometimes explicit hydrogen bond (HB) components. Each atom has associated with it an atomic charge, q_I, leading to an electrostatic energy of the form

$$E_{\text{elec}} = \sum_{I,J} \frac{q_I q_J}{\epsilon R_{IJ}}. \tag{6.2}$$

Here q_I is the charge on atom I, R_{IJ} is the distance between atoms I and J, and ϵ accounts for the units and dielectric constant of the medium. This raises the issue of how to determine the charges. For modern FF, the charges are determined either directly from QM or from charge equilibration (QEq), a general semiempirical scheme that allows charges to depend on instantaneous structure. Sometimes the dielectric constant is used to replace some effects of the solvent and sometimes the solvent is included explicitly or by a continuous Poisson–Boltzmann approximation.

For biological systems, hydrogen bonding (e.g., between the amide hydrogen and the carbonyl oxygen) is particularly important in determining structure and energetics. Thus, some FFs include specific special HB interaction terms in the VDW part of (6.2) (Levitt 1983, Brooks et al. 1983, Weiner et al. 1986, Mayo et al. 1990, Cornell et al. 1995). However since QM shows that electrostatics dominates hydrogen bond interactions, most modern FFs account for hydrogen bonding through the electrostatics (van Gunsteren et al. 1987, Jorgensen and Tirado-Tives 1988, Hermans et al. 1984).

The two most common forms for the VDW energy are the Lennard-Jones 6–12

$$E_{LJ} = AR_{IJ}^{-12} - BR_{IJ}^{-6} \tag{6.3}$$

and the Buckingham exponential, –6

$$E_{\exp 6} = Ae^{-CR_{IJ}} - BR_{IJ}^{-6},\tag{6.4}$$

where the *A, B,* and *C* parameters are usually defined by a comparison with experimental data or accurate QM calculations on small molecules. Morse functions

$$E_{IJ} = A\{[e^{-C(R-Ri)} - 1]^2 - 1\}\tag{6.5}$$

are also used (Gerdy 1995, Brameld et al. 1997).

Because the nonbond terms require calculation for all pairs of atoms (scaling as the square of the number of atoms), this is a bottleneck for simulations of very large systems. To reduce these costs, it is common to ignore interactions longer than some cutoff radius (using a spline function to smooth the potential at the cutoff radius). More recently, fast multipole techniques [the cell multipole method (CMM); Ding 1992, Lim et al. 1997, Figueirido et al. 1997] have been used to obtain accurate nonbond energies without cutoffs but scaling linearly with the size of the system. These fast and accurate methods are being used in MD simulations for large-scale biological systems (Vaidehi and Goddard 1997).

The valence energy is usually described as

$$E_{\text{valence}} = E_{\text{bond}} + E_{\text{angle}} + E_{\text{torsion}} + E_{\text{inversion}},\tag{6.6}$$

where bond describes the interaction between two bonded atoms; angle describes the interaction between two bonds sharing a common atom; torsion describes the interaction between a bond *IJ* and a bond *KL* connected through a bond *JK;* and inversion is used to describe nonplanar distortions at atoms with three bonds.

Since the covalent bonds are expected to remain near equilibrium, the bond stretching and angle bending are taken as harmonic:

$$E_{\text{bond}} = \frac{1}{2} K_{IJ}(R_{IJ} - R_0)^2\tag{6.7}$$

$$E_{\text{angle}} = \frac{1}{2} K_{IJK}(\theta - \theta_0)^2 \text{ or } (1/2)C(\cos\theta - \cos\theta_0)^2.\tag{6.8}$$

Here R_0 is the equilibrium bond distance, K_{IJ} is the bond force constant, θ is the bond angle between bonds *IJ* and *JK*, θ_0 is the equilibrium bond angle, and K_{IJK} (or *C*) is the angle force constant.

The torsion energy is described in terms of the dihedral angle, ϕ, between bonds *IJ* and *KL* along bond *JK*. This is periodic and can be written as

$$E_{\text{torsion}} = \frac{1}{2} \sum_{n=1}^{6} K_{\phi,n}[1 - d\cos(n\phi)],\tag{6.9}$$

where $K_{\phi,n}$ is the torsion energy barrier for periodicity n and $d = \pm 1$ describes whether the torsion angle ($\phi = 0$) is a minimum ($d = +1$) or a maximum. Torsion potentials are essential in describing the dependence of the energies on conformation.

Finally, an inversion term is needed to describe the distortions from planarity of atoms making three bonds (e.g., in aromatic amines or amides). For cases where the equilibrium geometry is planar (e.g., amine N or C), we use

$$E_{inversion} = K_{inv}(1 - \cos\omega). \tag{6.10}$$

For cases where the equilibrium geometry is nonplanar (e.g., an amine), we use

$$E_{inversion} = \frac{1}{2}C_I(\cos\omega - \cos\omega^0)^2, \tag{6.11}$$

where

$$C_I \frac{K_{inv}}{(\sin\omega^0)^2}. \tag{6.12}$$

Standard Force Fields   Given the functional forms as described above, the FF is defined by the particular choices for the parameters in the FF (force constants and equilibrium geometries). Three strategies have been used for biological systems.

One is to develop the FFs for a specific class of molecules. The most popular FFs for proteins and DNA are AMBER (Weiner et al. 1984, Cornell et al. 1995) and CHARMM (Brooks 1983, Mackrell et al. 1995), which are parameterized to describe the naturally occurring amino acids and nucleic acids. These parameterizations include the atomic charges required to describe the electrostatics. Such FFs have been quite useful, and the majority of simulations on natural protein and DNA systems use these FF. However, unusual ligands such as drug molecules, cofactors, substrates, or their modifications are difficult to incorporate, as are non-natural amino acids or bases. Useful here are FFs developed for organic systems [OPLS (Pranata et al. 1991) and MM3 (Allinger and Schleyer 1996)] which can describe parameters of non-natural amino acids or bases, along with most molecules that bind to biosystems.

The second strategy is to develop rule-based FFs based on the character and connectivity of the molecules. The simplest such generic FF is Dreiding (Mayo et al. 1990): Equilibrium bond distances are based on atomic radii, and the bond angles, inversion angles, and torsion periodicities are derived from simple rules based on fundamental ideas of bonding. There is only one bond force constant (bond order times 700 kcal/mol Å), one angle force constant 100 kcal/mol rad), and simple rules based on fundamental ideas of bonding. To obtain generic charges, the charge equilibrium method (Rappe and Goddard 1991) was developed in which all charges of all molecules are determined from three parameters per atom (radius, electronegativity, and hardness). Despite its simplicity, Dreiding gives accurate structures for the main group elements (the B, C, N, O, and F columns) most prevalent in biology.

A more general generic FF that also treats transition metals is the universal force field (UFF; Rappe et al. 1992), which treats all elements of the periodic table (through Lr, element 103). UFF includes simple rules in which the force constants of molecules are derived from atomic parameters. Such generic FFs are most useful for systems with unusual arrangements of atoms or for new molecules for which there are no experimental or QM data. For applications in which it is necessary to have the exact molecular structure, such generic FFs may not be sufficiently accurate. Hence, there is a need for a generic FF that incorporates just enough specificity for accurate simulations of biomolecules while providing the flexibility to model all other organic molecules.

Accurately predicting the vibrational spectra of the molecule, in addition to the geometry and energy, requires the third strategy for FFs in which the parameters are optimized for a specific class. This requires cross-terms coupling different bonds and angles. A general procedure for developing such spectroscopic FFs from QM is the biased Hessian method, which has been used for many systems (Dasgupta et al. 1996). Usually the spectroscopic quality FFs are useful only for limited classes of molecules.

Effect of Solvents   The role of solvents (particularly water) is critical in biological simulations, since the secondary and tertiary structures of proteins are determined by the nature of the solvent. Several levels of simulation have been used.

The earliest studies ignored the solvent entirely, usually replacing the effect of solvent polarization by using a dielectric constant larger than one (which is often distance dependent). At this level of approximation, it is important to include counterions to represent the effects of solvent on charged groups. Such simulations were useful in understanding the gross properties of systems.

The most accurate MD treatments include an explicit description of the water using an FF adjusted to describe the bulk properties of water (Jorgensen et al. 1983; Levitt et al. 1997; Rahman and Stillinger 1971; Berendsen et al. 1981, 1987). Although they are accurate, such calculations usually require very long time scales in order to allow the hydrogen-bonding network in the water to equilibrate as the biomolecule undergoes dynamic motion. In addition, for an accurate treatment of solvent effects, the number of solvent atoms may be ten times that of the biomolecule.

An excellent compromise for attaining most of the accuracy of explicit water, while eliminating the atoms and time scale of the solvent, is the dielectric continuum model. Here the electrostatic field of the protein is allowed to polarize the (continuum) solvent, which then acts back on the protein, leading to the Poisson–Boltzmann equation (Sitkoff et al. 1993). Recent developments have led to computationally efficient techniques (Tannor et al. 1994) that can accurately account for effects of solvation on the forces (geometry) and free energy. For example, solvation energies of neutral molecules are accurate to better than 1 kcal/mol.

6.2.2   Molecular Dynamics Methods

The Fundamental Equations   Given the FF, the dynamics are obtained by solving Newton's equations of motion:

$$-\mathbf{F}_i = m_i \ddot{x}_i,\tag{6.13}$$

where $\mathbf{F}_i$ is the force vector, $\ddot{x}_i$ denotes the acceleration, and m_i is the mass of atom i. Solving equation (6.13) leads to $3N$ coordinates and $3N$ velocities that describe the trajectory of the system as a function of time. Often this dynamic trajectory provides valuable information about a system. Thus, MD methods have been useful for the exploration of structure–activity relationships in biological molecules (McCammon 1987).

However, more often it is the ensemble of conformations near equilibrium that is required to calculate accurate properties. Assuming that the barriers between different relevant structures are sufficiently small that they can be sampled in the time scale of the simulations, we think that the collection of conformations sampled in the dynamics can be used as the ensemble for calculating properties.

The steps in MD simulations are as follows:

1. Start with the structure ($3N$ coordinates), which may be obtained from a crystal structure or from Build software using standard rules of bonding. In addition, it is necessary to have an initial set of velocities, which are chosen statistically to describe a Maxwell–Boltzmann distribution.

2. At each timestep, calculate the potential energy and its derivative to obtain the force on every atom in the molecule. Equation 6.14 is then solved to obtain the $3N$ accelerations at timestep t.

$$\ddot{x}_i = f_i / m_i = -\nabla E_{\text{tot}} / m_i,\tag{6.14}$$

where ∇E_{tot} is the gradient of the potential energy.

3. To obtain the velocities and coordinates of each atom as a function of time (the trajectory), we consider a timestep δ and write the acceleration at the nth timestep as

$$\ddot{x}_n = \frac{\dot{x}_{n+\frac{1}{2}} - x_{n-\frac{1}{2}}}{\delta},\tag{6.15}$$

leading to

$$\dot{x}_{n+\frac{1}{2}} = \dot{x}_{n+\frac{1}{2}} - \frac{\delta}{m_i} \nabla E_n.$$ (6.16)

Integrating (6.16) then leads to the coordinates at the next timestep,

$$x_{n+1} = x_n + \delta x_{n+\frac{1}{2}}.$$ (6.17)

Equations (6.16) and (6.17) are the fundamental equations for dynamics (the Verlet velocity leapfrog algorithm). The first segment of MD is used to equilibrate the system, removing any bias from the initial conditions.

4. The acceleration and then the velocities are integrated to determine the new atomic positions. This integration is usually performed using the Verlet leapfrog algorithm.

The timestep of integration, δ, must be short enough to provide several points during the period of the fastest vibration. If the hydrogen atoms are described explicitly, the timestep is usually 1 to 2 fs.

The trajectory of the molecular systems may require time scales ranging from picoseconds to hundreds of nanoseconds, depending on the application and the size of the system. Thus, a computationally efficient MD algorithm must allow fast and accurate calculation for atomic forces and use the longest possible timesteps compatible with accurate Verlet integration to simulate molecular motions on the longest time scale.

NPT and NVT Dynamics   Newton's equations of motion (6.13) describe a closed system. Thus the total energy (kinetic energy plus potential energy) of the system cannot change, and the system is adiabatic. If the volume is held constant, this simulation generates the microcanonical ensemble of statistical mechanics (denoted NVE for constant number of molecules, volume, and energy). However, most experiments deal with systems in equilibrium with temperature and pressure baths, leading to a Gibbs ensemble (Allen 1987). For the trajectory to generate a Gibbs ensemble, it is necessary to allow the internal temperature of the molecule to fluctuate in the same way it would if it were in contact with a temperature bath, and the volume must fluctuate in the same way it would in contact with a pressure bath.

Several methods (Woodcock 1971, Nose 1984, Hoover 1985, Vaidehi et al. 1996) are used to control the temperature of an MD simulation. The most rigorous method (Nose 1984) introduces into the equations of motion a new dynamic degree of freedom ζ, which is associated with energy transfer to the temperature bath (friction). If the volume is kept fixed, the Nose dynamics generates a Helmholtz canonical ensemble, giving rise to the correct partition function for an NVT system. These partition functions can be used to calculate macroscopic properties of the system. We consider NVT canonical dynamics as adequate for most biochemical problems [such as calculating binding energy and other molecular properties to be used in deriving quantitative structure–activity relationships (QSAR) that are useful for predicting enzyme activity].

Some interesting biological applications consider the response of proteins to external pressure (Floriano et al. 1998). In this case, we use periodic boundary conditions (described later), placing the molecule and solvent in a periodic cell that can be acted upon by external stresses. Examining the structural deformations of proteins under pressure or under external stress requires that the MD allow the internal pressure of a molecule to fluctuate in the same way as for a system in a constant-pressure environment (Parrinello and Rahman 1981). The modified Newton equations lead to NPT dynamics (constant number of particles, pressure, and temperature).

Constrained Internal Coordinates   The short time scale of 1 fs for MD is required to describe the very rapid oscillations involved in bond stretching and angle bending motions. However, for proteins and nucleic acids, it is the low-frequency motions involved in conformational changes that are of most interest. Several algorithms (Ryckaert et al. 1977, van Gunsteren et al. 1990, Mazur and Abagayan 1989, Jain et al. 1993, Rice and Brunger 1994) have been developed for fixing the bonds and angles in order to focus on the conformational motions. Such constrained dynamics algorithms lead to coupled equations of motion:

$$M(\theta)\ddot{\theta} + C(\theta,\dot{\theta}) = T(\theta) \tag{6.18}$$

for P degrees of freedom (torsions). Here $\ddot{\theta}$ is the angular acceleration; M is the $P \times P$ mass matrix (moment of inertia tensor), which depends on the internal coordinates θ; T is the vector of general forces (tensor) on the atoms; and C is the velocity- dependent Coriolis force. At each timestep we know θ, M, T, and C, and we must solve the matrix equation (6.18) to obtain the acceleration,

$$\ddot{\theta} = M^{-1}(T - C). \tag{6.19}$$

Integration of $\ddot{\theta}$ gives the velocity $\dot{\theta}$ and further integration leads to the torsion or the dihedral angles θ from which the coordinates can be obtained. The problem is that solving equation (6.19) requires inverting the $P \times P$ dense mass matrix M at every timestep. For a system with, say, 10,000 atoms, there might be 3000 torsional degrees of freedom, making the solution of (6.19) for every timestep impractical (the cost of inverting M scales as P^3). Recently we developed the Newton–Euler inverse mass operator (NEIMO) method (Jain et al. 1993, Mathiowetz et al. 1994, Vaidehi and Goddard 1996), which solves (6.19) at a computational cost proportional to P. The NEIMO method considers the molecule to be a collection of rigid "clusters" connected by flexible "hinges." A rigid cluster can be a single atom, a group of atoms (a peptide bond), or a secondary structure (a helix or even an entire domain of a protein). Such constrained models allow much larger timesteps.

Table 6.1
Time steps obtained in the hierarchical modeling of two proteins

Protein	MD method	Degrees of freedom	Time step (ps)
Protein A[1]	Newtonian	1062	0.001
	NEIMO (all torsions)	219	0.010
	H-NEIMO	92	0.020
PGK	Newtonian (all atom)	12525	0.001
	NEIMO (all torsions)	2210	0.005
	H-NEIMO	80	0.010

[1]Protein A is a helix-coil-helix segment from staphylococcus aureus

With fine-grain all-torsion NEIMO dynamics, we normally treat double bonds, terminal single bonds, and rings (benzene) as rigid bodies. In hierarchical or H-NEIMO dynamics (Vaidehi et al. 2000), we allow higher levels of coarseness, keeping various segments or parts of a domain of a protein rigid during the dynamics. This allows larger timesteps, as illustrated in table 6.1.

Using hierarchical NEIMO simulations on the glycolytic enzyme phosphoglycerate kinase (PGK), we are able to follow the long time-scale domain motions in PGK responsible for its function. PGK catalyzes an essential phosphorylation step in the glycolytic pathway. Under physiological conditions, PGK facilitates the phosphoryl transfer from adenosine diphosphate (ADP) to adenosine triphosphate (ATP). PGK consists of two major domains (1400 atoms in each domain) denoted as the C-domain and the N-domain. In most crystal structures (McPhillips et al. 1996, Bernstein et al. 1997), the substrates are found bound to the opposite domains at a distance of ~13 Å. Thus, Blake (1997) proposed a hinge-bending mechanism by which the protein brings the substrates together to react. Using H-NEIMO (figure 6.2; see also plate 19), we found low-frequency domain motions that take the open structure examined by yeast PGK (McPhillips 1996) to the closed structure examined by *Trypanosoma brucei* PGK (Bernstein et al. 1997). The NEIMO dynamics suggest that PGK undergoes long time-scale motions that put the substrate binding sites together, then take them apart, then put them together again. The extent and rate of the domain motions depend on the nature of the substrates bound.

MPSim   In order to allow long time simulations on very large systems (up to a million atoms), we developed the massively parallel simulation (MPSim) program (Lim et al. 1997) to operate efficiently with massively parallel computers. MPSim includes important algorithm developments such as CMM (for calculating long-range nonbond interactions) (Ding et al. 1992), NEIMO (Jain et al. 1993, Mathiowetz et al. 1994, Vaidehi et al. 1996), and the Poisson–Boltzmann solution (Tannor et al. 1994). It is compatible with massively parallel high-performance computers (SGI-Origin-2000, IBM-SP2, Cray T3D/T3E, HP/Convex-Exemplar, and Intel Paragon). MPSim has been used to understand the action of drugs on

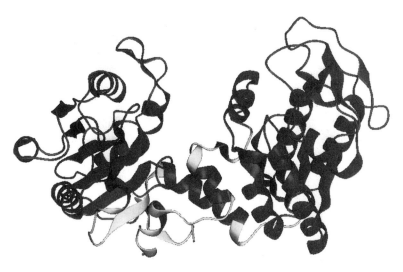

Figure 6.2
The H-NEIMO model for phosphoglycerate kinase. Here the parts of two domains that bind the substrates (shown in yellow) are kept rigid while the loops (shown in green) are allowed to have full torsional freedom. H-NEIMO MD simulations exhibit large-scale domain motions that bring the substrates close together and then apart. (See plate 19.)

the human rhinovirus (Vaidehi et al. 1996) as summarized in section 6.3.2. It is also being used for materials science problems (Miklis et al. 1997, Demiralp et al. 1999).

Periodic Boundary Conditions   In order to control the pressure on a system and to include explicit solvents without introducing complications of free surfaces, it is convenient to place the molecule in a large box (much larger than the molecule) and then reproduce it periodically to fill space. In addition, it is particularly useful to describe some systems, such as DNA, as periodic and repeating in one direction. The computational box containing the molecular system is surrounded by an infinite number of copies of itself (see figure 6.3).

Because periodic systems involve an infinite number of atoms, some care must be taken in calculating the long-range forces. Otherwise, singularities or wild oscillations can occur. The most common accurate method is Ewald summation (Ewald 1921, de Leeuw et al. 1980, Heyes 1981, Allen 1987, Karasawa and Goddard 1989, Chen et al. 1997), which

Figure 6.3
Molecules in a periodic system. (a) Two-dimensional view of a tripeptide in a periodic box. (b) The simulation unit cell is the box outlined in the center. It is surrounded by an infinite array of equivalent boxes to that there is no free surface. Molecules are allowed to move from box to box, but the number in the unit cell is always constant.

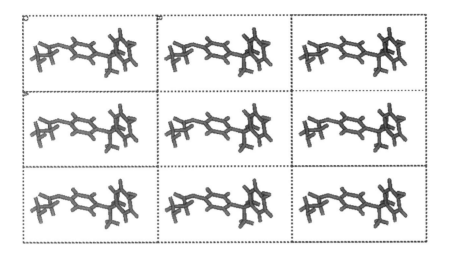

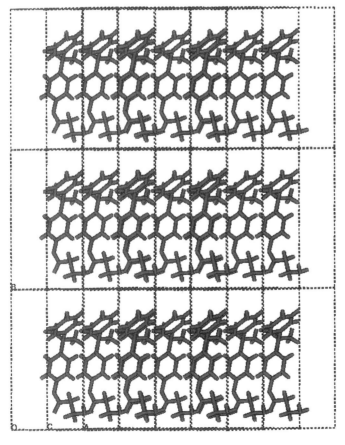

considers point charges smeared over a region of finite size $1/\eta$, chosen to converge rapidly (in real space), and then Fourier transforms the difference between the smeared charges and the point charges to obtain rapid convergence (in the reciprocal space sums). Karasawa and Goddard (1989) showed how to choose the optimum η to minimize computational cost for a given level of accuracy. This leads to costs scaling as $N^{3/2}$ for systems with N atoms per unit cell. Ding et al. (1992b) showed how to use the reduced CMM method to achieve linear scaling.

Monte Carlo Methods   For many systems, the barriers between low-lying structures may be too large for MD to sample all the structures. In such cases, we often use Monte Carlo or statistical sampling techniques, using a random search algorithm such as Monte Carlo metropolis. With a sufficiently large number of samples, the occurrence of each conformation is proportional to the Boltzmann factor, leading to a canonical distribution. The steps involved in the Monte Carlo simulation procedure are:

1. Starting from a given molecular conformation, a new conformation is generated by random displacement of one or more atoms. The random displacements should be such that in the limit of a large number of successive displacements, the available conformation space is uniformly sampled.

2. The newly generated conformation is accepted or rejected based on the change in the potential energy of the current step compared with the previous step. The new conformation is accepted if the change in potential energy $\Delta V = V$ (present step) $- V$ (previous step) ≤ 0, or if $\Delta V > 0$ when the Boltzmann factor is greater than a random number R.

Upon acceptance, the new conformation is counted and used as a starting point for the generation of the next random displacement. If the criteria are not met, then the new conformation is rejected and the previous conformation is counted again as a starting point for another random displacement. This method thus generates a Boltzmann ensemble of conformations. Many Monte Carlo methods are available (Allen 1987) and have been used as a fast conformational search tool in protein folding (Sternberg 1996). A recent advance, continuous conformation Boltzmann biased direct Monte Carlo (Sadanobu and Goddard 1997) has been used to determine the complete set of folding topologies for proteins with up to 100 residues (Debe et al. 1999). Such methods show considerable promise for solving the protein-folding problem (predicting the tertiary structure from a primary sequence).

6.3   Application to Biological Problems

In this section, we summarize some recent applications of quantum chemistry and molecular dynamics to problems in structural biology.

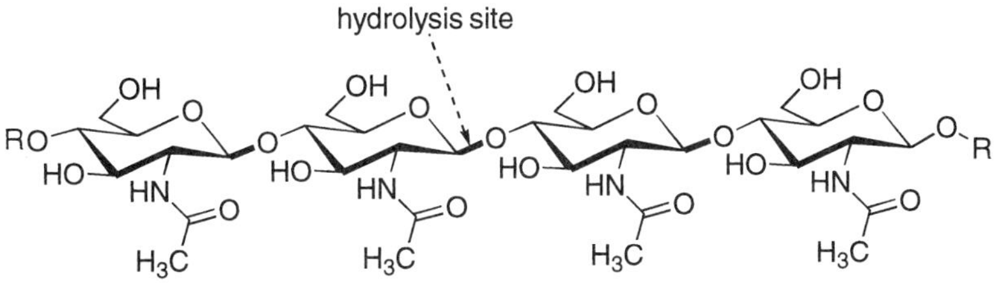

Figure 6.4
An arrow marks the hydrolysis site of chitin, the Symbol"b (1,4)-*N*-acetylglucosamine (GlcNAc) polysaccharide substrate of chitinases.

6.3.1 Study of Enzyme Reaction Mechanisms

The reaction mechanisms for many enzymes have been studied using a combination of QM and MD methods (McCammon 1987, Cunningham and Bach 1997). Examples include the hydrolysis of a peptide bond by serine proteases, and hydrolysis by the metalloenzyme, staphylococcal nuclease of both DNA and RNA, etc. (Warshel 1991). We describe here a recent QM and MD study for the elucidation of the mechanism of family 18 (Brameld and Goddard 1998a) and family 19 chitinases (Brameld and Goddard 1998b). This application relies heavily on a combination of QM and MD methods, demonstrating the feasibility of solving difficult problems using modern computational methods.

Chitin (see figure 6.4) is a β (1,4)-linked *N*-acetylglucosamine (GlcNAc) polysaccharide that is a major structural component of fungal cell walls and the exoskeletons of invertebrates (including insects and crustaceans). This linear polymer may be degraded through the enzymatic hydrolysis action of chitinases. Chitinases have been found in a wide range of organisms, including bacteria (Watanabe et al. 1990; plants (Collinge et al. 1993), fungi (Bartinicki-Garcia 1968), insects (Kramer et al. 1985), and crustaceans (Koga et al. 1987). For organisms that utilize the structural properties of chitin, chitinases are critical for the normal life-cycle functions of molting and cell division (Fukamizo and Kramer 1985, Kuranda and Robbins 1991). Because chitin is not found in vertebrates, inhibition of chitinases is a promising strategy for treatment of fungal infections and human parasitosis (Robertus 1995).

Based on amino acid sequence, the glycosyl hydrolases have been classified into 45 families. Using this classification method, the chitinases form families 18 and 19, which are unrelated, differing both in structure and in mechanism. Sequence analysis shows little homology between these classes of chitinases. Family 19 chitinases (found in plants) share the bilobal α + β folding motif of lysozyme, which forms a well-defined substrate binding cleft between the lobes. In contrast, family 18 chitinases share two short sequence motifs, which form the catalytic (βα) 8-barrel active site. Family 18 chitinases with diverse

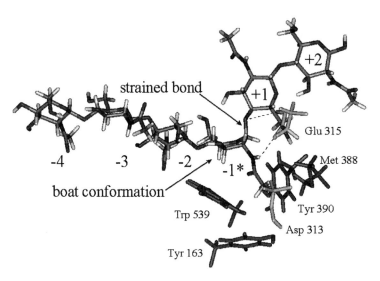

Figure 6.5
The minimum energy structure for the −1−boat hexaNAG conformation. A boat geometry for GlcNAc residue −1 and the twist between residues −1 and +1 strains the linking glycosidic bond. This distortion observed in the simulations should be included when designing new inhibitors.

sequences have been isolated from a wide range of eukaryotes and prokaryotes. The hydrolysis site of chitin is shown in figure 6.4.

Brameld investigated the hydrolysis mechanisms of the chitinases by examining the reactivity of the chitin substrate alone and in the presence of the enzyme. This was done using *ab initio* quantum mechanical calculations on three possible reaction intermediates for the enzymatic hydrolysis of chitin. He found that anchimeric assistance from the neighboring *N*-acetyl group of the chitin is critical in stabilizing the resulting oxazoline ion intermediate.

MD simulations of the complete enzyme with bound substrate led to further insights into the mechanisms of family 18 and 19 chitinases, which differ substantially. All MD simulations were carried out using the MSC-PolyGraf program using the Dreiding FF (Mayo et al. 1990). QEq charges (Rappe and Goddard 1991) were used for all GlcNAc residues. All nonbond interactions were considered explicitly (using a distance-independent dielectric constant), with a cutoff of 9.5 Å for MD simulations and 13.5 Å for energies. Solvation energies were estimated using the continuum solvent model in the Delphi program (Tannor et al. 1994). Two possible intermediates, namely, the oxazoline ion and the oxocarbenium ion, were proposed. The formation of the oxazoline ion intermediate results from substrate distortion (see figure 6.5) induced within the active site of family 18 chitinases. This substrate distortion was observed in the MD simulations and is necessary to design inhibitors for the family 18 chitinase. Yet surprisingly, the family 19 chitinases do not utilize an oxazoline ion intermediate and undergo a considerable change in enzyme conformation to stabilize the resulting oxocarbenium ion intermediate.

The oxazoline transition state serves as a target for the rational design of more potent glycosidase inhibitors specific to family 18 chitinases. Simple analogs of allosamidin that incorporate the key features of a delocalized positive charge while maintaining a chairlike sugar conformation may prove to be synthetically more accessible than allosamidin. Such analogs could lead to a new generation of chitinase transition-state inhibitors.

6.3.2 A Model for Drug Action on Rhinovirus-1A and Rhinovirus-14

Human rhinovirus (HRV) belongs to the picornavirus family. It has over 100 serotypes, providing a challenge to drug design. The serotypes of HRV are classified into two groups. The major receptor group (including HRV-14 and HRV-16) binds to the intercellular adhesion molecule 1 (ICAM-1) receptors. The minor receptor group (including HRV-1A) binds to low-density lipoprotein-type receptor molecules. The protein capsid of HRV consists of an icosahedral shell with 60 copies of the four viral proteins (VP1, VP2, VP3, and VP4) totaling 480,000 atoms (300 Å in diameter). Figure 6.6 (see also plate 20) shows the icosahedral shell, its elements such as the pentamers that make up the icosahedron, the asymmetric unit that makes up the pentamer, and the basic four viral proteins of the asymmetric unit. A single-stranded RNA is enclosed in the protein shell. The sequence of events involved in the endocytosis is not clear yet, but circumstantial evidence suggests that the RNA is released through the pentamer in the virus coat (Rueckert 1991).

There are several known isoxazole-derived drugs for HRV-1A and HRV-14 (Couch 1990). It is known that binding of these drugs to HRV-14 prevents binding of the virus to the ICAM-1 receptors. However for HRV-1A, binding of thee drugs does not block receptor attachment; rather, it prevents uncoating of the virus. One speculation is that this binding leads to stiffening of the viral capsid. Based on this speculation and the fact that the RNA is released through the pentamer channel, Vaidehi and Goddard (1997) proposed the pentamer channel stiffening Model (PCSM). Drug action on HRV-1A constricts or stiffens the pentamer channel sufficiently that the RNA cannot exit, thus preventing uncoating.

Using MPSim (see the section on this program) on the KSR-64 processor parallel computer, we showed a strong correlation (see figure 6.7) of drug effectiveness (minimum inhibitory concentration, MIC) with the strain energy increase calculated for various drugs. Here we defined the strain energy as the energy required to expand the pentamer channel to 25 Å. The strain energy was calculated using MPSim with the Amber FF for the viral proteins and the Dreiding FF for the drugs.

Figure 6.7 shows that all effective drugs cause an increase in the strain energy required to open the pentamer channel. The best drug, WIN56291 (MIC = 0.1 μM), shows the sharpest increase in the strain energy compared with the native HRV-1A or an ineffective

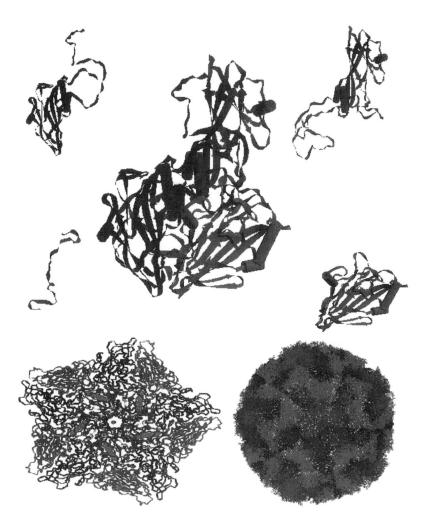

Figure 6.6
Structure of a rhinovirus. The complete icosahedral viral capsid (300 Å diameter) is shown at the bottom right. The four viral proteins making up the asymmetric unit of the virus are shown in the top half. The five asymmetric units forming the pentamer are shown in the bottom left. Twelve such pentamers form the complete viral capsid. (See plate 20.)

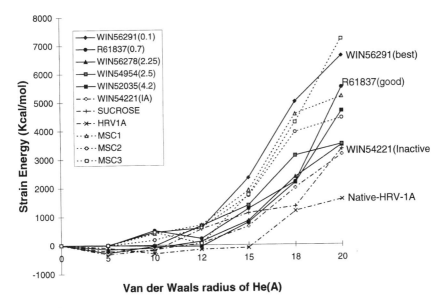

Figure 6.7
Pentamer stifness for HRV-1A with various Winthrop drugs. The MIC values for the drugs are given in parenthesis. MSC1, MSC2, and MSC3 denote new candidates for drug molecules.

drug, WIN54954 (MIC = 2.5 μM). This suggests that the PCSM can be used to predict the efficacy of a drug before synthesis and testing. Three drugs—MSC1, MSC2, and MSC3—were tested in this way.

6.3.3   Calculation of Binding Energy Using Free Energy Perturbation Theory

A major application of MD simulations is for drug design, where high binding energy is expected to be a necessary condition for a good drug. MD provides microscopic-level information (atomic and molecular positions, velocities, etc.) about the dynamics of macromolecules; this information can be averaged over time (using appropriate formulas from statistical mechanics) to obtain the macroscopic thermodynamic properties of a system, such as the free energy, temperature, and heat capacity. Thus, one assumes that the macroscopic property G_{obs} can be expressed as an average over the ensemble of conformation from the MD, $G_{obs} = <G>_{ens}$, and that this can be expressed as the time average of $G(r_i)$ over the trajectory over a sufficiently long time interval. This is written as

$$G_{obs} = <G(r_i)>_{time} = \lim_{t\to\infty} \frac{1}{t} \int_0^t G(r)dt, \tag{6.20}$$

where $G(r_i)$, r_i is a function of time and is generated as the trajectory of the MD simulations.

The problem is that the convergence of (6.20) is sufficiently slow that the error may be large compared with the difference in bond energies for the drugs. This problem is solved by transforming the system slowly from drug A ($\lambda = 0$) to drug B ($\lambda = 1$) and integrating the difference in G (λ) along the trajectory. This is called *free energy perturbation* (FEP) theory. It has been demonstrated to give a quantitative estimate of the relative free energy of binding of various drug molecules or inhibitors to its receptors. The integration in (6.20) is replaced by finite sum over timesteps. The free energy of molecular systems can be calculated (van Gunsteren and Weiner 1989) using equation (6.20). The free energy difference given by

$$G(\lambda) - G(\lambda_i) = -RT \ln < \exp\left[-\left(\frac{V_\lambda - V_{\lambda i}}{RT}\right)\right] > \lambda_i \qquad (6.21)$$

provides a means of calculating the free energy difference any two states $A(\lambda)$ and $A(\lambda_i)$. However, unless the states share a significant fraction of conformation space, convergence is very slow. The convergence time can be reduced by calculating the relative free energy difference between closely related states, using the thermodynamic cycle. To visualize the method (see figure 6.8), we consider the relative binding of two ligands, L_1 and L_2, to a receptor that is a protein or DNA. The appropriate thermodynamic perturbation cycle (Zwanzig 1956) for obtaining the relative binding constant is given in figure 6.8.

The ratio of the binding constants for L_1 and L_2 can be calculated from the equation

$$\frac{K_{L2}}{K_{L1}} = \exp[-(\Delta G_{L2} - \Delta G_{L1})/RT] \qquad (6.22)$$

where R denotes the gas constant and K_{L1} and K_{L2} are the binding constants for L1 and L2. Simulation of steps 1 and 3 involves replacement of solvent molecules in the binding pocket of the protein (receptor) by the ligand and removing the ligand from the solvent and binding it to the protein. These are very slow processes for the time scale of MD simulations. However, it is possible to simulate steps 2 and 4 using MD if the chemical composition and the structure of the ligands are very closely related, as described in the example below. Hence the free energy difference for steps 1 and 2 is related to the free energy difference between steps 2 and 4 by (figure 6.8):

$$\Delta G_3 - \Delta G_1 = \Delta G_2 - \Delta G_4$$

The crucial factor in the simulations of steps 2 and 4 is in the adequacy of the sampling in the configuration space of the system. Longer MD simulations are required for accurate free energies.

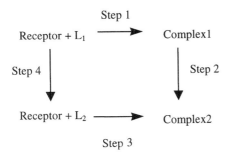

Figure 6.8
Thermodynamic cycle for the calculation of the relative free energies of binding.

This method has been used extensively to study the relative binding energies of drugs and inhibitors for enzymes (van Gunsteren and Weiner 1989; Marrone et al. 1997; Plaxco and Goddard 1994) and inhibitors. Here we describe briefly one of the recent applications (Marrone et al. 1996) of the free energy perturbation simulations to the study of inhibition of the enzyme adenosine deaminase by 8R-coformycin and (8R)-deoxycoformycin. The inhibition of the enzyme adenosine deaminase, which deaminates the base adenosine, could provide an effective treatment of some immunological disorders. Thus MD simulations can play a critical role in designing inhibitors for this enzyme since it gives a good model for the binding site of the inhibitors in the enzyme and also provides an estimate of their binding energies. The coformycin and deoxycoformycin molecules differ in the sugar moiety attached to them, which is ribose in the case of coformycin and deoxyribose for deoxycoformycin. The molecular structure of these two substrates is shown in figure 6.9.

It is clear that these two inhibitors differ only by only a small functional group and hence provide a case well suited for free energy perturbation calculations. Molecular dynamics and free energy simulations of coformycin and deoxycoformycin and their complexes with adenosine deaminase show a difference of -1.4 kcal/mol in binding energy between deoxycoformycin and coformycin. Deoxycoformycin and coformycin differ by a hydroxyl group, but the relative binding energy is small enough, showing that this hydroxyl group is buried near a flexible hydrophilic region of the enzyme conformation rather than being sequestered in a hydrophobic pocket. Thus, detailed structural aspects of the transition state analog have been derived in this study (Marrone et al. 1996).

6.3.4 Quantitative Structure–Activity Relationships

Following the calculation of relative binding constants, one of the most important steps in the process of drug discovery is the study of QSAR (Franke 1984; Fanelli et al. 1998). The aim is to explain the experimental data obtained for binding of various ligands to a recep-

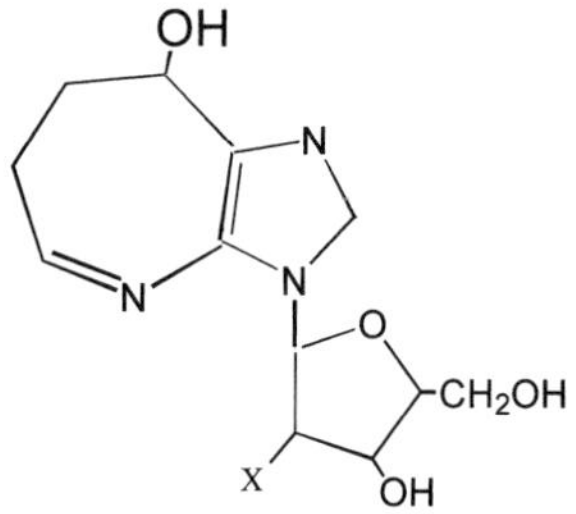

Figure 6.9
Molecular structure of adenosine deaminase inhibitors X = H for deoxycoformycin and X = OH for coformycin.

tor, at the molecular level, in terms of physiochemical properties of the ligands, and to predict or estimate similar biodata for new analogs. QSARs are equations that relate an
observed experimental quantity for ligand binding, for example, the minimum inhibitory
concentration measured routinely in drug industries, to the calculated molecular properties
of various regions of a drug molecule. QSARs are widely used today in the pharmaceutical
industry to design new drugs or inhibitors. Figure 6.10 shows the major steps involved in a
QSAR analysis.

Today computational chemistry using QM–MD simulations allows us to define and
compute ad hoc shape and size descriptors for the different conformations assumed by
drugs in biotest solutions. Together with the statistically sound experimental data measured on well-identified target receptors, these descriptors are essential elements for obtaining simple, consistent, comparable, and easily interpretable theoretical QSAR models
based on ligand similarity–target receptor complementarity paradigms. The molecular
properties calculated can be of varied natures. For example, based on the structural formula of the drugs, they can be broken into fragments, and the properties of the subsystem
or fragments that contribute to the drug activity can be calculated. Intermolecular interaction properties such as polarizability or the hydrophilic or hydrophobic nature of groups
can be computed. Thus, QM–MD simulations are used extensively in pharmaceutical com-
panies to derive QSARs (Fanelli et al. 1998).

6.4 Summary

Atomistic simulations constitute a powerful tool for elucidating the dynamic behavior of
biological systems. They can be used in a variety of problems in biological research and
drug design, as suggested in this chapter. The major thrust in this area is the simulation of
large systems (hundreds of angstroms) for long time scales (microseconds).

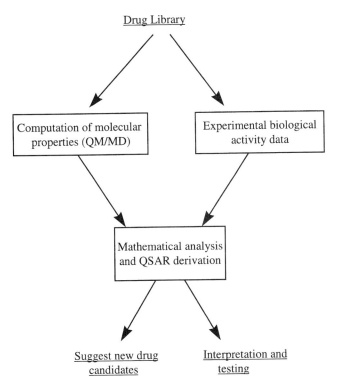

Figure 6.10
Principal steps involved in QSAR analysis.

The next challenge is to find accurate ways to describe the dynamics of biological systems at a coarse grain level while retaining the accuracy of atomic-level MD in order to examine dynamic behavior for very long times and large distance scales. The dream here would be eventually to simulate the processes of an entire cell, where the coarse grain description would describe the elastic properties of the cell membrane plus the chemical nature of the receptors, but without explicit atoms. In this approach, the grid points of the coarse grain would embody chemical and physical properties so that expanding the grid would lead automatically to a description of binding some new molecule or the active transport of a molecule down a channel. Thus, one can focus down to the atomic level to determine the molecular properties and then back to the coarse grain for the large-scale motions.

Probably the most important challenge in biology is the prediction of accurate three-dimensional structures and the functions of proteins (tertiary structure) entirely from the gene sequence (primary structure). Enormous progress is being made here (Debe et al. 1999), but there is not yet success in *ab initio* predictions.

Acknowledgments

This research was funded by the National Science Foundation (grants CHE-95-22179, GCAG CHE 95-2217, and SGER DBI-9708929), U.S. Department of Energy (DOE)-BCTR (David Boron), and the National Institute of Child Health and Development (National Institutes of Health) grant HD3638502.

The Materials Simulation Center is supported by grants from Army Research Office (DURIP), DOE-ASCI, British Petroleum Chemical, the Beckman Institute, Seiko-Epson, Exxon, Owens-Corning, Dow Chemical, Avery Dennison, Chevron Petroleum Technology, Chevron Chemical Co., Chevron Research Technology, and Asahi Chemical. Some calculations were carried out on the National Center for Supercomputing Applications (L. Smarr) at the University of Urbana.

References

Allen, M. P., and Tildesley, D. J. (1987). *Computer Simulations of Liquids.* Clarendon, Oxford.

Allinger, N. L., and Schleyer, P. V. (1996). Special issue on molecular mechanics. *J. Comp. Chem.* 17: 385–385.

Berendsen, H. J. C., Postma, J. P. M., van Gunsteren, W. F., and Hermans, J. (1981). Interaction models for water in relation to protein hydration. *Jerusalem Symp. Quantum Chem. Biochem.* 14: 331–342.

Berendsen, H. J. C., Grigera, J. R., and Straatsma, T. P. (1987). The missing term in effective pair potentials. *J. Phys. Chem.* 91: 6269–6271.

Bernstein, B. E., Michels, P. A. M., and Hol, W. G. J. (1997). Synergistic effects of substrate-induced conformational changes in phosphoglycerate kinase activation. *Nature* 385: 275–278.

Blake, C. F. (1997). Glycolysis-phosphotransfer hinges in PGK. *Nature* 385: 204–205.

Brameld, K. A., Dasgupta, S., and Goddard III, W. A. (1997). *Ab initio* derived spectroscopic quality force fields for molecular modeling and dynamics. *J. Phys. Chem.* B 101: 4851–4859.

Brameld, K. A., and Goddard III, W. A. (1998a). Substrate distortion to a boat conformation at subsite-1 is critical in the mechanism of family 18 chitinases. *J. Am. Chem. Soc.* 120: 3571–3580.

Brameld, K. A., and Goddard III, W. A. (1998b). The role of enzyme distortion in the single displacement mechanism of family 19 chitinases. *Proc. Natl. Acad. Sci.* 95: 4276–4281.

Brooks, B. R., Bruccoleri, R. E., Olafson, B. D., States, D. J., Swaminathan, and Karplus, M. (1983). CHARMM—a program for macromolecular energy, minimization, and dynamics calculations, *J. Comput. Chem.* 4: 187–217.

Che, J., Cagin, T., and Goddard III, W. A. (1999). Studies of fullerenes and carbon nanotubes by an extended bond order potential. *Nanotechnology* 10: 263–268.

Chen, Z. M., Cagin, T., and Goddard III, W. A. (1997). Fast Ewald sums for general van der Waals potentials. *J. Comp. Chem.* 18: 1365–1370.

Cornell, W. D., Cieplak, P., Bayly, C. I., Gould, I. R., Merz, K. M., Ferguson, D. M., Spellmeyer, D. C., Fox, T., Caldwell, J. W., and Kollman, P. A. (1995). A second generation force-field for the simulation of proteins, nucleic-acids, and organic-molecules. *J. Am. Chem. Soc.* 117: 5179–5197.

Collinge, D. B., Kragh, K. M., Mikkelsen, J. D., Nielsen, K. K., Rasmussen, U., and Vad, K. (1993). Plant kinases. *Plant J.* 3: 31–40.

Couch, R. B. (1990). Rhinoviruses. In *Virology,* B. N. Fields, and D. M. Knipe, eds., pp. 607–629. Raven, New York.

Cunningham, M. A., Ho, L. L., Nguyen, D. T., Gillilan, R. E., and Bash, P. A. (1997). Simulation of the enzyme reaction mechanism of malate dehydrogenase. *Biochemistry* 36: 4800–4816.

Dasgupta, S., Yamasaki, T., and Goddard III, W. A. (1996). The Hessian biased singular value decomposition method for optimization and analysis of force fields. *J. Chem. Phys.* 104: 2898–2920.

Debe, D., Carlson, M. J., and Goddard III, W. A. (1999). The topomer-sampling model of protein folding. *Proc. Natl. Acad. Sci. USA* 96: 2596–2601.

Demiralp, E., Cagin, T., Huff, N. T., and Goddard III, W. A. (1998). New interatomic potentials for silica. *XVIII Intl. Congress on Glass Proc.*, M. K. Choudhary, N. T. Huff, and C. H. Drummond III, ed., pp. 11–15.

Demiralp, E., Cagin, T., and Goddard III, W. A. (1999). Morse stretch potential charge equilibrium force field for ceramics: application to the quartz-stishovite phase transition and to silica glass. *Phys. Rev. Lett.* 82: 1708–1711.

Ding, H. Q., Karasawa, N., and Goddard III, W. A. (1992). Atomic level simulations on a million particles—the cell multipole method for Coulomb and London nonbond interactions. *J. Chem. Phys.* 97: 4309–4315.

Ding H. Q., Karasawa N., and Goddard III, W. A. (1992b). The reduced cell multipole method for coulomb interactions in periodic-systems with million-atom unit cells. *Chem. Phys. Lett.* 196: 6–10.

De Leeuw, S. W., Perram, J. W., and Smith, E. R. (1980). Simulation of electrostatic systems in periodic boundary conditions. *Proc. Roy. Soc. London* A373: 27–56.

Ewald, P. (1921). Die Berechnung optischer und elektrostatischer gitterpotentiale. *Ann. Phys.* (Leipzig) 64: 253–287.

Fanelli, F., Menziani, C., Scheer, A., Cotecchia, S., and De Benedetti, P. D. (1998). Ab initio modeling and molecular dynamics simulation of the alpha(lb)-adrenergic receptor activation). *Methods* 14: 302–317.

Figueirido, F., Levy, R. M., Zhou, R. H., and Berne, B. J. (1997). Large scale simulation of macromolecules in solution: Combining the periodic fast multipole method with multiple time step integrators. *J. Chem. Phys.* 107: 7002–700.

Floriano, W. B., Nascimento, M. A. C., Domont, G. B., and Goddard III, W. A. (1998). Effects of pressure on the structure of metmyoglobin: molecular dynamics predictions for pressure unfolding through a molten globule intermediate. *Protein Science* 7: 2301–2313.

Franke, A. R., (1984). *Theoretical Drug Design Methods*. Elsevier, Amsterdam.

Fujita, T. (1990). The extrathermodynamic approach to drug design. In *Comprehensive Medicinal Chemistry*, Hansch, C., ed., vol. 4, pp. 497–560. Pergamon, Oxford.

Fukamizo, T., and Kramer, K. J. (1985). mechanism of chitin hydrolysis by the binary chitinase system in insect molting fluid. *Insect. Biochem.* 15: 141–145.

Gerdy, J. J. (1995). Accurate Interatomic Potentials for Simulations. Ph.D thesis, California Institute of Technology.

Henrissat, B., and Bairoch, A. (1993). New families in the classification of glycosyl hydrolases based on amino-acid-sequence similarities. *Biochem. J.* 293: 781–788.

Hermans, J., Berendsen, H. J. C., van Gunsteren, W. F., and Postma, J. P. M. (1984). A consistent empirical potential for water-protein interactions. *Biopolymers* 23: 1513–1518.

Heyes, D. M. (1981). Electrostatic potentials and fields in infinite point charge lattices. *J. Chem. Phys.* 74: 1924–1929.

Hoover W. G. (1985). Canonical dynamics—equilibrium phase-space distributions. *Phys. Rev. A* 31: 1695–1697.

Jain, A., Vaidehi, N., and Rodriguez, G. (1993). A fast recursive algorithm for molecular dynamics simulation. *J. Comp. Phys.* 106: 258–268.

Jorgensen, W. L., Chandrasekar, J., Madura, J. D., Impey, R. W., and Klein. M. L. (1983). Comparison of simple potential functions for simulating liquid water. *J. Chem. Phys.* 79: 962.

Jorgensen, W. L., and Tirado-Tives, J. (1988). The OPLS potential functions for proteins—energy minimizations for crystals of cyclic-peptides and crambin. *J. Am. Chem. Soc.* 110: 1666–1671.

Karasawa, N., and Goddard III, W. A. (1989). Acceleration of convergence for lattice sums. *J. Phys. Chem.* 93: 7320–7327.

Kier, L. B., and Testa, B. (1995). Complexity and emergence in drug research In *Advances in Drug Research,* B. Testa, and U. A. Meyer eds., vol. 26, pp. 1–43. Academic Press, London.

Koga, D., Isogai, A., Sakuda, S., Matsumoto, S., Suzuki, A., Kimura, S., and Ide, A. (1987). Specific-inhibition of bombyx-mori chitinase by allosamidin. *Agri. Biol. Chem.* 51: 471–476.

Kramer, K. J., Dziadik-Turner, C., and Koga, D. (1985). In *Comprehensive Insect Physiology, Biochemistry and Pharmacology: Integument, Respiration and Circulation.* Pergamon Press, Oxford.

Kuranda, M. J., and Robbins, P. W. (1991). Chitinase is required for cell-separation during growth of saccharomyces-cerevisiae. *J. Biol. Chem.* 266: 19758–19767.

Levitt, M. (1983). Molecular dynamics of native protein. 1. Computer simulation of trajectories. *J. Mol. Biol.* 168: 595–620.

Levitt, M., Hirschberg, M., Sharon-R., Laidig, K. E., and Dagett, V. (1997). Calibration and testing of a water model for simulation of the molecular dynamics of proteins and nucleic acids in solution. *J. Phys. Chem.* B101: 5051–5061.

Lim, K. T., Brunett, S., Iotov, M., McClurg, B., Vaidehi, N., Dasgupta, S. Taylor, S., and Goddard III, W. A. (1997). Molecular dynamics for very large systems on massively parallel computers: the MPSim program. *J. Comp. Chem.* 18: 501–521.

Mackerell, A. D., Wiorkiewicz-Kuczera, J., and Karplus, M., (1995). All-atom empirical potential for molecular modeling and dynamics studies of proteins. *J. Am. Chem. Soc.* 117: 11946–11975.

Marrone, T. J., Straatsma, T. P., Briggs, J. M., Wilson D. K., Quiocho, F. A., and McCammon, J. A. (1996). Theoretical study of inhibition of adenosine deaminase by (8R)-coformycin and (8R)-deoxycoformycin, *J. Med. Chem.* 39: 277–284.

Marrone, T. J., Briggs, J. M., and McCammon, J. A. (1997). Structure-based drug design: computational advances. *Ann. Rev. Pharmacol. Toxicol.* 37: 71–90.

Mathiowctz, A., Jain, A., Karasawa, N., and Goddard III W. A. (1994) Protein simulations using techniques suitable for very large systems: the cell multipole method for nonbond interactions and the newton-euler inverse mass operator method for internal coordinate dynamics. *Proteins* 20: 227–247.

Mayo, S. L., Olafson, B. D., and Goddard, W. A. (1990). DREIDING—a generic force field for molecular simulations. *J. Phys. Chem.* 94: 8897–8909.

Mazur, A., and Abagayan, R. J. (1989). New methodology for computer-aided modeling of biomolecular structure and dynamics. 1. Non-cyclic structures. *Biomol. Struct. Dyn.* 6: 815–832.

McCammon, J. A. (1987). In *Dynamics of Proteins and Nucleic Acids.* Cambridge University Press, Cambridge.

McPhillips, T. M., Hsu, B. T., Sherman, M. A., Mas, M. T., and Rees, D. C., (1996). Structure of the R65Q mutant of yeast 3-phosphoglycerate kinase complexed with MG-AMP-PNP and 3-phospho-D-glyceratefe. *Biochem.* 35: 4118–4127.

Miklis, P., Cagin, T., and Goddard III, W. A. (1997). Dynamics of bengal rose encapsulated in the meijer dendrimer box. *J. Am. Chem. Soc.* 119: 7458–7462.

Montorsi, M., Menziani, M. C., Cocchi, M., Fanelli, F., and De Benedetti, P. G. (1998). Computer modeling of size and shape descriptors of alpha(1)-adrenergic receptor antagonists and quantitative structure-affinity/selectivity relationships. *Methods* 145: 239–254.

Nose, S. (1984). A unified formulation of the constant temperature molecular-dynamics methods. *Mol. Phys.* 52: 255–268.

Parr, R. G., and Yang, W. (1989). in *Density Functional Theory of Atoms and Molecules.* Clarendon Press, Oxford.

Parinello, M., and Rahman, A. (1981). Polymorphic transitions in single crystals: a new molecular dynamics method. *J. Appl. Phys.* 52: 7182–7190.

Plaxco, K. W., and Goddard III, W. A. (1994). Contributions of the thymine methyl-group to the specific recognition of polynucleotides and mononucleotides—an analysis of the relative free-energies of solvation of thymine and uracil. *Biochem* 33: 3050–3054.

Pople, J. A., and Beveridge, D. L. (1970). In *Approximate Molecular Orbital Theory.* McGraw-Hill, New York.

Pranata, J., Wierschke, S. G., and Jorgensen, W. L. (1991). OPLS potential functions for nucleotide bases—Relative association constants of hydrogen-bonded base-pairs in chloroform. *J. Am. Chem. Soc.* 113: 2810–2819.

Qi, Y., Cagin, T., Kimura, Y., and Goddard III, W. A. (1999). Molecular-dynamics simulations of glass formation and crystallization in binary liquid metals: Cu-Ag and Cu-Ni. *Phys. Rev. B* 59: 3527–3533.

Rahman, A., and Stillinger, F. H. (1971). Molecular dynamics study of liquid water. *J. Chem. Phys.* 55: 3336–3359.

Rappé, A. K., and Goddard III, W. A. (1991). *J. Chem. Phys.* 95: 3358–3363.

Rappé, A. K., Casewit, C. J., Colwell, K. S., Goddard, W. A., III, and Skiff, W. M. (1992). UFF, A full periodic-table force-field for molecular mechanics and molecular-dynamics simulations. *J. Am. Chem. Soc.* 114: 10024–10035.

Rice, L. M., and Brunger, A. T. (1994). Torsion angle dynamics—Reduced variable conformational sampling enhances crystallographic structure refinement. *Proteins* 19: 277–290.

Ringnalda, M. N., Langlois, J.-M., Greeley, B. H., Murphy, R. B., Russo, T. V., Cortis, C., Muller, R. P., Marten, B., Donnelly, R. E., Mainz, D. T., Wright, J. R., Pollard, W. T., Cao, Y., Won, Y., Miller, G. H., Goddard, W. A., III, and Friesner, R. A. Jaguar 3.0 from Schrödinger, Inc. Portland, OR.

Robertus J. D., Hart P. J., Monzingo A. F., Marcotte E., and Hollis, T. (1995). The structure of chitinases and prospects for structure-based drug design. *Can. J. Bot.* 73: S1142–S1146, Suppl. 1 E–H.

Rueckert, R. R. (1991). Picornaviridae and their replication. In *Fundamental Virology,* B. N. Fields, and D. M. Knipe, eds., pp. 409–450. Raven, New York.

Ryckaert, J. P., Cicotti, G., and Berendsen, H. J. C. (1977). Numerical Integration of Cartesian equations of motion of a system with constraints: Molecular dynamics of n-alkanes. *J. Comp. Phys.* 23: 327–341.

Sadanobu, J., and Goddard III, W. A. (1997). The continuous configurational Boltzmann biased direct Monte Carlo method for free energy properties of polymer chains. *J. Chem. Phys.* 106: 6722–6729.

Schaefer, H. F. (1984). In *Quantum Chemistry: The Development of Ab Initio Methods in Molecular Electronic Structure Theory.* Clarendon Press, Oxford.

Sitkoff D., Sharp K. A., and Honig B. (1993), A rapid and accurate route to calculating solvation energy using a continuum model which includes solute polarizability. *Biophys. J.* 64: A65–A65.

Sternberg, M. J. E. (1996). *Protein Structure Prediction.* Oxford University Press, Oxford.

Tannor, D. J., Marten, B., Murphy, R., Friesner, R. A., Sitkoff, D., Nicholls, A., Ringnalda, M., Goddard III, W. A., and Honig, B. (1994). Accurate first principles calculation of molecular charge-distributions and solvation energies from ab-initio quantum-mechanics and continuum dielectric theory. *J. Am. Chem. Soc.* 116: 11875–11882.

Vaidehi, N., Jain, A., and Goddard III, W. A. (1996). Constant temperature constrained molecular dynamics: the Newton-Euler inverse mass operator method. *J. Phys. Chem.* 100: 10508–10517.

Vaidehi, N., and Goddard III, W. A. (1997). The pentamer channel stiffening model for drug action on human rhinovirus HRV-1A. *Proc. Natl Acad. Sci. USA* 94: 2466–2471.

Vaidehi, N., and Goddard III, W. A. (2000) Domain motions in phospho glycerate kinase using hierarchical NEIMO simulations. *J. Phys. Chem. A* 104: 2375–2383.

van Gunsteren, W. F., and Berendsen, H. J. C. 1987. *Groningen Molecular Simulation (GROMOS) Library Manual.* Biomos, Groningen.

van Gunsteren, W. F., and Weiner, P. K. (1989). In *Computer Simulation of Biomolecular Systems.* Leiden.

van Gunsteren, W. F., and Berendsen H. J. C. (1990). Computer-simulation of molecular-dynamics—methodology, applications, and perspectives in chemistry. *Ang. Chem. Int. Ed. Engl.* 29: 992–1023.

Warshel, A. (1991). In *Computer Modeling of Chemical Reactions in Enzymes and Solutions.* Wiley, New York.

Watanabe, T., Suzuki, K., Oyanagi, W., Ohnishi, K., and Tanaka, H. (1990). Gene cloning of chitinase-A1 from bacillus-circulans W1-12 revealed its evolutionary relationship to serratia chitinase and to the type-iii homology units of fibronectin. *J. Biol. Chem.* 265: 15659–15665.

Weiner, S. J., Kollman, P. A., Case, D. A., Singh, U. C., Ghio, C., Alagona, G., Profeta, S., and Weiner, P. J. (1984). A new force-field for molecular mechanical simulation of nucleic-acids and proteins. *J. Am. Chem. Soc.* 106: 765–784.

Weiner, S. J., Kollman, P. A., Nguyen, D. T., and Case, D. A. (1986). An all atom force-field for simulations of proteins and nucleic-acids. *J. Comput. Chem.* 7: 230–252.

Woodcock, L. V. (1971). Isothermal molecular dynamics calculations for liquid salts. *Chem. Phys. Lett.* 10: 257–261.

7 Diffusion

Guy Bormann, Fons Brosens, and Erik De Schutter

7.1 Introduction

7.1.1 Why Model Diffusion in Cells?

Physical proximity is an essential requirement for molecular interaction to occur, whether it is between an enzyme and its substrate and modulators or between a receptor and its ligand. Cells have developed many mechanisms to bring molecules together, including structural ones [e.g., calcium-activated ionic channels often cluster with calcium channels in the plasma membrane (Gola and Crest 1993, Issa and Hudspeth 1994] or specific processes like active transport (Nixon 1998). In many situations, however, concentration gradients exist that will affect the local rate of chemical reactions. Such gradients can be static at the time scale of interest [e.g., the polarity of cells (Kasai and Petersen 1994)], or very dynamic [e.g., the intra- and intercellular signaling by traveling calcium waves (Berridge 1997)].

Diffusion is the process by which random Brownian movement of molecules or ions causes an average movement toward regions of lower concentration, which may result in the collapse of a concentration gradient. In the context of cellular models, one distinguishes—experimentally and functionally—between free diffusion and diffusion across or inside cell membranes (Hille 1992). We will only consider the first case explicitly, but methods similar to those described here can be applied to the latter.

Historically, diffusion has often been neglected in molecular simulations, which are then assumed to operate in a "well-mixed pool." This assumption may be valid when modeling small volumes, for example, inside a spine (Koch and Zador 1993), but should be carefully evaluated in all other contexts. As we will see in this chapter, the introduction of diffusion into a model raises many issues that otherwise do not apply: spatial scale, dimensionality, geometry, and which individual molecules or ions can be considered to be immobile or not. The need for modeling diffusion should be evaluated for each substance separately. In general, if concentration gradients exist within the spatial scale of interest, it is highly likely that diffusion will have an impact on the modeling results unless the gradients change so slowly that they can be considered stationary compared with the time scale of interest.

Simulating diffusion is in general a computationally expensive process, which is a more practical explanation of why it is often not implemented. This may very well change in the near future. A growing number of modeling studies (Markram et al. 1998, Naraghi and Neher 1997) have recently emphasized the important effects diffusion can have on molecular interactions.

7.1.2   When is Diffusion Likely to Make a Difference?

A reaction–diffusion system is classically described as one where binding reactions are fast enough to affect the diffusion rate of the molecule or ion of interest (Wagner and Keizer 1994, Zador and Koch 1994). A classic example that we will consider is the interaction of calcium with intracellular buffers, which cause calcium ions to spread ten times slower than one would expect from the calcium diffusion constant (Kasai and Petersen 1994).

Depending on which region of the cell one wants to model, this slow, effective diffusion can be good or bad news. In fact, a slow, effective diffusion rate will dampen rapid changes caused by calcium influx through a plasma membrane so much that in the center of the cell it can be modeled as a slowly changing concentration without a gradient. Conversely, below the plasma membrane, the slower changes in concentration gradient may need to be taken into full account.

This is, however, a static analysis of the problem, while the dynamics of the reaction–diffusion process may be much more relevant to the domain of molecular interactions. In effect, as we will discuss in some detail, binding sites with different affinities may compete for the diffusing molecule, which leads to highly nonlinear dynamics. This may be a way for cells to spatially organize chemical reactions in a deceptively homogeneous intracellular environment (Markram et al. 1998, Naraghi and Neher 1997).

Before dealing with reaction–diffusion systems, however, we will consider the diffusion process itself in some detail. In this context we will consider especially those conditions under which the standard one-dimensional diffusion equation description used in most models is expected to fail.

7.2   Characteristics of Diffusion Systems

Part I of this volume provided a concise introduction to the modeling of chemical kinetics based on mass action, expressed as reaction rate equations. It also discussed the shortcomings of the rate equation approach for describing certain features of chemical reactions and provided alternatives, for example, Markov processes, to improve models of chemical kinetics.

These improvements stem from the realization that under certain circumstances the stochastic nature of molecular interactions can have a profound effect on dynamic properties. The resulting models still share a common assumption with rate equation models in that they assume a well-mixed homogeneous system of interacting molecules. This ignores the fact that spatial fluctuations may introduce yet another influence on the dynamics of the system. For example, in many cases it is important to know how a spatially localized disturbance can propagate in the system.

From statistical mechanics it follows that adding spatial dependence to the molecule distributions calls for a transport mechanism that dissipates gradients in the distribution. This mechanism is called *diffusion* and it is of a statistical nature. The effects on the dynamics of the chemical system are determined by the relative time scales of the diffusion process and the chemical kinetics involved.

At the end of this section we will describe a number of examples of how diffusion interacts with a chemical system to create new dynamics. First, however, we will consider diffusion by itself as an important and relatively fast transport mechanism in certain systems. We will also discuss how to relate diffusion to experimental data and show that such data mostly provide a warped view on diffusion. Therefore it is usually more appropriate to talk about "apparent diffusion." The appearance of this apparent diffusion, that is, the degree of warpedness compared with free diffusion, is determined both by the nature of the chemical reactions and by the specific spatial geometry or the system under investigation.

This chapter provides methods of different types to solve the equations that govern these phenomena and will give guidelines concerning their applicability. This should enable the reader to make initial guesses about the importance of diffusion in the system under consideration, which can help in deciding whether to pursue a more detailed study of the full reaction–diffusion system.

7.2.1   What Is Diffusion?

Diffusional behavior is based on the assumption that molecules perform a random walk in the space that is available to them. This is called *Brownian motion*. The nature of this random walk is such that the molecules have an equal chance of going either way (in the absence of external fields). To see how this can generate a macroscopic effect (as described by Fick's law) let's consider a volume element in which there are more molecules than in a neighboring volume element. Although the molecules in both volume elements have an equal chance of going either way, more molecules happen to go to the element with the low initial number of molecules than to the element with the high initial number simply because they outnumber the molecules in the element with the lower number.

To describe this process mathematically, we assume the molecules perform a discrete random walk by forcing them to take a step of fixed length Δx in a random direction (i.e., left or right) every fixed Δt, independent of previous steps. The average change in number of molecules Δn_i^k at discrete position i in a timestep Δt from t_k to t_{k+1} for a high enough number of molecules is then given by (see, for instance, Schulman 1981).

$$n_i^{k+1} - n_i^k = \Delta n_i^k = p_l n_{i+1}^k - (p_l + p_r)n_i^k + p_r n_{i-1}^k, \tag{7.1}$$

where p_l and p_r are the probabilities of going either to the left or to the right. In the absence of drift we have for the probabilities (using the sum rule for independent events)

$$\begin{cases} p_l + p_r & = 1 \\ \quad p_l & = p_r \end{cases} \tag{7.2}$$

so that $p_l = p_r = \frac{1}{2}$.

Filling in these numbers in equation (7.1) gives

$$\Delta n_i^k = \frac{1}{2} n_{i+1}^k - n_i^k + \frac{1}{2} n_{i-1}^k$$

$$= \frac{1}{2}(n_{i+1}^k - 2n_i^k + n_{i-1}^k) \tag{7.3}$$

$$= \frac{\Delta x^2}{2} \frac{n_{i+1}^k - 2n_i^k + n_{i-1}^k}{\Delta x^2},$$

resulting in

$$\frac{\Delta n_i^k}{\Delta t} = \frac{\Delta x^2}{2\Delta t} \frac{n_{i+1}^k - 2n_i^k + n_{i-1}^k}{\Delta x^2}. \tag{7.4}$$

Letting $\Delta x \to 0$ and $\Delta t \to 0$ while keeping

$$D = \frac{(\Delta x)^2}{2\Delta t} \tag{7.5}$$

constant (and multiplying both sides with a constant involving Avogadro's number ($6.022 \cdot 10^{23}$ mol^{-1}) and volume to convert to concentrations), we get the following parabolic partial differential equation (PDE)[1]:

$$\frac{\partial C(x,t)}{\partial t} = D \frac{\partial^2 C(x,t)}{\partial x^2}. \tag{7.6}$$

This is generally known as the one-dimensional diffusion equation along a Cartesian axis. Other one-dimensional equations are used more often, but they arise from symmetries in a multidimensional formulation, for example, the radial component in a spherical or cylindrical coordinate system (see later discussion). The random walk can be easily extended to more dimensions by using the same update scheme for every additional (Cartesian) coordinate.

The random walk process described here generates a Gaussian distribution of molecules over all grid positions when all molecules start from a single grid position x_0 (i.e., a Dirac delta function):

$$C(x,t) = \frac{C_0}{\sqrt{4\pi Dt}} \exp\left[-\frac{(x-x_0)^2}{4Dt} \right]. \tag{7.7}$$

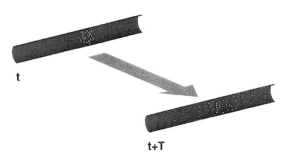

t

t+T

Figure 7.1
The initial condition is the distribution of molecules on a fixed time t (chosen as $t = 0$ in most cases). It determines the final shape of the distribution at a later time $t + T$. In general, a partial differential equation (in space and time) has a set of possible solutions for a fixed set of parameters and (spatial) boundary conditions. By setting the initial condition, one selects a specific solution.

It is left as an exercise for the reader to proof this.[2] From this, and the fact that every function can be written as a series of delta functions, it follows that the same solution can be generated using a continuous random walk by drawing the size of the step each molecule moves every timestep Δt from a Gaussian of the form

$$N_{\Delta t}(\Delta x) = \frac{1}{\sqrt{4\pi D\Delta t}} \exp\left[-\frac{(\Delta x)^2}{4D\Delta t}\right],\qquad(7.8)$$

which has a mean-square step size of $(\Delta x)^2 = 2D\Delta t$, relating the two random walk methods. The choice of method is determined by the following: Discrete random walks allow a fast evaluation of the local density of molecules, while continuous random walks allow easy inclusion[3] of complex boundary conditions.

The above method is a quick way to determine the solution of equation (7.6) for the initial condition $C(x,0) = C_0\delta(x-x_0)$ (figure 7.1). This is one of the few cases that can also be solved analytically. See Crank (1975) for a general and extensive analytical treatment of the diffusion equation.

A final remark concerning the scale on which the random walk approach applies should be made. The movements the molecules make in the diffusion process are related to the thermal movement of and collisions with the background solvent molecules. The path a single molecule follows is erratic, but each step is deterministic when the molecules are considered to be hard spheres. However, if the environment of the molecule is random, the resulting path can be considered the result of a random walk where each step is independent of the molecule's history (Hille 1992).

Nevertheless, even in a fluid, the close-range environment of a molecule has some short-term symmetry in the arrangement of the surrounding molecules so that on the scale of the mean free path (and the corresponding pico- or femtosecond time scale) there is some correlation in the steps of the molecule—that is, the molecule appears to be trapped in its environment for a short time. At this scale, using random walks is not appropriate and one has to resort to molecular dynamics (MD) instead. At a larger distance (and thus at a longer time scale), the environment appears random again in a fluid and the correlations disappear, so that a random walk approach is valid again.

7.2.2   A Macroscopic Description of Diffusion

Many discussions about diffusion start out with an empirical law: Fick's law (Fick 1885). This law states that a concentration gradient for a dissolved substance gives rise to a flux of mass in the (diluted) solution, proportional to but in the opposite direction of the concentration gradient. The constant of proportionality is called the (macroscopic) diffusion constant. It is valid under quite general assumptions. The corresponding equation in one dimension is of the form

$$J_F(x,t) = -D\frac{\partial C(x,t)}{\partial x} \tag{7.9}$$

where D is the macroscopic diffusion coefficient and $C(x,t)$ is the mass distribution function expressed as a concentration. The minus sign is there to ensure that the mass will flow downhill on the concentration gradient (otherwise we get an unstable system and the thermodynamic law of entropy would be violated).

The diffusion coefficient is usually treated as a fundamental constant of nature for the species at hand, determined from diffusion experiments. It is, however, not a real fundamental constant since it can be calculated, using kinetic gas theory, from thermodynamic constants, absolute temperature and molecule parameters. In fact, it can even be position dependent if the properties of the medium (such as viscosity) vary strongly with position. So, including diffusion in an accurate model could become a messy business in some cases.

Fick's law looks very much like an expression for a flux proportional to a conservative force on the molecules, with the "force" depending on the gradient of a field potential (the concentration). But, as can be concluded from section 7.2.1, that force is entirely of a statistical nature. The flux of mass changes the concentration at all points of nonzero net flux. The rate of change is given by a mass balance (i.e., a continuity) equation as follows:

$$\frac{\partial C(x,t)}{\partial t} = -\frac{\partial J_F(x,t)}{\partial x} \tag{7.10}$$

in the absence of other non-Fickian fluxes. Replacing $J_F(x,t)$ by the right-hand side of equation (7.9) gives equation (7.4) if we identify the diffusion coefficient with the former coefficient D. For more details on kinetic gas theory and a derivation of this macroscopic diffusion coefficient from microscopic considerations, see, for example, Feynman et al. (1989). A classic is of course Einstein's original treatment of Brownian motion (Einstein 1905).

Equations (7.9) and (7.10) can be easily extended to higher dimensions by using the following vector notations:

$$\begin{cases} \mathbf{J}_F(\mathbf{r},t) & = -D \, \boldsymbol{\nabla} \, C(\mathbf{r},t) \\ \dfrac{\partial C(\mathbf{r},t)}{\partial t} & = - \, \boldsymbol{\nabla} \cdot \mathbf{J}_F(\mathbf{r},t) \end{cases} \qquad (7.11)$$

where $\boldsymbol{\nabla}$ is the gradient operator and $\boldsymbol{\nabla} \cdot$ is the divergence operator. Combining the two equations then gives the general multidimensional diffusion equation:

$$\frac{\partial C(\mathbf{r},t)}{\partial t} = D\nabla^2 C(\mathbf{r},t) = D\Delta C(\mathbf{r},t), \qquad (7.12)$$

with Δ the Laplace operator (Laplacian). The explicit form of the Laplacian depends on the coordinate system chosen. In a Cartesian system in ordinary space it looks like this:

$$\Delta C(\mathbf{r},t) = \Delta C(x,y,z,t) = \left(\frac{\partial^2}{\partial x^2} + \frac{\partial^2}{\partial y^2} + \frac{\partial^2}{\partial x^2} \right) C(x,y,z,t). \qquad (7.13)$$

Sometimes the nature of the problem is such that it is simpler to state the problem in cylindrical[4]

$$\Delta C(\rho,\phi,z,t) = \left[\frac{1}{\rho} \frac{\partial}{\partial \rho} \left(\rho \frac{\partial}{\partial \rho} \right) + \frac{1}{\rho^2} \frac{\partial^2}{\partial \phi^2} + \frac{\partial^2}{\partial z^2} \right] C(\rho,\phi,z,t) \qquad (7.14)$$

or spherical[5] coordinates

$$\Delta C(r,\theta,\phi,t) = \left[\frac{1}{r} \frac{\partial^2}{\partial r^2} (r\cdot) + \frac{1}{r^2 \sin\theta} \frac{\partial}{\partial \theta} \left(\sin\theta \frac{\partial}{\partial \theta} \right) + \frac{1}{r^2 \sin^2\theta} \frac{\partial^2}{\partial \phi^2} \right] C(r,\theta,\phi,t). \qquad (7.15)$$

In the case of axial and spherical symmetry, respectively, this reduces to the following more generally used (one-dimensional) radial equations:

$$\begin{cases} \dfrac{\partial C(\rho,t)}{\partial t} & = & \dfrac{D}{\rho} \dfrac{\partial}{\partial \rho} \left[\rho \dfrac{\partial C(\rho,t)}{\partial \rho} \right] & = & D\dfrac{\partial^2 C(\rho,t)}{\partial \rho^2} + \dfrac{D}{\rho} \dfrac{\partial C(\rho,t)}{\partial \rho} \\ \dfrac{\partial C(r,t)}{\partial t} & = & \dfrac{D}{r} \dfrac{\partial^2 (rC(r,t))}{\partial r^2} & = & D\dfrac{\partial^2 C(r,t)}{\partial r^2} + \dfrac{2D}{r} \dfrac{\partial C(r,t)}{\partial r} \end{cases}. \qquad (7.16)$$

The solution of the spherical radial equation for an instantaneous point source, that is, for an initial condition of the form $C(r,0) = C_0\delta(r)/r$, is (for $r_0 = 0$):

$$C(r,t) = \frac{C_0}{2Dt} \exp\left(-\frac{r^2}{4Dt}\right). \tag{7.17}$$

This is the radial equivalent of equation (7.7) and again this is actually one of the few solutions that can be found analytically. Both solutions describe one-dimensional diffusion in an infinite space and the last one is generally referred to as free diffusion.

7.2.3 Importance of Boundary Conditions and Dimensionality

Often, properties of diffusion are derived from free diffusion, following equation (7.6), and proposed as generic diffusion properties, while the actual boundary conditions shape the concentration profiles in significantly different ways. Knowing these shapes can be important when interpreting experimental results or when selecting curves—that is, solutions of model problems—for fitting experimental data to estimate (apparent) diffusion parameters.

Choosing the appropriate boundary conditions is also important from a modeling point of view. Free diffusion is of use in modeling interactions in infinite space or in a bulk volume on appropriate time scales, that is, situations where diffusion is used as a simple passive transport or dissipative mechanism. However, when diffusion is used as a transport mechanism to link cascading processes that are spatially separated, or when diffusion is used in conjunction with processes that are triggered at certain concentration thresholds, a detailed description of the geometry of the system may become very important. In that case, the complex geometry can impose the majority of the boundary conditions.

To give an example, let's consider a cylinder of infinite length, with a reflecting mantle, in which there is an instantaneous point source releasing a fixed amount of a substance, A, while further down the cylinder there is a store of another substance, B. The store starts releasing substance B when the concentration of substance A reaches a certain threshold level. For cylinders of smaller radii, the time of release from the store will be much shorter. This is because the concentration of A at the store will rise sooner to the threshold level for smaller radii, although the intrinsic diffusion rate of A does not change.

When the concentration level of A has to be sustained for a longer period, consider what happens when we seal off the cylinder with a reflecting wall near the point source. Since mass can now only spread in one general direction, the concentration of A can become almost twice as high as in the unsealed case, the exact factor depending on the distance between the point source and the sealed end. Furthermore, the concentration near the store and point source will stay elevated for a longer period of time. Eventually, the concentration will drop to zero in this case since the cylinder is still semi-infinite. When we seal off the other end too (at a much larger distance), then the concentration of A will, not surprisingly, drop to a fixed rest level—above or below the threshold level, depending on the volume of the now finite closed cylinder—when no removal processes are present.

Boundary conditions also play a major role in the dimensionality of the diffusion equation one has to choose. Equation (7.16) is an example of equations that arise from certain symmetries in the boundary conditions. Often the dimensionality is also reduced by neglecting asymmetries in some components of the boundary conditions or by averaging them out for these components in order to reduce computation time or computational complexity. In some cases, though, it is necessary to keep the full dimensionality of the problem to be able to give an accurate description of the system. This is, for example, the case in cell models where compartmentalization is obtained through spatial colocalization of related processes. In order to simulate such models, one has to resort to fast multidimensional numerical methods. The methods section provides two such methods: one finite-differencing method called *the alternating direction implicit method* (ADI) and one Monte Carlo (MC) method called *Green's function Monte Carlo.*

7.2.4   Interactions among Molecules and with External Fields

In the reaction rate equation approach, a system of chemical reactions can be described in general by a set of coupled first-order ordinary differential equations (ODEs):

$$\begin{cases} \dfrac{dC_1(t)}{dt} &= f_1(C_1, C_2, \ldots, C_N) \\[2mm] \dfrac{dC_2(t)}{dt} &= f_2(C_1, C_2, \ldots, C_N) \\[1mm] \vdots \qquad \vdots & \qquad \vdots \\[1mm] \dfrac{dC_N(t)}{dt} &= f_N(C_1, C_2, \ldots, C_N), \end{cases} \tag{7.18}$$

where the forms of the functions f_s are determined by the reaction constants and structures of the M chemical reaction channels. The functions f_s are generally nonlinear in the C_s's. If the reaction mechanisms and constants are known, one can use the following normal form for the f_s's:

$$f_s(C_1, C_2, \ldots, C_N) = \sum_{c=1}^{M} K_{s,c} \prod_{j=1}^{N} C_j^{v_{j,c}^L} \tag{7.19}$$

with $K_{s,c} = k_c(v_{s,c}^R - v_{s,c}^L)$, where $(v_{s,c}^L, v_{s,c}^R)$ are the stoichiometric coefficients for substance s on the left-hand and right-hand sides, respectively, of the reaction equation for reaction c[6] and k_c is the reaction constant. Most of the $v_{s,c}^L$ and/or $v_{s,c}^R$ are zero. $R_c = \sum_{j=1}^{N} v_{j,c}^L$ is known as the reaction order and is related to the order of the collisions between the reaction molecules. Since collisions of an order higher than 4 are extremely improbable, $R_c \leq 4$ for most reactions. A reaction of higher order is most likely a chain of elementary reaction steps.

Combining the rate equations with a set of diffusion equations now results in a set of coupled nonlinear parabolic PDEs:

$$
\begin{cases}
\dfrac{\partial C_1(x,y,z,t)}{\partial t} &= D_1 \Delta C_1(x,y,z,t) + f_1(C_1,\dots,C_N) \\[2mm]
\dfrac{\partial C_2(x,y,z,t)}{\partial t} &= D_2 \Delta C_2(x,y,z,t) + f_2(C_1,C_2,\dots,C_N) \\
\ \ \vdots & \quad\quad \vdots \\
\dfrac{\partial C_N(x,y,z,t)}{\partial t} &= D_N \Delta C_N(x,y,z,t) + f_N(C_1,C_2,\dots,C_N)
\end{cases}
\tag{7.20}
$$

The first term on the right-hand side will be zero for all immobile substances. Systems governed by a set of equations like those in (7.18) can exhibit a broad array of complex temporal patterns. Including a transport mechanism to introduce spatial dependence as described by equation (7.20) can introduce additional complex spatial patterns.

Only very simple cases can be solved analytically for both types of problems, so one has to resort to numerical methods, which can become very involved for the second type of problem. At low molecular densities, classic integration methods like the ones introduced previously, cannot grasp the full dynamics of the system because they don't include the intrinsic fluctuations. One has to resort to stochastic approaches for the chemical kinetics like those discussed in chapter 2 and for diffusion the Monte Carlo method discussed in section 7.2.5.

Reaction partners can be electrically charged. Although this can have an important influence on the reaction kinetics, it is usually handled by adjusting the rate constants (in the case of rate models) or the reaction probabilities (in the case of stochastic models). However, one can think of problems where it's important to know the effects of local or external electric fields on diffusion. One can start studying these problems by rewriting equation (7.11) to

$$
\begin{aligned}
\mathbf{J}(\mathbf{r},t) &= \mathbf{J}_F(\mathbf{r},t) + \mathbf{J}_{el}(\mathbf{r},t) \\
&= \mathbf{J}_F(\mathbf{r},t) + \mathbf{J}_{ext}(\mathbf{r},t) + \mathbf{J}_C(\mathbf{r},t) \\
&= -D\left[\boldsymbol{\nabla} C(\mathbf{r},t) + \frac{zF}{RT} C(\mathbf{r},t)\, \boldsymbol{\nabla} V \right] + \mathbf{J}_C(\mathbf{r},t) \\
\frac{\partial C(\mathbf{r},t)}{\partial t} &= - \boldsymbol{\nabla} \cdot \mathbf{J}(\mathbf{r},t) \\
&= D\left\{ \Delta C(\mathbf{r},t) + \frac{zF}{RT} \boldsymbol{\nabla} \cdot [C(\mathbf{r},t)\, \boldsymbol{\nabla} V] \right\} - \boldsymbol{\nabla} \cdot \mathbf{J}_C(\mathbf{r},t) \quad,
\end{aligned}
\tag{7.21}
$$

where $\mathbf{J}_C$ is the net flux induced by the Coulomb interactions between the charged molecules and $\mathbf{J}_{ext}$ the net flux induced by the interaction of the charged molecules with an external field of potential V; z is the valence of the ionic species, F is the Faraday constant, R

is the gas constant, and T is the absolute temperature. The second equation is called the *electrodiffusion equation*. In the case of $\mathbf{J}_C = 0$, it is called the Nernst–Planck equation (Hille 1992). Special complicated numerical algorithms have to be used to solve the general electrodiffusion equation because of the long range and the $1/r$ divergence of Coulomb interactions. For an application to synaptic integration in dendritic spines, see, for example, Qian and Sejnowski (1990).

7.2.5 An Example of a Reaction–Diffusion System: Calcium Buffering in the Cytoplasm

Reaction–diffusion systems have been studied most extensively in the context of buffering and binding of calcium ions. Calcium is probably the most important intracellular signaling molecule. Its resting concentration is quite low (about 50 nM), but it can increase by several orders of magnitude (up to hundreds of micromoles) beneath the pore of calcium channels (Llinás et al. 1992). Cells contain an impressive variety of molecules with calcium binding sites. They include pumps that remove the calcium (e.g., calcium adenosine triphosphatase (ATPase) (Garrahan and Rega 1990), buffers that bind the calcium (e.g., calbindin, calsequestrin, and calretinin) (Baimbridge et al. 1992), enzymes activated by calcium (e.g., phospholipases) (Exton 1997) or modulated by calcium (e.g., through calmodulin activation) (Farnsworth et al. 1995). The interaction of the transient steep calcium concentration gradients with all these binding sites creates complex reaction–diffusion systems, the properties of which are not always completely understood. For example, it remains a point of discussion whether calcium waves, which are caused by inositol 1,4,5-triphosphate (IP$_3$)-activated release of calcium from intracellular stores, propagate because of diffusion of IP$_3$ (Bezprozvanny 1994, Sneyd et al. 1995) itself or of calcium (Jafri and Keizer 1995), which has a complex modulatory effect on the IP$_3$ receptor (Bezprozvanny et al. 1991).

We will focus here on the macroscopic interaction of calcium with buffer molecules. The main function of buffers is to keep the calcium concentration within physiological bounds by binding most of the free calcium. Almost all models of the interactions of calcium with buffers use the simple second-order equation:

$$Ca^{2+} + B \underset{b}{\overset{f}{\rightleftharpoons}} CaB \qquad (7.22)$$

Equation (7.22) is only an approximation because most buffer molecules have multiple binding sites with different binding rates (Linse et al. 1991) and require more complex reaction schemes. In practice, however, experimental data for these different binding rates are often not available. Nevertheless, it is important to express buffer binding with rate equations, even simplified ones, instead of assuming the binding reactions to be in a steady state described by the buffer dissociation constant $K_D = b/f$ as is often done (e.g., Goldbeter

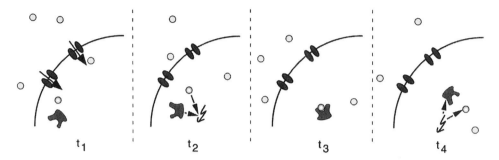

Figure 7.2
A pictorial description of the effect of buffer binding on calcium diffusion for slowly diffusing buffers. The yellow dots represent calcium ions, the green blobs buffer molecules. The purple blobs are calcium channels and the lightning symbol represents chemical interaction. The reacting dot is gradient colored to distinguish it from the other intracellular dot. Note that the figure only shows the net effect for the respective populations. In reality, individual molecules follow erratic paths that do not directly show the buffering effect. (See plate 21.)

et al. 1990, Sneyd et al. 1994). Even if the time scale of interest would allow for equilibrium to occur, the use of steady-state equations neglects the important effects of the forward binding rate constant f on diffusion and competition between buffers (figure 7.2; see also plate 21).

While the properties of buffered calcium diffusion have been mostly studied using analytical approaches (Naraghi and Neher 1997, Wagner and Keizer 1994, Zador and Koch 1994), it is fairly simple to demonstrate these concepts through simulation of the system corresponding to the set of differential equations described in the next section (see also Nowycky and Pinter 1993) for a similar analysis). The simulations of the simple models[7] that we used were run with the GENESIS software (the simulation scripts can be downloaded from the Web at the URL=http://www.bbf.uia.ac.be/models/models.shtml). For more information on simulating calcium diffusion with the GENESIS software (Bower and Beeman 1995), we refer the reader to De Schutter and Smolen (1998). We will focus on the transient behavior of the reaction–diffusion system, that is, the response to a short pulsatile influx of calcium through the plasma membrane. Also, since we are interested mostly in molecular reactions, we will stay in the nonsaturated regime ($[Ca^{2+}] \ll K_D$) though of course interesting nonlinear effects can be obtained when some buffers start to saturate. Finally, for simplicity we do not include any calcium removal mechanisms [calcium pumps (Garrahan and Rega 1990) and the Na^+Ca^{2+} exchanger (DiFrancesco and Noble 1985], although they are expected to have important additional effects by limiting the spread of the calcium transients (Markram et al. 1998, Zador and Koch 1994).

The diffusion coefficient of calcium (D_{Ca}) in water is $6 \cdot 10^{-6}$ cm^2 s^{-1}. In cytoplasm, however, the measured diffusion coefficient is much lower because of the high viscosity of this medium. Albritton et al. (1992) measured a D_{Ca} of $2.3 \cdot 10^{-6}$ cm² s⁻¹ in cytoplasm, and other ions and small molecules can be expected to be slowed down by the same factor. As shown in figure 7.3A (see also plate 22), with this diffusion coefficient calcium rises rapidly everywhere in a typical cell, with only a short delay of about 10 ms in the deepest regions. After the end of the calcium influx, the calcium concentration reaches equilibrium throughout the cell in less than 200 ms. Down to about 3.0 μm, the peak calcium concentration is higher than the equilibrium one because influx rates are higher than the rate of removal by diffusion.

This diffusion coefficient is attained only if no binding to buffers occurs. Albritton et al. (1992) completely saturated the endogenous buffers with calcium to obtain their measurements. Figure 7.3 (B–E) demonstrates the effect of stationary buffers on the effective calcium diffusion rate. They decrease both the peak and equilibrium calcium concentration, slow the diffusion process, and limit the spread of calcium to the submembrane region (that is, close to the site of influx). The total buffer concentration $[B]_T = [B] + [CaB]$ and its affinity, which is high in these simulations ($K_D = 0.2$ μM), will determine the amount of free calcium at equilibrium. This is called the *buffer capacity* of the system and is also the ratio of the inflowing ions that are bound to buffer to those that remain free (Augustine et al. 1985, Wagner and Keizer 1994):

$$\kappa = \frac{d[CaB]_i}{d[Ca^{2+}]_i} = \frac{K_D[B]_T}{(K_D + [Ca^{2+}]_i)^2}.$$

(7.23)

In the case of low calcium concentration, equation (7.23) reduces to $[B]_T/K_D$. Buffer capacities in the cytoplasm are high, ranging from about 45 in adrenal chromaffin cells (Zhou and Neher 1993) to more than 4000 in Purkinje cells (Llano et al. 1994).

While the buffer capacity determines the steady-state properties of the system, the dynamics are highly dependent on the forward binding rate f (compare figure 7.3B with figure 7.3C). In effect, for the same amount of buffer and the same high affinity, the calcium profiles are very different. For the slow buffer as in figure 7.3B, calcium rises rapidly everywhere in the cell, with delays similar to those as in figure 7.3A, but reaches much lower peak concentrations. Only about 50% of the buffer gets bound below the plasma membrane. After the end of the calcium influx, the decay to equilibrium has a very rapid phase, comparable to that in figure 7.3A, followed by a much slower relaxation that takes seconds. The peak calcium concentrations are comparatively much higher than the equilibrium concentration of 95 nM down to a depth of 5.0 μm. Note the slow relaxation to equilibrium of bound buffer in figure 7.3D. Conversely, in the case of a rapidly binding buffer (figure 7.3C), calcium rises slowly below the plasma membrane, but does not start to rise 1.0 μm deeper until 50 ms later. Deeper, the calcium concentration does not start to rise until after the end of the calcium influx. Much more calcium gets bound to buffer (figure

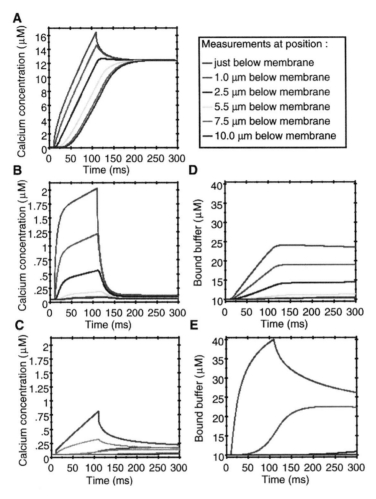

Figure 7.3
Effect of immobile buffer binding rates on calcium diffusion. The changes in calcium concentration at several distances from the plasma membrane due to a short calcium current into a spherical cell are shown. (A) Calcium concentration for simulation with no buffers present. (B) Same with buffer with slow forward rate f (similar to EGTA) present. (C) Same with buffer with fast f (similar to BAPTA) present. Fainter colors: approximation using D_{app} [equation (7.24)]. (D) Bound buffer concentration for simulation shown in (B). (E) Same for simulation shown in (C). (See plate 22.)

7.3E) below the plasma membrane than in figure 7.3D. The decay to equilibrium is monotonic and slower than in figure 7.3A.

Both the differences during the influx and equilibration processes compared with the unbuffered diffusion are due to the binding and unbinding rates of the buffer, which dominate the dynamics. The buffer binding not only determines the speed with which the transients rise, but also that with which they extend toward deeper regions of the cell. This is often described as the effect of buffers on the apparent diffusion coefficient D_{app} of calcium (Wagner and Keizer 1994):

$$D_{app} \approx \frac{1}{1+\kappa} D_{Ca}. \tag{7.24}$$

This equation gives a reasonable approximation for distances significantly far from the source of influx provided f is quite fast (Wagner and Keizer 1994). For a fast buffer similar to BAPTA, it does not capture the submembrane dynamics at all [fainter colors in figure 7.3C[8]]. In conclusion, in the presence of fast immobile buffers, experimental measurements of calcium diffusion can be orders of magnitude slower than in their absence. This severely limits the spatial extent to which calcium can affect chemical reactions in the cell compared with other signaling molecules with similar diffusion constants that are not being buffered, such as IP$_3$ (Albritton et al. 1992, Kasai and Petersen 1994). Conversely, for a slower buffer (figure 7.3B), the initial dynamics (until about 50 ms after the end of the current injection) are mostly determined by the diffusion coefficient.

When the calcium influx has ended, the buffers again control the dynamics while the concentrations equilibrate throughout the cell. Because much more calcium is bound than is free (buffer capacity of 250 in these simulations), the bulk of this equilibration process consists of shuffling of calcium ions from buffers in the submembrane region to those in deeper regions. Since this requires unbinding of calcium from the more superficial buffer, the dynamics are now dominated by the backward binding rate b, which is faster for the fast f buffer in figure 7.3C (both buffers having the same dissociation constant). As a consequence, the rapidly binding buffer concentration mimics the calcium concentration better than the slowly binding one. This is an important point when choosing a fluorescent calcium indicator dye; the ones with fast (un)binding rates will be much more accurate (Regehr and Atluri 1995). Finally, calcium removal mechanisms can also affect the equilibration process to a large degree. If the plasma membrane pumps remove the calcium ions fast enough, they will never reach the lower regions of the cell, especially in the case of the buffer with fast f (where initially more calcium stays close to the plasma membrane).

The above system changes completely when the buffer itself becomes diffusible (see schematic explanation in figure 7.2 and simulation results in figure 7.4; see also plate 23). Now the buffer itself becomes a carrier for calcium ions. Bound buffer diffuses to the re-

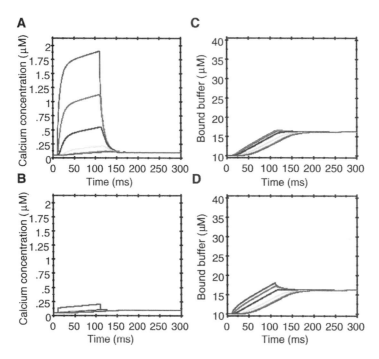

Figure 7.4
Effect of diffusible buffer binding rates on calcium diffusion. Simulation results similar to those in figure 7.3 (B–E), but the buffer is now diffusible. (A) Calcium concentration for simulation with diffusible buffer with slow forward binding rate f. (B) Same with diffusible buffer with fast f. (C) Bound buffer concentration for simulation shown in (B). (D) Same for simulation shown in (C). (See plate 23.)

gions of lower bound buffer concentration, which are also regions of lower calcium concentration so that the bound buffer will unbind its calcium. Again this effect can be described by how the apparent diffusion coefficient D_{app} of calcium changes (Wagner and Keizer 1994):

$$D_{app} \approx \frac{1}{1+\kappa} D_{Ca} + \kappa D_{B}. \tag{7.25}$$

This equation has all the limitations of a steady-state approximation. It demonstrates, however, that diffusion of buffers [which in the case of ATP have a D_B comparable to D_{Ca} in the cytoplasm (Naraghi and Neher 1997)] can compensate for the effect of (stationary) buffers on the apparent diffusion rate. In the case of figure 7.4B, it does not capture the submembrane transient at all (not shown), but reproduces the transients at deeper locations well. The diffusive effect of mobile buffers is an important artifact that fluorescent calcium indicator dyes (e.g., fura-2) can introduce in experiments because they will change the calcium dynamics themselves (Blumenfeld et al. 1992, De Schutter and Smolen 1998, Sala and Hernandez-Cruz 1990).

The actual effect on the apparent diffusion coefficient, however, will depend again on the f rate of the buffer. In the case of a slow f rate, the effects of buffer diffusion during the period of calcium influx are minor because the submembrane peak calcium concentration is only a bit lower compared with simulations with immobile buffers (compare figure 7.4A with figure 7.3B). At the same time, the diffusible buffer saturates much less than the immobile one (figure 7.3C). The biggest difference is, however, after the end of the calcium influx; the system reaches equilibrium within 100 ms! As in the case of an immobile fast buffer (figure 7.3C), a diffusible fast buffer (figure 7.4B) is much more effective in binding calcium than a slow one (compare figure 7.4A), but even more so than the immobile one. Now only the submembrane concentration shows a peak; at all other depths, the concentrations increase linearly to equilibrium [and can be approximated well by equation (7.25) for D_{app}] and this is reached almost immediately after the end of the calcium influx.

In general, it is mainly during the equilibration period after the calcium influx that the buffer diffusion causes a very rapid spread of calcium to the deeper regions of the cell. In other words, the apparent diffusion coefficient will actually increase under these situations because the reaction–diffusion system moves from initially being dominated by the slow calcium diffusion (during influx the apparent diffusion coefficient approaches the diffusion coefficient of calcium) to being dominated by the buffer diffusion (during equilibration where now only one unbinding step is needed instead of the shuffling needed with the immobile buffer). A similar phenomenon can be found in the case of longer periods of calcium influx. In this case the mobile buffer will start to saturate so that the unbinding process becomes more important and again the apparent diffusion coefficient rises, but now during the influx itself (Wagner and Keizer 1994).

When the buffer has a fast f rate, the effects of buffer diffusion become more apparent during the calcium influx itself also (compare figure 7.4B with figure 7.3C)! One sees indeed an almost complete collapse of the calcium gradient because the buffer binds most of the calcium flowing into the submembrane region without being affected by saturation (it gets continuously replenished by free buffer diffusing upward from deeper regions of the cell).

We end this section with a consideration of how competition among buffers in a reaction–diffusion system can cause localization of binding reactions. This is demonstrated in figure 7.5 (see also plate 24), which shows a simulation in which two buffers with identical affinities but different homogeneous concentrations are present.[9] While the calcium concentrations (figure 7.5A) are somewhat similar to those in figure 7.4B, the peak concentrations are much smaller and the reaction–diffusion system takes a much longer time to

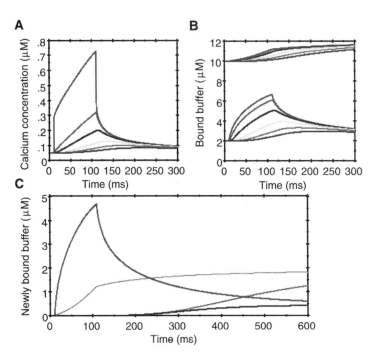

Figure 7.5
Effect of competition between two diffusible buffers. Simulations similar to those in figure 7.3 (B–E), with both
diffusible buffers present. (A) Calcium concentrations. (B) Bound buffer concentrations: upper traces, slow buff-
er; lower traces, fast buffer. (C) Superposition of bound buffer traces for fast buffer and slow buffer (fainter colors; traces have been superimposed by subtracting the steady-state bound buffer concentration). Concentrations
just below the plasma membrane (red curves) and at a 30-μm depth (blue curves) are shown. (See plate 24.)

reach equilibrium after the end of the calcium influx. More important are the large differences in the bound buffer profiles (figure 7.5B). The ones for the slow buffer rise continuously, while the ones for the fast buffer peak at low depths in the cell.

These differences are shown in more detail in figure 7.5C. A peak in a fast binding buffer concentration can be observed below the plasma membrane during the calcium influx.
Initially much more calcium is bound to the fast buffer than to the slow one (which has a
five times higher concentration). Conversely, in deeper regions of the cell, the slow buffer
always dominates (figure 7.5C). This effect, which has been called "relay race diffusion"
by Naraghi and Neher (1997) shows that under conditions of continuous inflow, buffers
have a characteristic length constant (depending on their binding rates, concentration, and
diffusion rate), which corresponds to the distance from the site of the influx at which they
will be most effective in binding calcium. If one replaces in the previous sentence the word

"buffer" with "calcium-activated molecule," it becomes obvious that relay race diffusion can have profound effects on the signaling properties of calcium because it will preferentially activate different molecules, depending on their distance from the influx or release site. The competition between buffers also slows down the dynamics of the reaction–diffusion system because calcium is again being shuttled between molecules (compare figure 7.4B and figure 7.5A).

Markram et al. (1998) have studied similar systems where calcium removal systems were included in the model. They show that if calcium removal is fast, only the fast-binding molecules may be able to react with calcium ions during transients. Conversely, in a case of repetitive pulses of calcium influx, the slower binding molecules will keep increasing their saturation level during the entire sequence of pulses, while the faster buffers will reach a steady saturation after only a few pulses (Markram et al. 1998), demonstrating the sensitivity to temporal parameters of such a mixed calcium-binding system. This is very reminiscent of recent experimental results demonstrating the sensitivity of gene expression systems to calcium spike frequency (Dolmetsch et al. 1998, Li et al. 1998).

7.2.6   Numerical Methods of Solving a Reaction–Diffusion System

This section presents examples of two distinct classes of numerical methods to approximately solve the full reaction–diffusion system. The first class of methods tries to solve the set of reaction–diffusion equations by simulating the underlying chemical and physical stochastic processes and/or by integrating the equations using stochastic techniques. Because of their stochastic nature, they are generally referred to as Monte Carlo methods. The second class of methods tries to solve the set of reaction–diffusion equations directly by numerically integrating the equations. The methods are called the *Crank–Nicholson* method and the *alternating direction implicit* method.

Monte Carlo Methods   The term "Monte Carlo method" is used for a wide range of very distinct methods that have one thing in common: They make use of the notion of randomness (through the use of computer-generated random numbers[10]), either to quickly scan a significant part of a multidimensional region or to simulate stochastic processes. The first type is used for efficiently integrating multivariate integrals. Reaction–diffusion problems are more naturally solved[11] using the second type of MC method. These MC methods can be subdivided once more in two classes: methods that follow the fate of individual molecules and those that follow the fate of mass elements. We will give an introduction to the first class of methods, called *diffusion MC,* and work out in more detail an example of the second class, called *Green's function MC.*

Diffusion MC Methods   These methods use random walks to simulate the diffusional motion of individual molecules. Successful applications include the study of synaptic transmission by simulating diffusion of neurotransmitter molecules across the synaptic cleft and including their reactions with postsynaptic receptors (Bartol et al. 1991, Wahl et al. 1996) and the study of G-protein activation in the submembrane region for a small membrane patch (Mahama and Linderman 1994). Although their usefulness is proven, they suffer from two major problems.

The first problem is related to the inclusion of chemical reactions. These have to be simulated by detecting collisions between the walking molecules, and current detection algorithms tend to be very inefficient. In addition, it is very hard to determine the reaction probabilities from the rate constants and/or molecule parameters since the molecule can experience more real, chemically effective collisions than the random walk would allow for. Therefore these methods are only useful for simulating chemical reactions of mobile molecules with immobile ones on a surface, like receptors, so that one can use the notion of effective cross-section to calculate the reaction probabilities.

The other problem is related to the scale of the system. Simulating individual molecules is only feasible for a small number of molecules, that is, on the order of ten thousand molecules,[12] as is the case in the examples mentioned. If one is interested, for instance, in simulating a large part of a dendritic tree or when the concentration of a substance has a large dynamic range, as is the case for calcium, these methods will make severe demands on computer memory and computation time.

A way to solve the first problem is to use a molecular dynamics approach which, however, suffers from the second problem as well, thereby eradicating the advantages of using random walks. A popular implementation of this kind of MC methods is MCell (Bartol et al. 1996). The authors of MCell pioneered the use of ray-tracing techniques for boundary detection in general MC simulators of biological systems.

Green's Function MC   We introduce an MC method that doesn't suffer from these problems, as can be seen from the following discussion. It is Green's function MC and it is based on a quantum MC method from solid state physics (De Raedt and von der Linden 1995). Technically, this method depends strongly on a result from function theory. Every function can be expanded in a series of Dirac delta functions, or

$$\tilde{C}(\mathbf{r}, t_0) = \sum_{j=1}^{N_\delta} C_i(t_0) \delta(\mathbf{r} - \mathbf{r}_i), \tag{7.26}$$

where $C_i(t_0) = C(\mathbf{r}_i, t_0)$ are N_δ samples of the function $C(\mathbf{r}, t_0)$ at fixed points. In our case, it denotes the distribution function (expressed in concentration units, for instance) of a substance at time t_0. Finding the evolution of the function under an evolution equation boils down to finding the evolution of the coefficients $C_i(t_0 \rightarrow t)$. We know from section 7.2.1 that the solution of the diffusion equation for a Dirac delta function[13] is a Gaussian and that this

Gaussian can be generated using a random walk, in this case a random walk in 3-space. In order to perform that random walk to solve the diffusion equation, we further subdivide the $C_i(t_0)$'s into n_i pieces of weight w_κ and assign this weight to a random walker for every piece, resulting in $N_w \sum_{i=1}^{N_\delta} n_i$ walkers in total.

Equation (7.26) now becomes

$$\tilde{C}(\mathbf{r}, t_0) = \sum_{i=1}^{N_\delta} \left(\sum_{\kappa=1}^{n_i} w_\kappa \right) \delta(\mathbf{r} - \mathbf{r}_i), \tag{7.27}$$

which is only valid for $t = t_0$ because at later times the successive random walk steps redistribute the walkers over neighboring positions so that we should write:

$$\tilde{C}(\mathbf{r}, t) = \sum_{i=1}^{N_\delta} \left(\sum_{\kappa=1}^{n_i(t)} w_\kappa \right) \delta(\mathbf{r} - \mathbf{r}_i) \tag{7.28}$$

instead for later times ($t > t_0$). $n_i(t)$ is the number of all walkers for which the position $\mathbf{r}_{w_k} \in \Delta\mathcal{V}_i$, $\Delta\mathcal{V}_i$ being a neighborhood of $\mathbf{r}_i$ and $\bigcup_{i=1}^{N_\delta} \Delta\mathcal{V}_i = \mathcal{V}$ ($\Delta\mathcal{V}_{i_1} \cap \Delta\mathcal{V}_{i_2} = \varnothing, i_1 \neq i_2$), the region in space in which the simulation takes place. When proper accounting of walkers is done, $N_w = \Sigma_i n_i(t)$ should be the initial number of walkers for every step.

The easiest way to perform the random walk, however, is to divide all $C_i(t_0)$'s into n_i equal pieces of weight w. At time t, $C(\mathbf{r}, t)$ can now be approximated by

$$\tilde{C}(\mathbf{r}, t) = w \sum_{i=1}^{N_\delta} n_i(t) \delta(\mathbf{r} - \mathbf{r}_i), \tag{7.29}$$

with $n_i(t)$ the number of walkers at or near $\mathbf{r}_i$. As a consequence, simulating diffusion has now become as simple as taking N_w samples of equal size w of the initial distribution function and keeping track of all the samples as they independently perform their random walks. In a closed region V, choose weights $w = (C_0 \mathcal{V})/N_w$ where C_0 is, for example, the final concentration at equilibrium and $V = \text{vol}(V)$. For pure diffusion, all the mass that was present initially will be preserved so that the weights will stay unchanged. Note that N_w determines the variance of the fluctuations around the expected mean[14] and can be much smaller than the number of molecules present in $\mathcal{V}$. Therefore this Green's function MC method does not suffer from the same bad scaling properties as the diffusion MC methods.

Implementing chemical interactions (in fact, almost any interaction) has now become straightforward. To make the idea clear while keeping it simple, we will use a reaction-rate approach. Once this is understood, it should not be too hard to switch to a population-level description (as opposed to a mass or number density description) and use the stochastic methods of chapter 20, to implement the chemical interactions. There are two main classes of algorithms that implement the chemical reactions, of which the first one is the most popular. This first class consists of weight-updating schemes; the other class is described as birth/death schemes since these schemes manipulate the number of random walkers. Because of its popularity, we will first derive a weight-updating scheme by combining a form of equation (7.28) with a set of equations like those in equation (7.20). However, it is advisable not to use this scheme. The reason will become clear from the introduction to an algorithm of the second class.

Since the interactions can change the identity of the molecules of a substance and thereby change the total mass of that substance present in $\mathcal{V}$, we expect the interactions to change the weight of the walkers. Therefore we should write equation (7.28) as follows:

$$\tilde{C}(\mathbf{r},t) = \sum_{i=1}^{N_\delta} \left[\sum_{\kappa=1}^{n_i(t)} w_\kappa(t) \right] \delta(\mathbf{r} - \mathbf{r}_i). \tag{7.30}$$

Filling in equation (7.30) for all substances in equation (7.20), one can show that one will get the following update scheme for all the weights associated with substance s:

$$\frac{d}{dt}\left[\sum_{\kappa=1}^{n_{i,s}(t)} w_{s,\kappa}(t) \right] = f_s[\tilde{C}_1(\mathbf{r}_i,t),\tilde{C}_2(\mathbf{r}_i,t),\ldots,\tilde{C}_N(\mathbf{r}_i,t)] \quad \forall \kappa{:}\mathbf{r}_{w_\kappa} \in \Delta\mathcal{V}_i$$

$$\sum_{\kappa=1}^{n_{i,s}(t)} \frac{dw_{s,\kappa}(t)}{dt} = \sum_{c=1}^{M} K_{s,c} \prod_{j=1}^{N} \left[\sum_{\kappa=1}^{n_{t,j}(t)} w_{j,\kappa}(t) \right]^{v_{j,c}^L}, \tag{7.31}$$

where, in the second set of equations, we used the normal form for the f_s's [equation (7.19)]. (The diffusional parts are handled by the random walks and make the n_i time dependent.) Generally, this means solving $N_{w(s)}$ coupled nonlinear ODEs for every substance s, amounting to $\sum_{s=1}^{N} N_{w(s)}$ equations in total for every timestep. When the f_s's are linear in $\tilde{C}_s(\mathbf{r},t)$, the equations become as follows:

$$\frac{dw_{s,\kappa}(t)}{dt} = f_s[\tilde{C}_1(\mathbf{r}_i,t),\ldots,\tilde{C}_{s-1}(\mathbf{r}_i,t),\tilde{C}_{s+1}(\mathbf{r}_i,t),\ldots,\tilde{C}_N(\mathbf{r}_i,t)]w_{s,\kappa}(t) \tag{7.32}$$

$$\kappa{:}1,\ldots,n_{i,s}(t); \mathbf{r}_{w_\kappa} \in \Delta\mathcal{V}_i$$

with the formal solution:

$$w_{s,\kappa}(t+\Delta t) = w_{s,\kappa}(t)\exp\left(\int_t^{t+\Delta t} f_s\, d\tau \right) \approx w_{s,\kappa}(t)\exp(f_{s,t}\Delta t), \tag{7.33}$$

where the approximation is assumed valid when the integrandum can be considered (nearly) constant in $]t, t + \Delta t[$. This update scheme guarantees positive weight values for positive initial values. Also, at a fixed position, the same exponential factor can be used for all the walkers corresponding to the same substance, providing a significant optimization. The accuracy and stability of the solution will depend on the problem at hand, the degree of stiffness will especially be very important.

The term "stiffness" is generally used in the context of a coupled system of equations. It is a measure of the range of time scales in the simultaneous solution of the equations. If one is interested in fast time scales, it implies that one will lose a lot of time calculating the slowly changing components. If one is only interested in the slow components, it means that one runs the risk of numerical instability. Therefore special numerical algorithms should be used to solve stiff systems. Runge–Kutta–Fehlberg with Rosenbrock-type extensions are simple and robust examples of such algorithms (see, for example, Press et al. 1992).

Sometimes the term "stiffness" is also used in the context of certain PDEs. In this case it means that the spatial and temporal scales are linked by a relation like equation (7.5). However, in most cases one is interested in longer time scales for a fixed spatial scale than is prescribed by this relation. So, again, one is confronted with the dilemma of efficiency vs. numerical stability. Therefore also in this case one should use numerical algorithms specially suited for these equations (for example, the implicit schemes as described below).

Since the weights can change, nothing can stop them from becoming zero or very large. Walkers with zero (or arbitrary subthreshold) weights do not contribute (much) to the distribution function. Generating random walks for them is a waste of computation time. A popular practice is to disregard these small weight walkers altogether (and possibly reuse their data structures for redistributing large weight walkers, as described next). If the weights become very large, they can cause instability, excessive variance, and huge numerical errors. Therefore one could divide a walker with a large weight into smaller pieces. There are other methods to increase efficiency and reduce variance, but they depend on the specifics (such as relative time scales) of the problem at hand. Without these weight control mechanisms, however, weight-update schemes tend to be very inefficient and produce results with large variances.

Sherman and Mascagni (1994) developed a similar method, called *gradient random walk* (GRW). Instead of calculating the solution directly, this method calculates the gradient of the solution. Their work is a good source of references to other particle methods to solve reaction–diffusion equations and to the mathematical foundations and computational properties of particle methods in general.

When using the combination of weight control mechanisms as described above, an important principle of MC simulations is violated. This "correspondence principle" leads to the condition of "detailed balance"[15] at thermodynamic equilibrium. It guarantees global balance (at thermodynamic equilibrium) and meaningful, consistent statistics (in general). A prerequisite for the correspondence principle to be met is to preserve the close correspondence—hence the name "correspondence principle"—between local changes in the contribution of random walkers and the physical processes that cause these local changes. This means that local reshuffling of random walkers for performance reasons is out of the question. Algorithms of the second class, however, meet this principle in an efficient manner when correctly implemented.

Deriving a particular scheme for second-class algorithms starts from equation (7.29) instead of (7.28). Instead of making the weights dependent on time, as in equation (7.30), one can find a scheme for changing the number $n_i(t)$ while keeping w_s fixed (on a per substance basis). By substituting equation (7.29) in (7.20), one gets for every position $\mathbf{r}_i$:

$$
\begin{aligned}
w_s \frac{dn_{i,s}(t)}{dt} &= \sum_{c=1}^{M} K_{s,c} \prod_{j=1}^{N} [w_s n_{i,s}(t)]^{v_{j,c}^L} \\
\frac{dn_{i,s}(t)}{dt} &= \sum_{c=1}^{M} \frac{K_{s,c}}{w_s} \prod_{j=1}^{N} w_s^{v_{j,c}^L} \prod_{j=1}^{N} n_{i,s}^{v_{j,c}^L}(t) \\
&= \sum_{c=1}^{M} \Lambda_{s,c} \prod_{j=1}^{N} n_{i,s}^{v_{j,c}^L}(t)
\end{aligned}
\tag{7.34}
$$

(now with $\Lambda_{s,c} = (K_{s,c}/w_s) \prod_{j=1}^{N} w_j^{v_{j,c}^L} = [k_c(v_{s,c}^R - v_{s,c}^L)/w_s] \prod_{j=1}^{N} w_j^{v_{j,c}^L}$, which can be precomputed). Now one only has to solve $N[\ll N_{w(s)}(t)]$-coupled nonlinear ODEs for every position $\mathbf{r}_i$ that is occupied (by at least one substance), in the worst-case scenario, leading to at most $\sum_{s=1}^{N} N_{w(s)}(t) [\neq \sum_{s=1}^{N} N_{w(s)}(0)]$ sets of N coupled equations, of which most are trivial (i.e., zero change). Integrating this set of equations over Δt gives a change in the number of random walkers $\Delta n_{i,s}(t)[=n_{i,s}(t+\Delta t)-n_{i,s}(t)]$ at position $\mathbf{r}_i$. Generally, $\Delta n_{i,s}(t)$ is not an integer; therefore one uses the following rule: create or destroy $\lfloor \Delta n_{i,s}(t) \rfloor$[16] random walkers and create or destroy an extra random walker with probability frac[$\Delta n_{i,s}(t)$]. Note that now $N_w = \Sigma_i n_i(t)$ is not preserved! The correspondence principle is still met when globally scaling down the number of random walkers (and as a compensation renormalizing their weight) for a substance in case their number has grown out of the band as a result of non-Fickian fluxes.

Now that one can select a method to incorporate chemical interactions, it has to be combined with the discrete random walk generator for diffusion. To simplify the implementation, one chooses a common spatial grid for all substances. Since for diffusion the temporal scale is intrinsically related to the spatial scale (by equation (7.5)], this means a different Δt for every unique D. The simulation clock ticks with the minimal Δt, $\Delta t_{\min}$. Every clock tick, the system of equation (7.34) is solved to change the numbers of random walkers. Then a diffusion step should be performed for every substance s for which an interval Δt_s has passed since its last diffusion update at time t_s. Here we made the following implicit assumptions: The concentration of substance s in $\Delta \mathcal{V}_i$ is assumed not to change due to diffusion during a time interval Δt_s and all substances are assumed to be well mixed in very $\Delta \mathcal{V}_i$. The assumptions are justified if one chooses vol($\Delta \mathcal{V}_i$) in such as way that in every $\Delta \mathcal{V}_i$ the concentration changes every timestep, with at most 10% for the fastest-diffusing substance(s) [i.e., the one(s) with $\Delta t_s = Dt_{min}$).

Discretization in Space and Time   For a large enough number of molecules, it is sufficient to integrate the diffusion equations directly using, for instance, discretization methods. Their scaling properties are much better (although for multidimensional problems, this is questionable) and theoretically they are also much faster. However, as can be seen from the equations (see later discussion), their implementation can become more and more cumbersome for every additional interaction term and for growing geometrical complexity. So, although these methods are very popular because of their perceived superior performance, it's sometimes more feasible to use the Green's function MC method (or another equivalent particle method) as a compromise between performance and ease of implementation.

The Crank–Nicholson method is an example of a classical implicit finite-difference discretization method that is well suited for diffusion problems. The next subsection introduces the formulation for one-dimensional (reaction-)diffusion, and this discussion is followed by ADI, an extension to Crank–Nicholson for solving multidimensional problems.

One-Dimensional Diffusion   Because the diffusion equation (7.6) is a parabolic PDE, which makes it very stiff for most spatial scales of interest (Carnevale 1989, Press et al. 1992), its solution requires an implicit[17] solution method and its numerical solution is sensitive to the accuracy of the boundary conditions (Fletcher 1991). For the discretization, the volume will be split into a number of elements and the concentration computed in each of them. This approximation for a real concentration will be good if sufficient molecules are present in each element so that the law of large numbers applies. This implies that the elements should not be too small, though this may of course lead to the loss of accuracy in representing gradients. The problem is not trivial. Take, for example, the calcium concentration in a dendritic spine (Koch and Zador 1993). A resting concentration of 50 nM corresponds to exactly two calcium ions in the volume of a 0.5-μm diameter sphere, which is about the size of a spine head! The random walk methods described in the previous sections seem more appropriate for this situation.

A popular assumption is that of a spherical cell where diffusion is modeled using the spherical radial equation [the second equation in (7.16)]. The discretization is performed by subdividing the cell into a series of onion shells with a uniform thickness of about 0.1 μm (Blumenfeld et al. 1992) (see figures 7.3–7.5). The Crank–Nicholson method is then the preferred solution method for finding the concentration in each shell after every timestep Δt (Fletcher 1991, Press et al. 1992) because it is unconditionally stable and second-order accurate in both space and time. We will continue using our example of calcium diffusion, which leads to a set of difference equations of the form:

$$-\frac{\Delta t}{2} D_{Ca} C_{i-1,i} [Ca^{2+}]_{i-1,t+\Delta t} + \left(1 + \frac{\Delta t}{2} D_{Ca} (C_{i-1,i} + C_{i,i+1})\right) [Ca^{2+}]_{i,t+\Delta t}$$

$$-\frac{\Delta t}{2} D_{Ca} C_{i,i+1} [Ca^{2+}]_{i+1,t+\Delta t} \tag{7.35}$$

$$= \frac{\Delta t}{2} D_{Ca} C_{i-1,i} [Ca^{2+}]_{i-1,t} + \left(1 - \frac{\Delta t}{2} D_{Ca} (C_{i-1,i} + C_{i,i+1})\right) [Ca^{2+}]_{i,t}$$

$$+ \frac{\Delta t}{2} D_{Ca} C_{i,i+1} [Ca^{2+}]_{i+1,t}$$

Note that the equations express the unknown $\left([Ca^{2+}]_{i,t+\Delta t}\right)$ in terms of the knowns (the right-hand side terms), but also of the unknowns for the neighboring shells. Therefore, this requires an iterative solution of the equations, which is typical for an implicit method. The terms containing $i + 1$ are absent in the outer shell $i = n$ and those containing $i-1$ in the innermost shell $i = 0$. This system of equations can be applied to any one-dimensional morphology where the geometry of the problem is described by the coupling constants $C_{i,i+1}$. In the case of a spherical cell, the coupling constants are of the form:

$$C_{i,i+1} = \frac{3(i+1)^2}{(3i^2 + 3i + 1)\Delta r^2} \tag{7.36}$$

where Δr is the uniform thickness of the shells and the diameter of the cell is equal to $d = 2 \cdot n\Delta r$. See De Schutter and Smolen (1998) for practical advice on how to apply these equations to more complex geometries.

The system of algebraic equations (7.35) can be solved very efficiently because its corresponding matrix is tridiagonal with diagonal dominance (Press et al. 1992). An influx across the plasma membrane can be simulated by adding a term $(\Delta t/2)(J_{n,t+\Delta t/2}/v_n)$ to the right-hand side of the equation corresponding to the outer shell. Computing the flux at $t + \Delta t/2$ maintains the second-order accuracy in time (Mascagni and Sherman 1998). Note that the overall accuracy of the solution will depend on how exactly this boundary condition J_n is met (Fletcher 1991).

In the case of buffered diffusion, the buffer reaction equation (7.22) has to be combined with (7.35). This leads to a new set of equations with, for every shell, one equation for each diffusible buffer:

$$
-\frac{\Delta t}{2} D_B C_{i-1,i}[B]_{i-1,t+\Delta t} - \frac{\Delta t}{2} D_B C_{i,i+1}[B]_{i+1,t+\Delta t}
$$

$$
+\left(1 + \frac{\Delta t}{2}\left(D_B\left(C_{i-1,i} + C_{i,i+1}\right) + b + f[Ca^{2+}]_{i,t}\right)\right)[B]_{i,t+\Delta t} \tag{7.37}
$$

$$
= \frac{\Delta t}{2} D_B C_{i-1,i}[B]_{i-1,t} + \frac{\Delta t}{2} D_B C_{i,i+1}[B]_{i+1,t}
$$

$$
+\left(1 - \frac{\Delta t}{2}\left(D_B\left(C_{i-1,i} + C_{i,i+1}\right) + b\right)\right)[B]_{i,t} + \Delta t b[B]_{T,i}
$$

and an equation for buffered calcium diffusion:

$$
-\frac{\Delta t}{2} D_{Ca} C_{i-1,i}[Ca^{2+}]_{i-1,t+\Delta t} - \frac{\Delta t}{2} D_{Ca} C_{i,i+1}[Ca^{2+}]_{i+1,t+\Delta t}
$$

$$
+\left(1 + \frac{\Delta t}{2}\left(D_{Ca}\left(C_{i-1,i} + C_{i,i+1}\right) + f[B]_{i,t}\right)\right)[Ca^{2+}]_{i,t+\Delta t} \tag{7.38}
$$

$$
= \frac{\Delta t}{2} D_{Ca} C_{i-1,i}[Ca^{2+}]_{i-1,t} + \frac{\Delta t}{2} D_{Ca} C_{i,i+1}[Ca^{2+}]_{i+1,t}
$$

$$
+\left(1 - \frac{\Delta t}{2} D_{Ca}\left(C_{i-1,t} + C_{i,i+1}\right)\right)[Ca^{2+}]_{i,t} + \Delta t b\left([B]_{T,i} - \frac{1}{2}[B]_{i,t}\right).
$$

Equation (7.38) applies to a system with only one buffer. In the case of multiple buffers, one additional term appears at the left-hand side and two at the right-hand side for each buffer. We have also assumed that D_B is identical for the free and bound forms of the buffer, which gives a constant $[B]_{T,i}$. This is reasonable, taking into account the relative sizes of calcium ions compared with most buffer molecules. Equations (7.37) and (7.38) should be solved simultaneously, resulting in a diagonally banded matrix with three bands for the calcium diffusion and two bands extra for each buffer included in the model.

Two-Dimensional Diffusion   An elegant extension to the Crank–Nicholson method is the alternating direction implicit method, which allows for an efficient solution of two-dimensional diffusion problems (Fletcher 1991, Press et al. 1992). Each timestep is divided into two steps of size $\Delta t/2$, and in each substep the solution is computed in one dimension:

$$
-\frac{\Delta t}{4} D_{Ca} C_{i-1,i}[Ca^{2+}]_{i-1,j,t+\Delta t/2} + \left(1 + \frac{\Delta t}{4} D_{Ca}\left(C_{i-1,i} + C_{i,i+1}\right)\right)[Ca^{2+}]_{i,j,t+\Delta t/2}
$$

$$
-\frac{\Delta t}{4} D_{Ca} C_{i,i+1}[Ca^{2+}]_{i+1,j,t+\Delta t/2} \tag{7.39}
$$

$$
= \frac{\Delta t}{4} D_{Ca} C_{i-1,i}[Ca^{2+}]_{i-1,j,t} + \left(1 - \frac{\Delta t}{4} D_{Ca}\left(C_{i-1,i} + C_{i,i+1}\right)\right)[Ca^{2+}]_{i,j,t}
$$

$$+ \frac{\Delta t}{4} D_{Ca} C_{i,i+1} [Ca^{2+}]_{i+1,j,t},$$

respectively

$$-\frac{\Delta t}{4} D_{Ca} c_{j-1,j} [Ca^{2+}]_{i,j-1,t+\Delta t} + \left(1 + \frac{\Delta t}{4} D_{Ca}\left(c_{j-1,j} + c_{j,j+1}\right)\right)[Ca^{2+}]_{i,j,t+\Delta t}$$

$$-\frac{\Delta t}{4} D_{Ca} c_{j,j+1} [Ca^{2+}]_{i,j+1,t+\Delta t} \tag{7.40}$$

$$= \frac{\Delta t}{4} D_{Ca} c_{j-1,j} [Ca^{2+}]_{i,j-1,t+\Delta t/2} + \left(1 - \frac{\Delta t}{4} D_{Ca}\left(c_{j-1,j} + c_{j,j+1}\right)\right)[Ca^{2+}]_{i,j,t+\Delta t/2}$$

$$+\frac{\Delta t}{4} D_{Ca} c_{j,j+1} [Ca^{2+}]_{i,j+1,t+\Delta t/2},$$

where first diffusion along the i dimension is solved implicitly and then over the j dimension. The coupling constants $C_{i,i+1}$ and $c_{j,j+1}$ again depend on the geometry of the problem and are identical and constant in the case of a square grid. The two matrices of coupled equations (7.39) and (7.40) are tridiagonal with diagonal dominance and can be solved very efficiently. In two dimensions, the ADI method is unconditionally stable for the complete timestep and second-order accurate in both space and time, provided the appropriate boundary conditions are used (Fletcher 1991). Holmes (1995) used the two-dimensional ADI method with cylindrical coordinates to model glutamate diffusion in the synaptic cleft. The ADI method can also be used to simulate three-dimensional diffusion with substeps of $\Delta t/3$, but in this case it is only conditionally stable (Fletcher 1991).

7.3   Case Study

In general, we're interested in compartmental models of neurons using detailed reconstructed morphologies, in particular of the dendritic tree. These compartments are cylindrical idealizations of consecutive tubular sections of a dendritic membrane. Because of the cylindrical nature of the compartments, we prefer a description in cylindrical coordinates, giving us an accurate representation of the concentration in the submembrane region. This is important for a number of our other simulations. As can be seen from equation (7.15), using cylindrical coordinates introduces factors depending on $1/\rho$ and an additional first-order term in ρ. These are pure geometric effects and arise from the fact that an infinitesimal volume element in cylindrical coordinates is of the form $dV = \rho d\rho d\phi dz$, which is dependent on ρ.

Just as in the one-dimensional case, the second equation in (7.11) can be derived from a (discrete) random-walk process. That is, in Cartesian coordinates, for every second-order term on the right-hand side of (7.11) (the second equation), the corresponding coordinate of the random walker's position is updated according to the one-dimensional random walk scheme. However, this recipe does not work when writing out a random walk scheme for (7.14) because of the extra first-order term. A more general recipe for deriving the appropriate random-walk process can be obtained by considering more closely the consequences of the ρ dependence of dV.

In order to maintain a uniform distribution of mass at equilibrium, there should be less mass in the volume elements closer to the origin, since for smaller ρ, that is, nearer to the origin, the volume elements get smaller. Therefore the random walkers (which correspond to mass units) should tend to step more often away from the origin than toward the origin when equilibrating, creating a net drift outward. Eventually, in a closed volume with reflecting walls, this drift will come to a halt at equilibrium, while the outward tendency creates a gradient in ρ in the number (but not density!) of walkers. The tendency is introduced by adjusting the probability to choose the direction of a ρ-step.

The adjustment is derived as follows for a drifting one-dimensional random walk: Suppose the walkers have a drift speed c (negative to go to the left, positive to go to the right); then the probability to go to the left is

$$\left\{ \begin{array}{ll} p_1 & = \dfrac{1}{2}\left(1 - \dfrac{c\Delta t}{\Delta x}\right) \\ p_r = 1 - p_1 & = \dfrac{1}{2}\left(1 + \dfrac{c\Delta t}{\Delta x}\right) \end{array} \right. \quad (0 \le p_1, p_r \le 1 \tag{7.41}$$

(with obvious conditions on Δt and Δx for a given c). When filling in these expressions for $p_{l,r}$ in (7.1) and taking the appropriate limits, one gets the following equation:

$$\frac{\partial C(x,t)}{\partial t} = D\frac{\partial^2 C(x,t)}{\partial x^2} + c\frac{\partial C(x,t)}{\partial x}. \tag{7.42}$$

Note the correspondence between this equation and the radial part of equation (7.14). Therefore we extend the recipe as follows: Every first-order term corresponds to a drift, with the drift speed c given by the coefficient of the partial derivative and the probabilities adjusted according to (7.41). If we ignore the $1/\rho$ factors by defining new "constants" $c = D/r$ and $D' = D/r^2$, we get the following random walk scheme in cylindrical coordinates: For every timestep $t \to t + \Delta t$, for every walker $\kappa (1 \le \kappa \le N_w)$, the following coordinate update is performed:

$$\left\{ \begin{array}{llll} r_\kappa(t) = k\Delta r & \to & r_\kappa(t + \Delta t) = (k \pm 1)\Delta r & (k \in \mathbb{N}) \\ \phi_\kappa(t) & \to & \phi_\kappa(t + \Delta t) = \phi_\kappa(t) \pm \Delta\phi(r) & \\ z_\kappa(t) = l\Delta z & \to & z_\kappa(t + \Delta t) = (l \pm 1)\Delta z & (l \in \mathbb{N}), \end{array} \right. \tag{7.43}$$

where we choose

$$
\begin{cases}
(\Delta r)^2 = (\Delta z)^2 = 2D\Delta t \\
[\Delta\phi(r)]^2 = 2D'\Delta t = \left(\dfrac{\Delta r}{r}\right)^2 = \dfrac{1}{k^2} \quad (k > 0)
\end{cases}
\tag{7.44}
$$

with the following probabilities for the respective signs:

$$
\begin{cases}
p_\pm^r \quad = \quad \dfrac{1}{2}\left(1 \pm \dfrac{c\Delta t}{\Delta r}\right) = \dfrac{1}{2}\left(1 \pm \dfrac{1}{2k}\right) \quad k > 0 \\
p_\pm^\phi = p_\pm^z \quad = \qquad\qquad \dfrac{1}{2}
\end{cases}
\tag{7.45}
$$

and ($p_-^r = 0$, $p_+^r = 1$) for $k = 0$. This type of random walk is called a *Bessel walk*. From the $1/r_n$ factors in the diffusion equation, one expects problems of divergence in the origin. The way we handle the random walks in the origin (i.e., for $k = 0$) seems to work fine, as can be seen from simulation results. Of course, owing to the low number of walkers in the origin, the variance is quite large in the region around the origin.

The discrete natures of the geometry and the random walk introduce other specific problems, among which are connections of compartments with different radii, branch points, and regions of spine neck attachments to the compartmental mantle. We used minimalistic approaches to these problems by using probabilistic models. Concerning the connections, for example, we assigned a probability for the transition that the random walker could make between two connected cylinders in such a way that $p_\pm^z$ becomes

$$
p_\pm^z =
\begin{cases}
\dfrac{1}{2} & R_1 \le R_2 \\
\dfrac{1}{2}\left(1 - \left(\dfrac{R_2}{R_1}\right)^2\right) & R_1 > R_2
\end{cases}
,
\tag{7.46}
$$

(where the sign is chosen in accordance with the end of the cylinder at which the walker is located) when the walker is at the end of cylinder 1 that is connected to cylinder 2. When the walker makes the transition to cylinder 2, its ρ-coordinate is adjusted according to $\rho_{new} = (R_2/R_1)\rho_{old}$. This way no artificial gradients are created in the area around the discrete transition. Instead, the walkers behave more like walkers in a tapered cylinder, which is the more natural geometry for dendritic branches.

Recent caged compound-release experiments in neuronal spiny dendrites (Wang and Augustine 1995, Bormann et al. 1997) show a slowing in the apparent diffusion (see figure 7.6). The apparent diffusion coefficient is defined as the diffusion coefficient for the effective one-dimensional diffusion along the longitudinal axis of the dendritic shaft. The ex-

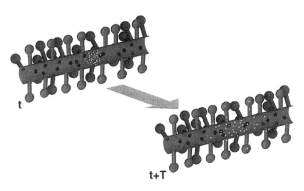

Figure 7.6
The effect of the presence of spines on the longitudinal component of diffusion in the dendritic shaft. Compare with figure 7.1 for the smooth case (i.e., without spines). The small dots represent small spherelike molecules. In the model, the spine heads (the small spheres on the spine necks shown here as sticks perpendicular to the dendritic shaft) are modeled as short, wide cylinders.

periments were performed with both relatively inert and highly regulated substances. We were interested in estimating the contribution of geometric factors to this slowing in apparent diffusion rate, especially the effect of "hidden" volume (mostly spines) and morphology (branching, structures in cytoplasm).

In order to isolate the contributing factors, we constructed simple geometrical models of dendritic structures, including spines and single branch points. Then we simulated idealized-release experiments by using square pulses as an initial condition and calculated the apparent diffusion coefficient $D_{app}(t)$ from the resulting concentration profiles as follows:

$$D_{app}(t) = \frac{\langle z^2 \rangle(t) - \langle z^2 \rangle(0)}{2t} \tag{7.47}$$

with

$$\langle z^2 \rangle(t) = \int_{-\infty}^{+\infty} (z - \langle z \rangle)^2 \, C(z,t) dz \,. \tag{7.48}$$

The integral is cut off at the end points of the (chain of consecutive) cylinders, which is all right when the distance between the end points is large enough so that only very few walkers will arrive at either of the end points in the simulated time period. The simulator is instructed to output $C(z,t) = \overline{C(\rho,\phi,z,t)}^{(\rho,\phi)}$, that is, the ρ and ϕ dependence are averaged out. Slowing is expressed by a slowing factor $\lambda(t)$, defined as $\lambda(t) = D_{app}(t)D \leq 1.0 \, \forall \, t \geq 0$. The apparent diffusion coefficient, D_{app}, at equilibrium is defined as $D_{app} = D_{app}(t \rightarrow \infty)$.

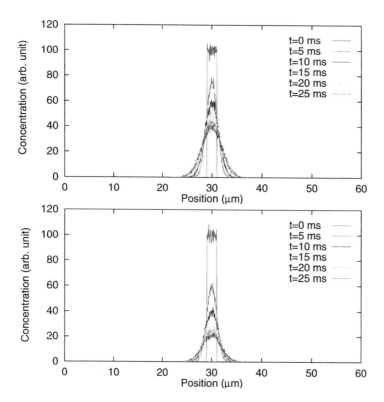

Figure 7.7
Concentration profiles along the long axis of the dendritic shaft. (Top) Smooth dendrite case (see text). (Bottom) Spiny dendrite case (see text). (See plate 25.)

Figure 7.7 (see also plate 25) shows the difference in time evolution of two concentration profiles. One is for a smooth dendrite (modeled as a simple cylinder), the other for a spiny dendrite of the same length, but with spines distributed over its plasma membrane. The spines are attached perpendicularly on the cylindrical surface and consist of two cylinders: one narrow for the neck (attached to the surface) and one wide for the head. The figure shows the results for the most common spine parameters (density, neck, and head lengths and radii) on Purkinje cell dendrites.

The corresponding $\lambda(t)$ curves can be found overlapped in figure 7.8. The small amount of slowing in the smooth dendrite case is due to reflection at the end points, which is negligible for the simulation time used. We found $\lambda(0)/\lambda(\infty)$'s of 1.0–3.0 for various geometric parameters of the spines. [We used $\lambda(0)/\lambda(\infty)$ to express total slowing instead of $\lambda^{-1}(\infty)$ be-

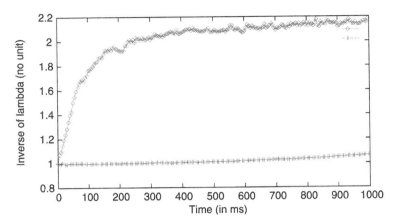

Figure 7.8
Slowing factor curves. Upper curve: spiny dendrite case (see text). Lower curve: smooth dendrite case (see text).

cause $D_{app}(0)$ is only equal to D for initial conditions that are the solution of the diffusion equation.]

Notes

1. The term "parabolic" describes equations of the form:

$$\frac{\partial u(\mathbf{r},t)}{\partial t} = \nabla \cdot (\mathbf{D}(\mathbf{r})\nabla u(\mathbf{r},t)) = \sum_j \frac{\partial}{\partial x_j}\left(\sum_i D_{i,j}(\mathbf{r})\frac{\partial u(\mathbf{r},t)}{\partial x_i}\right) \qquad (D_{i,j} \ge 0)$$

$$= D\sum_i \frac{\partial^2 u(\mathbf{r},t)}{\partial^2 x_i} \qquad\qquad (D_{i,j} = 0\forall i \ne j; D_{i,i} = D\forall i).$$

We only treat the last case.

2. Hint (Schulman 1981): Write down the binomial distribution for the number of molecules at a fixed position after $N=N_l+N_r$ steps. Then determine the proper conditions and apply the appropriate limit theorem of formal statistics to find the normal (i.e., Gaussian) distribution.

3. This is generally done by using boundary detection techniques based on the ray-tracing techniques of computer graphics (Foley et al., 1990). See section 7.2.5.

4. Definition:

$$\begin{cases} x &= \rho\cos\phi \\ y &= \rho\sin\phi & \rho \ge 0 \\ z &= z & 0 \le \phi \le 2\pi \end{cases}$$

5. Definition:

$$\begin{cases} x &= r\cos\phi\sin\theta & r \ge 0 \\ y &= r\sin\phi\sin\theta & 0 \le \phi \le 2\pi \\ z &= z\cos\theta & 0 \le \theta \le \pi \end{cases}$$

6. In this scheme, reversible reactions are written as a pair of irreversible reactions!

7. The spherical cell had a diameter of 20 μm and was simulated using a Crank–Nicholson solution method [equation (7.35)] with 0.1-μm-thick shells. Resting calcium concentration was 50 nM and the diffusion coefficient for calcium was $2 \cdot 10^{-6}$ cm^2 s^{-1}. Removal mechanisms were not implemented. Slow buffer (EGTA): concentration 50 μM, $K_d = 0.20$ μM, $f = 2.5 \cdot 10^6$ M^{-1}s^{-1}, $b = 0.5$ s^{-1}. Fast buffer (BAPTA): concentration 50 μM, $K_d = 0.20$ μM, $f = 4.0 \cdot 10^8$ M^{-1}s^{-1}, $b = 8$ s^{-1}. Diffusible buffers had a diffusion coefficient of $2 \cdot 10^{-6}$ cm^2 s^{-1}. The current was 100 ms long and had an amplitude of 50 pA.

8. The D_{app} was computed as $(1/1 + [B]_T/K_d)D_{Ca}$, and the calcium influx was scaled to achieve the same final free calcium concentration as in the case of a real buffer.

9. Concentrations were 50 μM for the slow buffer and 10 μM for the fast buffer. The cell is also larger (a diameter of 60 μm) and additional traces are shown for concentrations 20 and 30 μm below the plasma membrane.

10. Since computers are generally deterministic machines, it is better to use the term "pseudo-random" numbers. It is not the randomness of every number per se that's important, but the statistical properties of a sequence, which have to match the statistical properties of a sequence of "real" random numbers.

11. Of course, nothing keeps you from reformulating a set of reaction–diffusion equations into a set of integral equations and using the first type to solve them. Good luck!

12. That is, on today's workstations. More molecules (on the order of a hundred thousand to a million) can be tracked using workstation clusters or supercomputers.

13. Such a solution is called the *Green's function of the differential equation at hand*. Hence the name Green's function Monte Carlo.

14. It can be proven that var $\propto 1/\sqrt{N_w}$ when no additional information about the solution is used.

15. An exact definition of detailed balance strongly depends on the type of stochastic process involved (see Fishman 1996).

16. $\lfloor x \rfloor$ is the integer part of x.

17. A definition for implicit method is given after introducing the equations.

References

Albritton, M. L., Meyer, T., and Stryer, L. (1992). Range of messenger action of calcium ion and inositol 1,4,5-triphosphate. *Science* 258: 1812–1815.

Augustine, G. J., Charlton, M. P., and Smith, S. J. (1985). Calcium entry and transmitter release at voltage-clamped nerve terminals of squid. *J. Physiol.* 367: 163–181.

Baimbridge, K. G., Celio, M. R., and Rogers, J. H. (1992). Calcium-binding proteins in the nervous system. *Trends Neurosci.* 15: 303–308.

Bartol, T. M., Land, B R., Salpeter, E. E., and Salpeter, M. M. (1991). Monte Carlo simulation of miniature end-plate current generation in the vertebrate neuromuscular junction. *Biophys. J.* 59: 1290–1307.

Bartol, T. M. J., Stiles, J. R., Salpeter, M. M., Salpeter, E. E., and Sejnowski, T. J. (1996). MCELL: Generalised Monte Carlo computer simulation of synaptic transmission and chemical signaling. *Abstr. Soc. Neurosci.* 22: 1742.

Berridge, M. J. (1997). The AM and FM of calcium signaling. *Nature* 386: 759–760.

Bezprozvanny, I. (1994). Theoretical analysis of calcium wave propagation based on inositol (1,4,5)-triphosphate (InsP$_3$) receptor functional properties. *Cell Calcium* 16: 151–166.

Bezprozvanny, I., Watras, J. and Ehrlich, B. E. (1991). Bell-shaped calcium-dependent curves of Ins(1,4,5)P$_3$-gated and calcium-gated channels from endoplasmic reticulum of cerebellum. *Nature* 351: 751–754.

Blumenfeld, H., Zablow, L., and Sabatini, B. (1992). Evaluation of cellular mechanisms for modulation of calcium transients using a mathematical model of fura-2 Ca^{2+} imaging in *Aplysia* sensory neurons. *Biophys. J.* 63: 1146–1164.

Bormann, G., Wang, S. S. H., De Schutter, E., and Augustine, G. J. (1997). Impeded diffusion in spiny dendrites of cerebellar Purkinje cells. *Abstr. Soc. Neurosci.* 23: 2008.

Bower, J. M., and Beeman, D., (1995). *The book of GENESIS: Exploring realistic neural models with the GEneral NEural SImulation System.* TELOS, New York.

Carnevale, N. T. (1989). Modeling intracellular ion diffusion. *Abstr. Soc. Neurosci.* 15: 1143.

Crank, J. (1975). *The Mathematics of Diffusion.* Clarendon Press, Oxford.

De Raedt, H., and von der Linden, W. (1995). *The MC Method in Condensed Matter Physics,* 2nd ed. vol. 71. Springer-Verlag, Berlin.

De Schutter, E., and Smolen, P. (1998). Calcium dynamics in large neuronal models. In *Methods in Neuronal Modeling: From Ions to Networks,* 2nd ed. C. Koch and I. Segev, eds. pp. 211–250. MIT Press, Cambridge, Mass.

DiFrancesco, D., and Noble, D. (1985). *A model of cardiac electrical activity incorporating ionic pumps and concentration changes. Phil. Trans. Roy. Soc. Lond.* B 307: 353–398.

Dolmetsch, R. E., Xu, K., and Lewis, R. S. (1998). Calcium oscillations increase the efficiency and specificity of gene expression. *Nature* 392: 933–936.

Einstein, A. (1905). On the movement of small particles suspended in a stationary liquid demanded by the molecular kinetics of heat. *Ann. Phys.* 17: 549–560.

Exton, J. H. (1997). Phospholipase D: enzymology, mechanisms of regulation, and function. *Physiol. Rev.* 77: 303–320.

Farnsworth, C. L., Freshney, N. W., Rosen, L. B., Ghosh, A., Greenberg, M. E., and Feig, L. A. (1995). Calcium activation of Ras mediated by neuronal exchange factor Ras-GRF. *Nature* 376: 524–527.

Feynman, R. P., Leighton, R. B., and Sands, M. (1989). *The Feynman Lectures on Physics* (commemorative ed.). Addison-Wesley, Reading, Mass.

Fick, A. (1885). Ueber diffusion. *Ann. Phys. Chem.* 94: 59–86.

Fishman, G. S. (1996). *Monte Carlo: Concepts, Algorithms and Applications.* Springer-Verlag, New York.

Fletcher, C. A. J. (1991). *Computational Techniques for Fluid Dynamics.* vol. I. Springer-Verlag, Berlin.

Foley, J., Van Dam, A., Feiner, S., and Hughes, J. (1990). *Computer graphics: Principles and Practice,* 2nd ed. Addison-Wesley, Reading, Mass.

Garrahan, P. J., and Rega, A. F. (1990). Plasma membrane calcium pump. In *Intracellular Calcium Regulation,* F. Bronner, ed. pp. 271–303. Alan R. Liss, New York.

Gola, M., and Crest, M. (1993). Colocalization of active KCa channels and Ca^{2+} channels within Ca^{2+} domains in *Helix* neurons. *Neuron* 10: 689–699.

Goldbeter, A., Dupont, G., and Berridge, M. J. (1990). Minimal model for signal-induced Ca^{2+} oscillations and for their frequency encoding through protein phosphorylation. *Proc. Natl. Acad. Sci. U.S.A.* 87: 1461–1465.

Hille, B. (1992). *Ionic Channels of Excitable Membranes.* Sinauer Associates, Sunderland, Mass.

Holmes, W. R. (1995). Modeling the effect of glutamate diffusion and uptake on NMDA and non-NMDA receptor saturation. *Biophys. J.* 69: 1734–1747.

Issa, N. P., and Hudspeth, A. J. (1994). Clustering of Ca^{2+} channels and Ca^{2+}-activated K^+ channels at fluorescently labeled presynaptic active zones of hair cells. *Proc. Natl. Acad. Sci. U.S.A.* 91: 7578–7582.

Jafri, S. M., and Keizer, J. (1995). On the roles of Ca^{2+} diffusion, Ca^{2+} buffers, and the endoplasmic reticulum in IP_3-induced Ca^{2+} waves. *Biophys. J.* 69: 2139–2153.

Kasai, H., and Petersen, O. H. (1994). Spatial dynamics of second messengers: IP_3 and cAMP as long-range and associative messengers. *Trends Neurosci.* 17: 95–101.

Koch, C., and Zador, A. (1993). The function of dendritic spines: devices subserving biochemical rather than electrical compartmentalization. *J. Neurosci.* 13: 413–422.

Li, W., Llopis, J., Whitney, M., Zlokarnik, G., and Tsien, R. Y. (1998). Cell-permeant caged InsP$_3$ester shows that Ca^{2+} spike frequency can optimize gene expression. *Nature* 392: 936–941.

Linse, S., Helmersson, A., and Forsen, S. (1991). Calcium-binding to calmodulin and its globular domains. *J. Biol. Chem.* 266: 8050–8054.

Llano, I., DiPolo, R., and Marty, A. (1994). Calcium-induced calcium release in cerebellar Purkinje cells, *Neuron* 12: 663–673.

Llinás, R. R., Sugimori, M., and Silver, R. B. (1992). Microdomains of high calcium concentration in a presynaptic terminal. *Science* 256: 677–679.

Mahama, P. A., and Linderman, J. J. (1994). A Monte Carlo study of the dynamics of G-protein activation. *Biophys. J.* 67: 1345–1357.

Markram, H., Roth, A., and Helmchen, F. (1998). Competitive calcium binding: implications for dendritic calcium signaling. *J. Comput. Neurosci.* 5: 331–348.

Mascagni, M. V., and Sherman, A. S. (1998). Numerical methods for neuronal modeling. In *Methods in Neuronal Modeling: From Ions to Networks,* 2nd ed. C. Koch and I. Segev, eds. pp. 569–606. MIT Press, Cambridge, Mass.

Naraghi, M., and Neher, E. (1997). Linearized buffered Ca^{2+} diffusion in microdomains and its implications for calculation of [Ca^{2+}] at the mouth of a calcium channel. *J. Neurosci.* 17: 6961–6973.

Nixon, R. A. (1998). The slow axonal transport of cytoskeletal proteins. *Curr. Opin. Cell Biol.* 10: 87–92.

Nowycky, M. C., and Pinter, M. J. (1993). Time courses of calcium and calcium-bound buffers following calcium influx in a model cell. *Biophys. J.* 64: 77–91.

Press, W. H., Teukolsky, S. A., Vetterling, W. T., and Flannery, B. P. (1992). *Numerical Recipes in C: The Art of Scientific Computing,* 2nd ed. Cambridge University Press, Cambridge.

Qian, N., and Sejnowski, T. J. (1990). When is an inhibitory synapse effective? *Proc. Natl. Acad. Sci. U.S.A.* 87: 8145–8149.

Regehr, W. G., and Atluri, P. P. (1995). Calcium transients in cerebellar granule cell presynaptic terminals. *Biophys. J.* 68: 2156–2170.

Sala, F., and Hernandez-Cruz, A. (1990). Calcium diffusion modeling in a spherical neuron: relevance of buffering properties. *Biophys. J.* 57:313–324.

Schulman, L. S. (1981). *Techniques and Applications of Path Integration.* Wiley, New York.

Sherman, A., and Mascagni, M. (1994). A gradient random walk method for two-dimensional reaction-diffusion equations. SIAM *J. Sci. Comput.* 15: 1280–1293.

Sneyd, J., Charles, A. C., and Sanderson, M. J. (1994). A model for the propagation of intracellular calcium waves. *Am. J. Physiol.* 266: C293–C302.

Sneyd, J., Wetton, B. T. R., Charles, A. C., and Sanderson, M. J. (1995). Intercellular calcium waves mediated by diffusion of inositol triphosphate: a two-dimensional model. *Am J. Physiol.* 268: C1537–C1545.

Wagner, J., and Keizer, J. (1994). Effects of rapid buffers on Ca^{2+} diffusion and Ca^{2+} oscillations. *Biophys. J.* 67: 447–456.

Wahl, L. M., Pouzat, C., and Stratford, K. J. (1996). Monte Carlo simulation of fast excitatory synaptic transmission at a hippocampal synapse. *J. Neurophysiol.* 75: 597–608.

Wang, S. S. H., and Augustine, G. J. (1995). Confocal imaging and local photolysis of caged compounds: dual probes of synaptic function. *Neuron* 15: 755–760.

Zador, A., and Koch, C. (1994). Linearized models of calcium dynamics: formal equivalence to the cable equation. *J. Neurosci.* 14: 4705–4715.

Zhou, Z., and Neher, E. (1993). Mobile and immobile calcium buffers in bovine adrenal chromaffin cells. *J. Physiol.* 469: 245–273.

8 Kinetic Models of Excitable Membranes and Synaptic Interactions

Alain Destexhe

8.1 Introduction

Ion channels are transmembrane proteins containing a pore permeable specifically to one or several ionic species. This property of ionic selectivity is at the basis of the establishment of a voltage difference across the membrane (Hille 1992). This membrane potential is influenced by the opening or closing of ion channels, which can be gated by various factors, such as the membrane potential or the binding of extracellularly or intracellularly released messenger molecules (figure 8.1). In particular, the permeability of some ion channels may depend on voltage (figure 8.1A), which is a fundamental property explaining the electrical excitability of the membrane, as shown by the influential work of Hodgkin and Huxley (1952). Today, several types of voltage-dependent ion channels have been identified and are responsible for a rich repertoire of electrical behavior essential for neuronal function (Llinás 1988).

In addition to being responsible for intrinsic cellular behavior, ion channels also mediate cellular interactions. Transmitter-gated ion channels (figure 8.1B) provide the basis for "fast" synaptic interactions between many cell types, such as neurons. Slower types of synaptic interactions may also take place through the gating of ion channels by second messengers (figure 8.1C). In this case, an extracellularly released transmitter activates a separate receptor, which leads to the production of the second messenger. The second messenger may affect not only ion channels, but also the metabolism of the cell, up to gene regulation (Berridge and Irvine 1989, Tang and Gilman 1991, Gutkind 1998, Selbie and Hill 1998).

The biophysical properties of ion channels have been studied in depth using patch recording techniques (Sakmann and Neher 1995). Single-channel recordings have shown that ion channels display rapid transitions between conducting and nonconducting states. It is now known that conformational changes of the channel protein give rise to opening and closing of the channel, as well as changes in its voltage sensitivity. Conformational changes can be described by state diagrams and Markov models analogous to conformational changes underlying the action of enzymes.

Markov models involving multiple conformational states are usually required in order to accurately capture the dynamics of single-ion channels. However, this complexity may not be necessary in all cases, and simpler models can be adequate for simulating larger scale systems. For example, networks of neurons in the central nervous system involve several types of neurons, each containing multiple types of voltage-dependent ion channels, and synaptic interactions mediated by several types of synaptic receptors and their associated ion channels. Modeling such a system therefore requires modeling thousands of ionic cur-

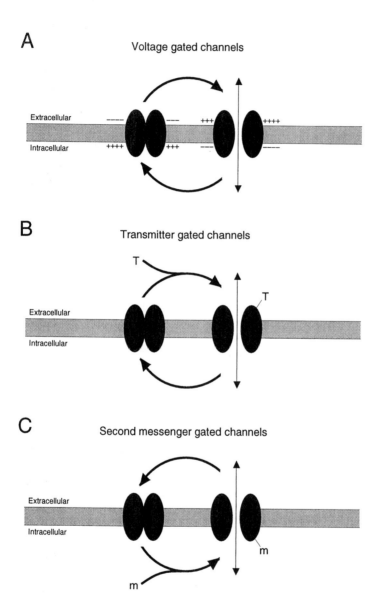

Figure 8.1
Different types of ion channels and their gating. The ion channels are represented in two configurations: closed (left) and open (right). Intermediate configurations may exist, but were not considered for simplicity. (A) Voltage-dependent ion channels. The ion channel undergoes a conformational change as a result of the action of the electric field and redistribution of charges around the channel, which may act to destabilize the closed conformation and stabilize the open configuration. A change of membrane potential then results in favoring the opening of the channel. (B) Transmitter-gated ion channels. In this case, the ion channel must bind a specific transmitter (T) molecule to open. The transmitter is released extracellularly by another cell. This type of interaction underlies

rents and representing each current by multistate Markov schemes. This is clearly not feasible with normal computing facilities.

In this chapter we review kinetic models of ion channels; the models range from multistate Markov models to highly simplified schemes. We show that kinetic models can be used to describe a large spectrum of physiological processes involved in cellular behavior, including voltage-dependent channels, synaptic transmission, and second-messenger actions.[1] We show that kinetic models provide a way to describe a wide range of cellular and subcellular interactions by a unique formalism and that this formalism naturally connects to kinetic equations used to model biochemical reactions.

8.2   Kinetic Models of Ion Channels

The molecular processes leading to opening and closing of ion channels may depend on various factors extrinsic to the channel, such as the electric field across the membrane or the binding of a ligand. The exact molecular mechanisms underlying the gating of ion channels are far from being understood (for a recent review, see Armstrong and Hille 1998). Nevertheless, it is possible to describe the electrical properties of ion channels accurately (Hille 1992). In this section, we review the general equations to describe the gating of ion channels and how to integrate them with membrane equations. Different classes of ion channels will be considered in later sections.

8.2.1   The Kinetic Description of Ion Channel Gating

Assuming that gating occurs following conformational changes of the ion channel protein, the sequence of conformations involved in this process can be described by state diagrams:

$$S_1 \rightleftharpoons S_2 \rightleftharpoons \ldots \rightleftharpoons S_n \, , \tag{8.1}$$

where $S_1 \ldots S_n$ represents distinct conformational states of the ion channel. Define $P(S_i, t)$ as the probability of being in a state S_i at time t and $P(S_i \rightarrow S_j)$ as the transition probability from state S_i to state S_j:

$$S_i \underset{P(S_j \rightarrow S_i)}{\overset{P(S_i \rightarrow S_j)}{\rightleftharpoons}} S_i \, . \tag{8.2}$$

most "fast" synaptic interactions in the central nervous system, through glutamate (AMPA and NMDA receptors) and GABA (GABA$_A$ receptors). (C) Second-messenger-gated channels. Some ion channels are gated intracellularly by binding to a "second messenger" (m). Examples of second messengers are intracellular Ca2+ ions or G-proteins. The latter are usually activated by a separate receptor and mediate slow "metabotropic" types of synaptic interactions, such as GABA$_B$ receptors. G-proteins may also interact with the metabolism of the cell and affect gene regulation, thus providing cellular interactions over a wide range of time scales.

The time evolution of the probability of state S_i is described by the master equation (see e.g., Colquhoun and Hawkes 1977, 1981):

$$\frac{dP(S_i,t)}{dt} = \sum_{j=1}^{n} P(S_j,t)\, P(S_j \rightarrow S_i) - \sum_{j=1}^{n} P(S_i,t)\, P(S_i \rightarrow S_j). \tag{8.3}$$

The left-hand term represents the "source" contribution of all transitions entering state S_i, and the right-hand term represents the "sink" contribution of all transitions leaving state S_i. In this equation, the time evolution depends only on the present state of the system and is defined entirely by knowledge of the set of transition probabilities. Such systems are called *Markovian systems*.

In the limit of large numbers of identical channels or other proteins, the quantities given in the master equation can be replaced by their macroscopic interpretation. The probability of being in a state S_i becomes the fraction of channels in state S_i, noted as s_i, and the transition probabilities from state S_i to state S_j become the rate constants, r_{ij}, of the reactions

$$S_i \overset{r_{ij}}{\underset{r_{ji}}{\rightleftharpoons}} S_j. \tag{8.4}$$

In this case, one can rewrite the master equation as

$$\frac{ds_i}{dt} = \sum_{j=1}^{n} s_j r_{ji} - \sum_{j=1}^{n} s_i r_{ij}, \tag{8.5}$$

which is a conventional kinetic equation for the various states of the system.

Stochastic Markov models [as in equation (8.3)] are adequate to describe ion channels as recorded using single-channel recording techniques (see Sakmann and Neher 1995). In other cases, where a larger area of membrane is recorded and large numbers of ion channels are involved, the macroscopic currents are more adequately described by conventional kinetic equations [as in equation (8.5)]. In the following discussion, only systems of the latter type will be considered.

8.2.2 Integration of Kinetic Models into Models of Excitable Membranes

Kinetic models of ion channels and other proteins are coupled through various types of interactions, which can be expressed by equations governing the electrical and chemical states of the cell. Using the equivalent circuit approach, the general equation for the membrane potential of a single isopotential compartment is

$$C_m \frac{dV}{dt} = \sum_{k=1}^{n} I_k, \tag{8.6}$$

Where V is the membrane potential, C_m is capacitance of the membrane, and I_k are the contributions of all channels of one type to the current across a particular area of membrane. Only single compartments were simulated, but the same approach could be extended to multiple compartments using cable equations (Rall 1995). In this case, a compartment may represent a small cylinder of dendritic or axonal processes. We assumed that (1) the membrane compartment contained a sufficiently large number of channels of each type k for equation (8.5) to hold, and (2) there was a single open or conducting state, O_k, for each channel type with maximum single-channel conductance, γ_k. Then, to a first approximation, each I_k can be calculated assuming a linear I–V relationship, giving the familiar Ohmic form:

$$I_k = \bar{g}_k o_k (V - E_k) , \tag{8.7}$$

where o_k is the fraction of open channels, $\bar{g}_k$ is the maximum conductance, and E_k is the equilibrium (reversal) potential. $\bar{g}_k$ is the product of the single-channel conductance and the channel density, $\bar{g}_k = \gamma_k \rho_k$.

Calcium acts as ligand for many channels and proteins. In order to track the concentration of ions such as these, standard chemical kinetics can also be used. The contribution of calcium channels to the free Ca^{2+} inside the cell was calculated as

$$\frac{d[Ca^{2+}]_i}{dt} = \frac{-I_{Ca}}{zF\,Ad} \tag{8.8}$$

where $[Ca^{2+}]_i$ is the intracellular submembranal calcium concentration, $z = 2$ is the valence of Ca^{2+}, F is the Faraday constant, A is the membrane area, and $d = 0.1$ μm is the depth of an imaginary submembrane shell. Removal of intracellular Ca^{2+} was driven by an active calcium pump obeying Michaelis–Menten kinetics (see Destexhe et al. 1993a):

$$Ca_i^{2+} + P \underset{r_2}{\overset{r_1}{\rightleftharpoons}} CaP \overset{r_3}{\longrightarrow} P + Ca_O^{2+} , \tag{8.9}$$

where P represents the calcium pump, CAP is an intermediate state, Ca_O^{2+} is the extracellular Ca^{2+} and r_1, r_2, and r_3 are rate constants as indicated. Ca^{2+} ions have a high affinity for the pump P, whereas extrusion of Ca^{2+} follows a slower process (Blaustein 1988). The extrusion process was assumed to be fast (milliseconds). Diffusion inside the cytoplasm was not included here, but is considered in detail in chapter 7.

8.3   Voltage-Dependent Ion Channels

Some ion channels may be gated by the electric field across the membrane. The opening of these channels may in turn influence the membrane potential. This interaction loop between membrane potential and ion permeability is at the basis of membrane excitability and the complex intrinsic firing properties of many types of neurons. In this section we discuss different types of kinetic models used to describe voltage-dependent channels and compare these models using the generation of a classical sodium-potassium action potential as an example.

8.3.1   The Kinetics of Voltage-Dependent Gating

Voltage-dependent ion channels can be described using Markov kinetic schemes in which transition rates between some pairs of states, i and j, are dependent on the membrane potential, V:

$$S_i \underset{r_{ji}(V)}{\overset{r_{ij}(V)}{\rightleftharpoons}} S_j . \tag{8.10}$$

According to the theory of reaction rates (e.g., Johnson et al. 1974), the rate of transition between two states depends exponentially on the free energy barrier between them:

$$r_{ij}(V) = \exp{-U_{ij}(V)/RT}, \tag{8.11}$$

where R is the gas constant and T is the absolute temperature. The free energy function $U_{ij}(V)$ is in general very difficult to evaluate and may involve both linear and nonlinear components arising from interactions between the channel protein and the membrane's electrical field. This dependence can be expressed without assumptions about underlying molecular mechanisms by a Taylor series expansion of the form:

$$U(V) = c_0 + c_1 V + c_2 V^2 + \dots , \tag{8.12}$$

giving a general transition rate function:

$$r(V) = \exp[-(c_0 + c_1 V + c_2 V^2 + \dots)/RT], \tag{8.13}$$

where $c_0, c_1, c_2 \dots$ are constants that are specific for each transition. The constant c_0 corresponds to energy differences that are independent of the applied field, the linear term $c_1 V$ corresponds to the translation of isolated charges or the rotation of rigid dipoles, and the higher-order terms to effects such as electronic polarization and pressure induced by V (Stevens 1978, Andersen and Koeppe 1992). In the "low field limit" (during relatively small applied voltages), the contribution of the higher-order terms may be negligible (Stevens 1978). Thus, a simple, commonly used voltage dependence results from the first-order approximation of equation (8.13) and takes the form:

$$r_{ij}(V) = a_{ij} \exp(-V/b_{ij}), \tag{8.14}$$

where a_{ij} and b_{ij} are constants.

This simple exponential form for the voltage dependence of rate constants is commonly used (e.g., Chabala 1984, Vandenberg and Bezanilla 1991, Perozo and Bezanilla 1990, Harris et al. 1981). However, the interactions of a channel protein with the membrane field might be expected to yield a rather more complex dependence (Stevens 1978, Neher and Stevens 1979, Andersen and Koeppe 1992, Clay 1989). Significantly, it has been shown that the number of states needed by a model to reproduce the voltage-dependent behavior of a channel may be reduced by adopting functions that saturate at extreme voltages (Keller et al. 1986, Clay 1989, Chen and Hess 1990, Borg-Graham 1991). In this case, the following form can be used for voltage-dependent transition rates:

$$r_i(V) = \frac{a_i}{1 + \exp[-(V - c_i)/b_i]},$$ (8.15)

which can be obtained from equation (8.13) by considering nonlinear as well as linear components of the voltage dependence of the free energy barrier between states. The constant a_i sets the maximum transition rate, b_i sets the steepness of the voltage dependence, and c_i sets the voltage at which the half-maximal rate is reached. Through explicit saturation, equation (8.15) effectively incorporates voltage-independent transitions that become rate limiting at extreme voltage ranges (Keller et al. 1986, Vandenberg and Bezanilla 1991, Chen and Hess 1990), eliminating the need for additional closed or inactivated states (see discussion by Chen and Hess 1990).

8.3.2   The Hodgkin–Huxley Model of Voltage-Dependent Channels

The most elementary Markov model for a voltage-gated channel is the first-order scheme:

$$C \underset{r_2(V)}{\overset{r_1(V)}{\rightleftharpoons}} O,$$ (8.16)

with voltage-gated rates $r_1(V)$ and $r_2(V)$ between a single open or conducting state, O, and a single closed state, C.

The model introduced by Hodgkin and Huxley (1952) described the permeability to Na^+ and K^+ ions in terms of gating particles described by open and closed transitions similar to equation (8.16). Interpreted in terms of ion channels, the Hodgkin–Huxley (H–H) model considers that the channel is composed of several independent gates and that each gate must be in the open state in order for the channel to conduct ions. Each gate has two states, with first-order kinetics described by equation (8.16).

Gates are divided into two types: usually several gates for activation, m, and a single gate for inactivation, h. All activation gates are assumed to be identical to one another, reducing the number of state transitions that must be calculated to two:

$$C_m \underset{\beta_m(V)}{\overset{\alpha_m(V)}{\rightleftharpoons}} O_m \tag{8.17}$$

$$C_h \underset{\beta_h(V)}{\overset{\alpha_h(V)}{\rightleftharpoons}} O_h \,, \tag{8.18}$$

where C_m, O_m and C_h, O_h are the closed or open states of each m and h gate, respectively. Because all gates must be open to allow the channel to conduct ions, the channel conductance is equal to the product of the fractions of gates in the open state, yielding

$$o = m^M h \,, \tag{8.19}$$

where $m = [O_m]/[O_m + C_m]$, $h = [O_h]/[O_h + C_h]$, and M is the number of identical m gates.

The Hodgkin–Huxley formalism is a subclass of the more general Markov representation. An equivalent Markov model can be written for any Hodgkin–Huxley scheme, but the translation of a system with multiple independent particles into a single-particle description results in a combinatorial explosion of states. Thus, the Markov model corresponding to the Hodgkin–Huxley sodium channel is

$$
\begin{array}{ccccccc}
C_3 & \underset{\beta_m}{\overset{3\alpha_m}{\rightleftharpoons}} & C_2 & \underset{2\beta_m}{\overset{2\alpha_m}{\rightleftharpoons}} & C_1 & \underset{3\beta_m}{\overset{\alpha_m}{\rightleftharpoons}} & O \\
\alpha_h \updownarrow \beta_h & & \alpha_h \updownarrow \beta_h & & \alpha_h \updownarrow \beta_h & & \alpha_h \updownarrow \beta_h \\
I_3 & \underset{\beta_m}{\overset{3\alpha_m}{\rightleftharpoons}} & I_2 & \underset{2\beta_m}{\overset{2\alpha_m}{\rightleftharpoons}} & I_1 & \underset{3\beta_m}{\overset{\alpha_m}{\rightleftharpoons}} & I
\end{array} \tag{8.20}
$$

(Fitzhugh 1965). The states represent the channel with the inactivation gate in the open state (top) or closed state (bottom) and (from left to right) three, two, one, or none of the activation gates closed. To reproduce the m^3 formation, the rates must have a 3:2:1 ratio in the forward direction and a 1:2:3 ratio in the backward direction. Only the O state is conducting. The squid delayed rectifier potassium current modeled by Hodgkin and Huxley (1952) with four activation gates and no inactivation can be treated analogously (Fitzhugh 1965, Armstrong 1969), giving

$$C_4 \underset{\beta_m}{\overset{4\alpha_m}{\rightleftharpoons}} C_3 \underset{2\beta_m}{\overset{3\alpha_m}{\rightleftharpoons}} C_2 \underset{3\beta_m}{\overset{2\alpha_m}{\rightleftharpoons}} C_1 \underset{4\beta_m}{\overset{\alpha_m}{\rightleftharpoons}} O. \tag{8.21}$$

8.3.3 Markov Models of Voltage-Dependent Channels

In more general models using Markov kinetics, independent and identical gates are not assumed. Rather, a state diagram is written to represent the set of configurations of the channel protein. This relaxes the constraints on the form of the diagram and the ratios of rate constants imposed by the Hodgkin–Huxley formulation. One may begin with the elementary two-state scheme [equation (8.16)] augmented by a single inactivated state, giving

$$
\begin{array}{ccc}
C & \underset{r_2}{\overset{r_1}{\rightleftharpoons}} & O \\
{\scriptstyle r_6}\Big\Updownarrow{\scriptstyle r_5} & & {\scriptstyle r_4}\Big\Updownarrow{\scriptstyle r_3} \\
& I &
\end{array}
\qquad . \tag{8.22}
$$

All six possible transitions between the three states are allowed, giving this kinetic scheme a looped form. The transition rates may follow voltage-dependent equations in the general form of equation (8.13), or some of these rates may be taken as either zero or independent of voltage to yield simpler models (see later discussion).

To more accurately fit the time course of channel openings or gating currents, additional closed and inactivated states may be necessary. As an example of a biophysically derived multistate Markov model, we consider the squid sodium channel model of Vandenberg and Bezanilla (1991). The authors fit by least squares a combination of single-channel, macroscopic ionic, and gating currents using a variety of Markov schemes. The nine-state diagram

$$
\begin{array}{ccccccccc}
C & \underset{r_6}{\overset{r_5}{\rightleftharpoons}} & C_1 & \underset{r_6}{\overset{r_5}{\rightleftharpoons}} & C_2 & \underset{r_6}{\overset{r_5}{\rightleftharpoons}} & C_3 & \underset{r_2}{\overset{r_1}{\rightleftharpoons}} & C_4 & \underset{r_4}{\overset{r_3}{\rightleftharpoons}} & O \\
 & & & & & & {\scriptstyle r_{10}}\Updownarrow{\scriptstyle r_8} & & & & {\scriptstyle r_9}\Updownarrow{\scriptstyle r_7} \\
 & & & & & & I_1 & \underset{r_3}{\overset{r_4}{\rightleftharpoons}} & I_2 & \underset{r_1}{\overset{r_2}{\rightleftharpoons}} & I_3
\end{array} \tag{8.23}
$$

was found to be optimal by maximum likelihood criteria. The voltage dependence of the transition rates was assumed to be a simple exponential function of voltage [equation (8.14)].

To complement the sodium channel model of Vandenberg and Bezanilla, we also examined the six-state scheme for the squid delayed rectifier channel used by Perozo and Bezanilla (1990):

$$
C \underset{r_2}{\overset{r_1}{\rightleftharpoons}} C_1 \underset{r_4}{\overset{r_3}{\rightleftharpoons}} C_2 \underset{r_4}{\overset{r_3}{\rightleftharpoons}} C_3 \underset{r_4}{\overset{r_3}{\rightleftharpoons}} C_4 \underset{r_6}{\overset{r_5}{\rightleftharpoons}} O \, , \tag{8.24}
$$

where again rates were described by a simple exponential function of voltage [equation (8.14)].

8.3.4 Example: Models of Voltage-Dependent Sodium Currents

The different types of models reviewed above are characterized by different complexities, ranging from a two-state representation [equation (8.16)] to transition diagrams involving many states [equation (8.23)]. The two-state description is adequate to fit the behavior of some channels (see e.g., Labarca et al. 1985; Yamada et al. 1989; Borg-Graham 1991; Destexhe et al. 1994b, c, 1998b), but for most channels more complex models must be considered. Many models of sufficient complexity are capable of fitting any limited set of experimental data. To demonstrate this, we compared three alternative models of the fast sodium channel underlying action potentials (figure 8.2); see also plate 26.)

First, the original quantitative description of the squid giant axon sodium conductance given by Hodgkin and Huxley (1952) was reproduced (figure 8.2A). The Hodgkin–Huxley model has four independent gates, each undergoing transitions between two states with first-order kinetics as described by equations (8.17) and (8.18). Three identical m gates represent activation and one h gate represents inactivation, leading to the familiar form for the conductance, $g_{Na} \sim m^3h$ [equation (8.19)].

Since the work of Hodgkin and Huxley, the behavior of the sodium channel of the squid axon has been better described in studies using Markov kinetic models. To illustrate the nature of these studies, we simulated the detailed sodium channel model of Vandenberg and Bezanilla (1991) [equation (8.23); figure 8.2B]. This particular nine-state model was selected to fit not only the measurements of macroscopic ionic currents available to Hodgkin and Huxley but also recordings of single-channel events and measurements of currents resulting directly from the movement of charge during conformational changes of the protein (so-called gating currents).

We also used a simplified Markovian sodium channel model (figure 8.2C). The scheme was chosen to have the fewest possible number of states (three) and transitions (four) while still being capable of reproducing the essential behavior of the more complex models. The form of the state diagram was based on the looped three-state model [equation (8.22)], with several transitions eliminated to give an irreversible loop (Bush and Sejnowski 1991):

$$
\begin{array}{ccc}
 & r_1(V) & \\
C & \rightleftharpoons & O \\
 & r_2(V) & \\
 & \searrow \quad \swarrow & \\
 & r_4(V) \quad r_3 & \\
 & I &
\end{array}
\tag{8.25}
$$

This model incorporated voltage-dependent opening, closing, and recovery from inactivation, while inactivation was voltage independent. For simplicity, neither opening from the inactivated state nor inactivation from the closed state was permitted. Although there is clear evidence for occurrence of the latter (Horn et al. 1981), it was unnecessary under the

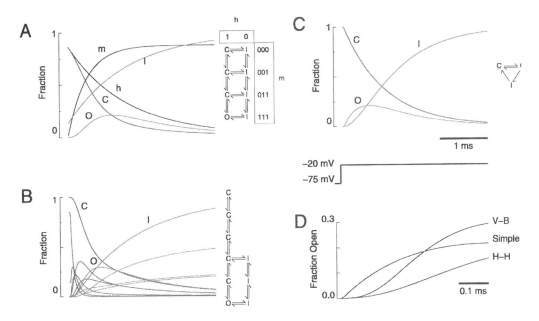

Figure 8.2

Three kinetic models of a squid axon sodium channel produce qualitatively similar conductance time courses. A voltage-clamp step from rest, V = −75 mV, to V = −20 mV was simulated. The fraction of channels in the open state (O, red), closed state (C, blue), and inactivated state (I, green) is shown for the Hodgkin-Huxley model, a detailed Markov model, and a simple Markov model. (A) An equivalent Markov scheme for the Hodgkin–Huxley model is shown [right insert, equation (8.20)]. Three identical and independent activation gates (m) give a form with three closed states (corresponding to 0, 1, and 2 activated gates) and one open state (3 activated gates). The independent inactivation gate (h) adds 4 corresponding inactivated states. Voltage-dependent transitions were calculated using the original equations and constants of Hodgkin and Huxley (1952). (B) The Markov model of Vandenberg and Bezanilla (1991) [equation (8.23)]. Individual closed (violet) and inactivated (yellow) states are shown, as well as the sum of all five closed states (C, blue), the sum of all three inactivated states (I, green) and the open state (red). (C) A simple 3-state Markov model (simple) fit to approximate the detailed model [equation (8.23)]. (D) A comparison of the time course of open channels for the three models on a faster time scale shows differences immediately following a voltage step. The Hodgkin–Huxley (H-H) and Vandenberg-Bezanilla (V-B) models give smooth, multiexponential rising phases, while the 3-state Markov model (Simple) gives a single exponential rise with a discontinuity in the slope at the beginning of the pulse. Modified from Destexhe et al. (1994c), where all parameters are given. (See plate 26.)

conditions of the present simulations. Rate constants were described by equation (8.15) with all $b_i = b$ and $c_1 = c_2$ to yield a model consisting of nine total parameters.

The response of the three sodium channel models to a voltage-clamp step from rest (-75 mV) was simulated (figure 8.2). For all three models, closed states were favored at hyperpolarized potentials. Upon depolarization, forward (opening) rates sharply increased while closing (backward) rates decreased, causing a migration of channels in the forward direction toward the open state. The three closed states in the Hodgkin–Huxley model and the five closed states in the detailed (Vandenberg–Bezanilla) model gave rise to the characteristic sigmoidal shape of the rising phase of the sodium current (figure 8.2D). In contrast, the simple model, with a single closed state, produced a first-order exponential response to the voltage step.

Even though the steady-state behavior of the Hodgkin–Huxley model of the macroscopic sodium current is remarkably similar to that of the microscopic Markov models (Marom and Abbott 1994), the relationship between activation and inactivation is different. First, in the Hodgkin–Huxley model, activation and inactivation are kinetically independent. This independence has been shown to be untenable on the basis of gating and ion current measurements in the squid giant axon (Armstrong 1981, Bezanilla 1985). Consequently, Markov models that are used to reproduce gating currents, such as the Vandenberg–Bezanilla model examined here, require schemes with coupled activation and inactivation. Likewise, in the simple model, activation and inactivation were strongly coupled owing to the unidirectional looped scheme [equation (8.25)], so that channels were required to open before inactivating and could not reopen from the inactivated state before closing.

Second, in the Hodgkin–Huxley and Vandenberg–Bezanilla models, inactivation rates are slow and activation rates are fast. In simple Markov models, the situation was reversed, with fast inactivation and slow activation. At the macroscopic level modeled here, these two relationships gave rise to similar time courses for open channels (figure 8.2A–C); see Andersen and Koeppe (1992). However, the two classes of models make distinct predictions when interpreted at the microscopic (single-channel) level. Whereas the Hodgkin–Huxley and Vandenberg–Bezanilla models predict the latency to first channel opening to be short and channel open times to be comparable to the time course of the macroscopic current, the simplified Markov model predicts a large portion of first channel openings to occur after the peak of the macroscopic current and to have open times much shorter than the duration of the current. Single-channel recordings have confirmed the latter prediction (Sigworth and Neher 1980, Aldrich et al. 1983, Aldrich and Stevens 1987).

Despite significant differences in their complexity and formulation, the three models of the sodium channel all produced very comparable action potentials and repetitive firing when combined with appropriate delayed-rectifier potassium channel models (figure 8.3).

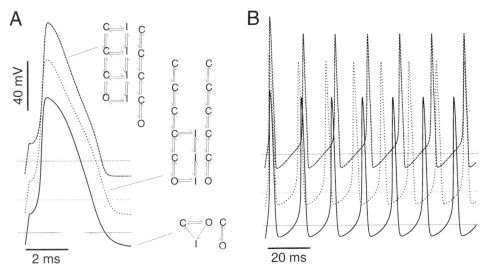

Figure 8.3
Similar action potentials produced using three different kinetic models of squid fast sodium and delayed rectifying potassium channels. (A) Single action potentials in response to a 0.2-ms, 2-nA current pulse are elicited at similar thresholds and produce similar wave forms using three different pairs of kinetic models: Hodgkin–Huxley (dashed line; Hodgkin and Huxley 1952), detailed Markov models (dotted line; Vandenberg and Bezanilla 1991; Perozo and Bezanilla 1990), and simplified kinetic models (solid line). (B) Repetitive trains of action potentials elicited in response to sustained current injection (0.2 nA) have slightly different frequencies. Sodium channels were modeled as described in figure 8.2. The detailed Markov potassium channel model had six states (Perozo and Bezanilla 1990) [equation (8.24)] and the simple model of potassium channel had two states [equation (8.16)]. Figure modified from Destexhe et al. (1994c), where all parameters are given.

None of the potassium channel models had inactivation. The main difference was in the number of closed states, from six for the detailed Markov model of Perozo and Bezanilla (1990) [equation (8.24)], to four for the original (Hodgkin and Huxley 1952) description of the potassium current, to just one for a minimal model [equation (8.16)] with rates of sigmoidal voltage dependence [equation (8.15)].

8.4 Transmitter Release

When action potentials reach the synaptic terminals of an axon, they trigger the release of a chemical messenger (transmitter) in the extracellular space. These transmitter molecules then bind to receptors and affect the voltage or the intracellular metabolism of the target cells. The release of transmitter involves complex calcium-dependent molecular mechanisms that link the arrival of an action potential in the synaptic terminal to the exocytosis of transmitter. This section describes the link between action potentials and transmitter release. Kinetic models incorporating various levels of detail are compared with simplified models of the release process.

8.4.1 Kinetic Model of Transmitter Release

The exact mechanisms whereby Ca^{2+} enters the presynaptic terminal, the specific proteins with which Ca^{2+} interacts, and the detailed mechanisms leading to exocytosis represent an active area of research (e.g., Neher 1998). It is clear that an accurate model of these processes should include the particular clustering of calcium channels, calcium diffusion and gradients, all enzymatic reactions involved in exocytosis, and the particular properties of the diffusion of transmitter across the fusion pore and synaptic cleft. For our present purpose, we use a kinetic model of calcium-induced release inspired by Yamada and Zucker (1992). This model of transmitter release assumed that (1) upon invasion by an action potential, Ca^{2+} enters the presynaptic terminal owing to the presence of a high-threshold Ca^{2+} current; (2) Ca^{2+} activates a calcium-binding protein that promotes release by binding to the transmitter-containing vesicles; (3) an inexhaustible supply of "docked" vesicles is available in the presynaptic terminal, ready to be released; (4) the binding of the activated calcium-binding protein to the docked vesicles leads to the release of n molecules of transmitter in the synaptic cleft. The latter process is modeled here as a first-order process with a stoichiometry coefficient of n (see details in Destexhe et al. 1994c).

The calcium-induced cascade leading to the release of transmitter is described by the following kinetic scheme:

$$4Ca^{2+} + X \underset{k_u}{\overset{k_b}{\rightleftharpoons}} X^* \tag{8.26}$$

$$X^* + V_e \underset{k_2}{\overset{k_1}{\rightleftharpoons}} V_e^* \tag{8.27}$$

$$V_e^* \xrightarrow{k_3} nT \tag{8.28}$$

$$T \xrightarrow{k_c} \ldots \tag{8.29}$$

Here, calcium ions bind to a calcium-binding protein, X, with a cooperativity factor of 4 (see Augustine and Charlton 1986 and references therein), leading to an activated calcium-binding protein, X^* [equation (8.26)]. The associated forward and backward rate constants are k_b and k_u. X^* then reversibly binds to transmitter-containing vesicles, V_e, with corresponding rate constants k_1 and k_2 [equation (8.27)]. The exocytosis process is then modeled by an irreversible reaction [equation (8.28)] where activated vesicles, V_e^*, release n molecules of transmitter, T, into the synaptic cleft with a rate rate constant k_3. The values of the parameters of these reactions were based on previous models and measurements (Yamada and Zucker 1992).

The concentration of the liberated transmitter in the synaptic cleft, [T], was assumed to be uniform in the cleft and cleared by processes of diffusion outside the cleft (to the extrajunctional extracellular space), uptake or degradation. These contributions were modeled by a first-order reaction [equation (8.29)] where k_c is the rate constant for clearance of T. The values of parameters have been detailed previously (Destexhe et al. 1994c).

Figure 8.4A shows a simulation of this model of transmitter release associated with a single-compartment presynaptic terminal containing mechanisms for action potentials, high-threshold calcium currents, and calcium dynamics (see Destexhe et al. 1994c for details). Injection of a short current pulse into the presynaptic terminal elicited a single action potential. The depolarization of the action potential activated high-threshold calcium channels, producing a rapid influx of calcium. The elevation of intracellular $[Ca^{2+}]$ was transient owing to clearance by an active pump. The time course of activated calcium-binding proteins and vesicles closely followed the time course of the transient calcium rise in the presynaptic terminal. This resulted in a brief (≈ 1 ms) rise in transmitter concentration in the synaptic cleft (figure 8.4A, bottom curve D). The rate of transmitter clearance was adjusted to match the time course of transmitter release estimated from patch clamp experiments (Clements et al. 1992, Clements 1996) as well as from simulations of extracellular diffusion of transmitter (Bartol et al. 1991, Destexhe and Sejnowski 1995).

8.4.2   Simplified Models of the Release Process

The above-described release model would be computationally very expensive if it had to be used in network simulations involving thousands of synapses. Therefore, for large-scale network models, simplification of the release process is needed. The first alternative is to use a continuous function to transform the presynaptic voltage into transmitter concentration. An expression for such a function was derived previously (Destexhe et al. 1994c) by assuming that all intervening reactions in the release process are relatively fast and can be considered at steady state. The stationary relationship between the transmitter concentration [T] and presynaptic voltage was well fit by the following relation (Destexhe et al. 1994c):

$$[T] = \frac{T_{max}}{1 + \exp[-(V_{pre} - V_p)/K_p]} , \qquad (8.30)$$

where T_{max} is the maximal concentration of transmitter in the synaptic cleft, V_{pre} is the presynaptic voltage, $K_p = 5$ mV gives the steepness, and $V_p = 2$ mV sets the value at which the function is half-activated. One of the main advantages of using equation (8.30) is that it provides a simple and smooth transformation between presynaptic voltage and transmitter

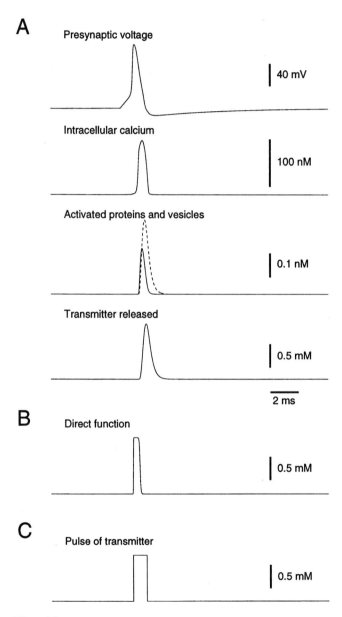

Figure 8.4
Models for the brief time course of a transmitter in the synaptic cleft. (A) Kinetic model of presynaptic reactions leading to a transmitter release. A presynaptic action potential elicited by injection of a 0.1-nA current pulse lasting 2 ms in the presynaptic terminal (top curve). The intracellular Ca^{2+} concentration in the presynaptic terminal increased (second curve) owing to the presence of a high-threshold calcium current that provided a transient cal-

concentration. Using this direct function provides a fair approximation of the transmitter concentration (figure 8.4B). This form, in conjunction with simple kinetic models of postsynaptic channels, provides a model of synaptic interaction based on autonomous differential equations with only one or two variables (see also Wang and Rinzel 1992).

The second alternative is to assume that the change in the transmitter concentration occurs as a brief pulse (Destexhe et al. 1994b). There are many indications that the exact time course of the transmitter in the synaptic cleft is not, under physiological conditions, a main determinant of the time course of postsynaptic responses at many synapses (e.g., Magleby and Stevens 1972, Lester and Sahr 1992, Colquhoun et al. 1992, Clements et al. 1992). Indeed, fast application techniques have shown that 1-ms pulses of 1 mM glutamate reproduced postsynaptic currents (PSCs) in membrane patches that were quite similar to those recorded in the intact synapse (Hestrin 1992, Colquhoun et al. 1992, Sakmann and Neher 1995, Standley et al. 1993). A pulse of 1 ms and 1 mM is shown in figure 8.4(C) and provides the simplest approximation of the time course of the transmitter.

In models, release simulated by a kinetic model or by a pulse of transmitter had a barely detectable influence on the time course of the synaptic current (see Destexhe et al. 1994c, 1998b). It is worth mentioning that pulse-based Markov models can be solved analytically, leading to very fast algorithms to simulate synaptic currents (Destexhe et al. 1994b, c, 1998b; Lytton 1996). Pulse-based models are considered in the next section.

8.5 Ligand-Gated Synaptic Ion Channels

An important class of ion channels remain closed until a ligand binds to the channel, inducing a conformational change that allows channel opening. This class of gating properties is referred to as *ligand gating*. A principal class of ligand-gated channels consists of those activated directly by transmitter. In this case, the receptor and the ion channel are part of the same protein complex, which is called the *ionotropic receptor*. Ligand gating also commonly occurs through secondary agonists, such as glycine activation of the glutamate N-methyl-D-aspartate (NMDA) receptor, or through second messengers, such as calcium, G-proteins, or cyclic nucleotides. In this section we review different kinetic models for synaptic ion channels gated by transmitter molecules. We illustrate the behavior of these models using fast glutamate-mediated transmission as an example.

---

cium influx during the action potential. Removal was provided by an active calcium pump. The relative concentration of activated calcium-binding protein X^* (third curve, solid line) and vesicles V_e^* (third curve, dotted line) also increased transiently, as did the concentration of transmitter in the synaptic cleft (bottom curve). (B) Approximation of the transmitter time course using a direct sigmoid function of the presynaptic voltage. (C) Pulse of 1 ms and 1 mM shown for comparison. Panel A was modified from Destexhe et al. (1994c), where all parameters were given.

8.5.1 Kinetic Models of Ligand-Gated Channels

In general, for a ligand-gated channel, transitions rates between unbound and bound states
of the channel depend on the binding of a ligand:

$$T + S \quad \overset{r_{ij}}{\underset{r_{ji}}{\rightleftharpoons}} \quad S_j \, , \tag{8.31}$$

where T is the ligand, S_i is the unbound state, S_j is the bound state (sometimes written $S_i T$),
and r_{ij} and r_{ji} are rate constants as defined before.

The same reaction can be rewritten as

$$S_i \quad \overset{r_{ij}([T])}{\underset{r_{ji}}{\rightleftharpoons}} \quad S_j , \tag{8.32}$$

where $r_{ij}([T]) = [T] \, r_{ij}$ and $[T]$ is the concentration of the ligand. Written in this form, equation (8.32) is equivalent to equation (8.10). Ligand-gating schemes are generally equivalent to voltage-gating schemes, although the functional dependence of the rate on $[T]$ is simple compared with the voltage dependence discussed earlier.

Thus, the most elementary state diagram for a ligand-gated channel is

$$C \quad \overset{r_1([T])}{\underset{r_2}{\rightleftharpoons}} \quad O, \tag{8.33}$$

where C and O represent the closed and open states of the channel and $r_1([T])$ and r_2 are the
associated rate constants.

8.5.2 Markov Models of Ligand-Gated Channels

Single-channel recording techniques have led to the formulation of multistate Markov kinetic models for most ionotropic synaptic receptors. Glutamate α-amino-3-hydroxy-5-methyl-4-isoxazolepropionic acid (AMPA) receptors mediate the prototypical fast excitatory synaptic currents in the brain. AMPA-mediated responses are thought to be due to a combination of rapid clearance of transmitter and rapid channel closure (Hestrin 1992). A Markov kinetic model of AMPA receptors was proposed by Patneau and Mayer (1991) (see also Jonas et al. 1993) and had the following state diagram:

$$C_0 \overset{r_1([T])}{\underset{r_2}{\rightleftharpoons}} C_1 \overset{r_1([T])}{\underset{r_3}{\rightleftharpoons}} C_2 \overset{r_6}{\underset{r_7}{\rightleftharpoons}} O \atop \qquad\quad r_4 \Updownarrow r_5 \qquad\quad r_4 \Updownarrow r_5 \atop \qquad\quad D_1 \qquad\qquad\quad D_2 \tag{8.34}$$

where the unbound form of the receptor C_0 binds to one molecule of transmitter T, leading to the singly bound form C_1, which itself can bind another molecule of T, leading to the doubly bound form C_2. r_1 is the binding rate and r_2 and r_3 are unbinding rates. Each form C_1 and C_2 can desensitize, leading to forms D_1 and D_2 with rates r_4 and r_5 for desensitization and resensitization, respectively. Finally, the doubly bound receptor C_2 can open, leading to the open form O, with opening and closing rates of r_6 and r_7, respectively.

The AMPA current is then given by

$$I_{AMPA} = \bar{g}_{AMPA}[O](V - E_{AMPA}), \tag{8.35}$$

where $\bar{g}_{AMPA}$ is the maximal conductance, $[O]$ is the fraction of receptors in the open state, V is the postsynaptic voltage, and $E_{AMPA} = 0$ mV is the reversal potential.

Another model for AMPA receptors was also proposed by Standley et al. (1993) to account for single-channel recordings in locust muscle. Several possible models were tested, and these workers found that their data were best fit by a six-state scheme:

$$D_2 \underset{r_9([T])}{\overset{r_{10}}{\rightleftharpoons}} D_1 \underset{r_7([T])}{\overset{r_8}{\rightleftharpoons}} C \underset{r_2}{\overset{r_1([T])}{\rightleftharpoons}} C_1 \underset{r_4}{\overset{r_3([T])}{\rightleftharpoons}} C_2 \underset{r_6}{\overset{r_5}{\rightleftharpoons}} O, \tag{8.36}$$

where C is the unbound closed state; C_1 and C_2 are, respectively, the singly and doubly bound closed states; O is the open state; and D_1 and D_2 are, respectively, the desensitized singly and doubly bound states.

Markov models were proposed for various other receptor types, including NMDA or γ-aminobutyric acid (GABA) type A receptors (reviewed in Sakmann and Neher 1995). For ionotropic receptor types, detailed Markov models as well as simplified kinetic models consisting of fewer states have been proposed and compared (Destexhe et al. 1994c, 1998b). We illustrate this approach in the next section by considering the example of AMPA receptors.

8.5.3 Glutamate AMPA Receptors

Glutamate AMPA receptors are used here to compare the behavior of different models of ligand-gated receptors. The behavior of the detailed, six-state model for AMPA receptors derived by Standley et al. (1993) [equation (8.36)] is illustrated in figure 8.5A in conjunction with the detailed model of transmitter release described before. The postsynaptic response showed a fast time to peak of about 1 ms and a decay phase lasting 5–10 ms, which is in agreement with Standley et al. (1993). In addition, the response to a series of presynaptic action potentials at a rate of approximately 20 Hz shows a progressive desensitization of the response due to an increase in the fraction of desensitized channels (states D_1 and D_2). A similar behavior was also found with another model of the AMPA current (Raman and Trussell 1992) (not shown in figure).

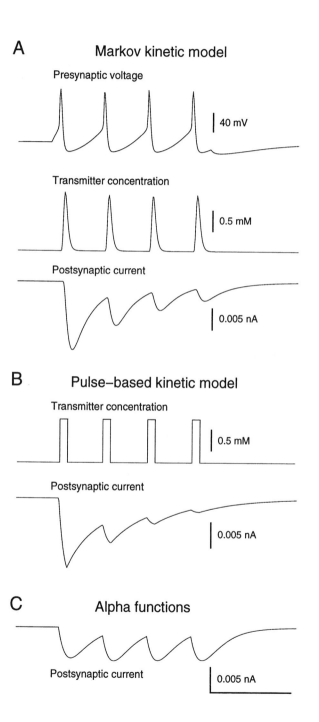

Simplified models were obtained by using pulses of transmitter and fewer states in the Markov scheme. The use of transmitter pulses was motivated by fast-perfusion experiments showing that 1-ms pulses of glutamate applied to patches containing AMPA receptors produced responses that closely matched the time course of synaptic currents (Colquhoun et al. 1992, Hestrin 1992). The simplified diagram was found by comparing all possible two- and three-state schemes with the detailed model.

The following three-state model

$$C \underset{r_5}{\overset{r_1([T])}{\rightleftarrows}} O \qquad \diagdown \diagup \qquad D \qquad r_3 \tag{8.37}$$

was found to be the best approximation for AMPA current (Destexhe et al. 1994c). Here, D represents the desensitized state of the channel and $r_1...r_5$ are the associated rate constants. This model reproduced the progressive desensitizing responses (figure 8.5B) as well as the time course of the AMPA current observed in the more accurate model. On the other hand, simpler two-state models [equation (8.33)] provided good fits of single PSCs, but did not account for desensitization (Destexhe et al. 1994c).

These models were compared with the alpha function, originally introduced by Rall (1967):

$$r(t - t_0) = \frac{(t - t_0)}{\tau_1} \exp[-(t - t_0) / \tau_1]. \tag{8.38}$$

This function gives a stereotyped wave form for the time course of the postsynaptic current following a presynaptic spike occurring at time $t = t_0$. τ_1 is the time constant of the alpha function. Alpha functions often provide approximate fits for many synaptic currents and have been widely used for computing synaptic currents in neural models (see, e.g., Koch and Segev 1998).

The summation behavior of alpha functions is illustrated in figure 8.5C. In this case, the fit of alpha functions to the time course of the AMPA PSCs was poor (the rise time was too slow compared with the decay time). More important, dynamic interactions between successive events, such as desensitization, are not present with alpha functions.

Figure 8.5
Comparison of three models for AMPA receptors. (A) Markov model of AMPA receptors. A presynaptic train of action potentials was elicited by current injection (presynaptic voltage). The release of glutamate was calculated using a kinetic model of synaptic release (transmitter concentration). The postsynaptic current from AMPA receptors was modeled by a six-state Markov model. (B) Same simulation with transmitter modeled by pulses (transmitter concentration) and AMPA receptors modeled by a simpler three-state kinetic scheme (postsynaptic current). (C) Postsynaptic current modeled by summed alpha functions. Modified from Destexhe et al. (1994c).

8.6   Second-Messenger-Gated Synaptic Channels

In contrast to ionotropic receptors, for which the receptor and ion channel are both part of the same protein complex, other classes of synaptic responses are mediated by an ion channel that is independent of the receptor. In this case, the binding of the transmitter to the receptor induces the formation of an intracellular second messenger, which in turn activates (or inactivates) ion channels. The advantage of this type of synaptic transmission is that a single activated receptor can lead to the formation of thousands of second-messenger molecules, thereby providing an efficient amplification mechanism.

These so-called *metabotropic* receptors not only act on ion channels but may also influence several key metabolic pathways in the cell through second messengers such as G-proteins or cyclic nucleotides. In this section, we review kinetic models for synaptic interactions acting through second messengers. We illustrate this type of interaction using the example of $GABA_B$ receptors, whose response is mediated by K^+ channels through the activation of G-proteins (Andrade et al. 1986).

8.6.1   Kinetic Models of Second-Messenger-Gated Channels

One particularity of metabotropic responses is that the receptor is part of a protein complex that also catalyzes the production of an intracellular second messenger. These steps can be represented by the general scheme:

$$R_0 + T \rightleftharpoons R \rightleftharpoons D \tag{8.39}$$

$$R + G_0 \rightleftharpoons RG \longrightarrow R + G \tag{8.40}$$

$$G \longrightarrow G_0 . \tag{8.41}$$

Here the transmitter, T, binds to the receptor, R_0, leading to its activated form, R, and desensitized form, D [equation (8.39)]. The activated receptor R catalyzes the formation of an intracellular second messenger, G, from its inactive form, G_0, through a Michaelis–Menten scheme [equation (8.40)]. Finally, the second messenger is degraded back into its inactive form [equation (8.41)].

The intracellular messenger G can affect various ion channels as well as the metabolism of the cell. The second messenger may act directly on its effector, as demonstrated for some K^+ channels that are directly gated by G-proteins (VanDongen et al. 1988). However, more complex reactions may be also be involved, as in the case of phototransduction (see Lamb and Pugh 1992).

We consider the simplest case of direct binding of the second messenger to the ion channel, leading to its activation:

$$C + G \rightleftharpoons \dots \rightleftharpoons O \tag{8.42}$$

or deactivation:

$$O + G \rightleftharpoons \ldots \rightleftharpoons C. \tag{8.43}$$

The first case [equation (8.42)] is analogous to ligand-gated channels; the channel opens following the binding of one or several molecules of second messenger. In the second case [equation (8.43)], the channel is open at rest but closes following the binding of G. These types of gating may be characterized by Markov schemes involving several states, analogously to schemes considered for ionotropic receptors in section 8.5.

Transmitters, including glutamate (through metabotropic receptors), GABA (through $GABA_B$ receptors), acetylcholine (through muscarinic receptors), noradrenaline, serotonin, dopamine, histamine, opiods, and others, have been shown to mediate slow intracellular responses. These transmitters induce the intracellular activation of G-proteins, which may affect ionic currents as well as the metabolism of the cell. One of the main electrophysiological functions of these modulators is to open or close K^+ channels (see Brown 1990, Brown and Birnbaumer 1990, McCormick 1992). This type of action is illustrated below for GABA acting on $GABA_B$ receptors.

8.6.2 $GABA_B$-mediated Neurotransmission

$GABA_B$ receptors mediate slow inhibitory responses mediated by K^+ channels through the activation of G-proteins (Andrade et al. 1986, Dutar and Nicoll 1988). There is strong evidence that direct G-protein binding mediates the gating process (Andrade et al. 1986, Thalmann 1988, Brown and Birnbaumer 1990). The typical properties of $GABA_B$-mediated responses in hippocampal, thalamic, and neocortical neurons can be reproduced, assuming that several G-proteins directly bind to the associated K^+ channels (Destexhe and Sejnowski 1995, Thomson and Destexhe 1999), leading to the following scheme:

$$R_0 + T \rightleftharpoons R \rightleftharpoons D \tag{8.44}$$

$$R + G_0 \rightleftharpoons RG \longrightarrow R + G \tag{8.45}$$

$$G \longrightarrow G_0 \tag{8.46}$$

$$C + nG \rightleftharpoons O, \tag{8.47}$$

where the symbols have the same meaning as above and n is the number of independent binding sites of G-proteins on K^+ channels.

The current is then given by

$$I_{GABA_B} = \bar{g}_{GABA_B} [O] (V - E_K),$$

where $[O]$ is the fraction of K^+ channels in the open state.

The behavior of this model is shown in figure 8.6. Very brief changes in transmitter concentration gave rise to an intracellular response of a much longer duration. The rate constants of enzymatic reactions were estimated based on pharmacological manipulations and recordings *in vivo* (Breitwieser and Szabo 1988, Szabo and Otero 1989). In this simulation, the release was obtained according to the kinetic model of synaptic release and the K^+ channel was gated by G according to a simple two-state scheme involving four binding sites for G. Based on whole-cell recordings of $GABA_B$ currents in dentate granule cells (Otis et al. 1993), the rate constants of this model were adjusted to experimental data using a simplex fitting procedure (Destexhe and Sejnowski 1995, Destexhe et al. 1998b, Thomson and Destexhe 1999).

This model reproduces a nonlinear dependence of $GABA_B$ responses to the number of presynaptic spikes. A single presynaptic action potential induces a relatively small increase of G-protein, which is insufficient to activate a significant postsynaptic response (figure 8.6, dashed lines). However, a train of 10 presynaptic spikes at high frequency induces higher levels of activated G-proteins and evokes a significant postsynaptic response (figure 8.6, solid lines). This type of dependence explains experimental observations that $GABA_B$ responses only appear under high stimulus intensities (Dutar and Nicoll 1988, Kim et al. 1997, Thomson and Destexhe 1999) and the absence of $GABA_B$-mediated miniature events (Thompson and Gahwiler 1992, Otis and Mody 1992).

A simplified two-state model of $GABA_B$-mediated responses must include the sensitivity to the number of spikes. A one-variable model (corresponding to an open/closed scheme) would be too simple in this case. The simplest two-variable model was obtained from the model above by considering equations (8.45) and (8.47) at steady state, by considering G_0 in excess, and by neglecting $GABA_B$ receptor desensitization. This simplified two-variable model of $GABA_B$-mediated currents was (Destexhe et al. 1998b)

$$\frac{dr}{dt} = K_1 T(1-r) - K_2 r \tag{8.48}$$

$$\frac{dg}{dt} = K_3 r - K_4 g \, , \tag{8.49}$$

where r *is* the fraction of receptors in the active form, T is the GABA concentration in the synaptic cleft, and g is the normalized concentration of G-proteins in the active state. The current is then given by

$$I_{GABA_B} = \bar{g}_{GABA_B} \frac{g^n}{g^n + K_D}(V - E_K) \, , \tag{8.50}$$

where $\bar{g}_{GABA_B}$ is the maximal conductance of K^+ channels, E_K is the potassium reversal potential, and K_D is the dissociation constant of the binding of G-proteins on K^+ channels.

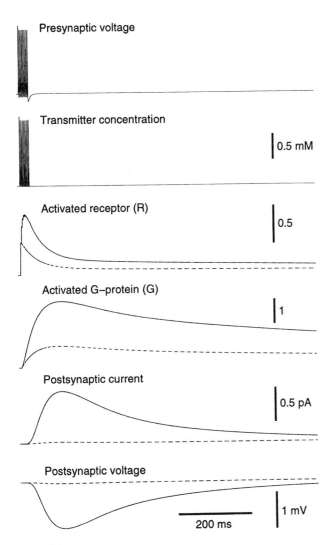

Figure 8.6
Model of synaptic interactions mediated by second messengers. The model simulates slow inhibitory responses mediated by GABA$_B$ receptors through the activation of G-proteins. The response is shown following a presynaptic train of 10 action potentials at high frequency (~300 Hz). The transmitter concentration was calculated using the kinetic model of release shown in figure 8.4A. The fractions of activated receptor (R) and G-protein (G), as well as the postsynaptic current and voltage are shown. Very brief synaptic events can evoke slow intracellular changes through the production of second messengers (G-proteins here). The response to a train of 10 presynaptic spikes (continuous lines) is compared with that of a single presynaptic spike (dashed lines).

These parameters were estimated by fitting the model to experimental data using a simplex procedure (see Destexhe et al. 1998b).

The behavior of the simplified model is shown in figure 8.7. Transmitter concentration was described by pulses of the transmitter as shown above and provided a time course and stimulus dependence similar to that of the detailed model. A single presynaptic action potential induced low intracellular levels of activated G-protein, which were insufficient to evoke a postsynaptic response (figure 8.7, dashed lines). However, a train of 10 presynaptic spikes at ~300 Hz induced G-protein concentrations that were high enough to activate a postsynaptic response (figure 8.7, solid lines), similarly to the detailed model (compare with figure 8.6). The two-variable model is therefore the minimal representation for simulating both the time course and the stimulus dependence of GABA$_B$-mediated responses.

8.7   Applications to Modeling Complex Neuronal Interactions

Modeling ion channels and synaptic interactions with the appropriate formalism is required in problems where the kinetics of ion channels are important. The case of the thalamus is a good example. Thalamic neurons have complex intrinsic firing properties owing to the presence of several types of voltage-dependent currents (Steriade and Llinás 1988) and synaptic connections between thalamic cells are mediated by different types of ionotropic and metabotropic receptors (McCormick 1992). Moreover, thalamic neurons and interconnectivity patterns have been well characterized by anatomists (Jones 1985), and the thalamus exhibits pronounced oscillatory properties that have been well characterized in vivo and in vitro (see Steriade et al. 1993). Despite the abundant anatomical and physiological information available for this system, the exact mechanisms leading to oscillatory behavior are still open because of the complexity of the interactions involved.

Thalamic oscillations have been explored by computational models (Andersen and Rutjord 1964, Destexhe et al. 1993b, Wang et al. 1995, Destexhe et al. 1996a, Golomb et al. 1996; reviewed in Destexhe and Sejnowski 2000). At the single-cell level, models investigated the ionic mechanisms underlying the repertoire of firing properties of thalamic neurons based on voltage-clamp data on the voltage- and calcium-dependent currents present in these cells (McCormick and Huguenard 1992; Destexhe et al. 1993a, 1996b). At the network level, synaptic interactions were simulated using the main receptor types identified in thalamic circuits (Destexhe et al. 1993b; Wang et al. 1995; Destexhe et al. 1996a; Golomb et al. 1996).

In network simulations, every cell is described by the membrane equation:

$$C_m \dot{V}_i = -g_L(V_i - E_L) - \sum_j I_{ji}^{\text{int}} - \sum_k I_{ki}^{\text{syn}} \tag{8.51}$$

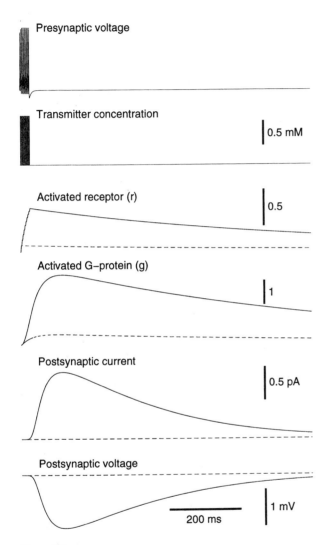

Figure 8.7
Simplified model of GABA$_B$-mediated responses. The same paradigm as in figure 8.6 is shown here using the simplified model. Presynaptic trains of action potentials were the same as in figure 8.6. Transmitter concentration was described by brief pulses (1 ms), and the fraction of activated receptor (r) and G-protein (g), as well as the postsynaptic current and voltage, are shown for the simplified two-variable model [equation (8.49)]. The responses to trains of 10 presynaptic spikes (continuous lines) or to a single presynaptic spike (dashed lines) are similar to those of the detailed model (compare with figure 8.6).

where V_i is the membrane potential, $C_m = 1$ μF/cm² is the specific capacity of the membrane, g_L is the leakage conductance, and E_L is the leakage reversal potential. Intrinsic and synaptic currents are respectively represented by I_{ji}^{int} and I_{ki}^{syn}.

Intrinsic currents are described by the generic form:

$$I_{ji}^{\text{int}} = \bar{g}_j m_j^M h_j^N (V_i - E_j) \, , \tag{8.52}$$

where the current is expressed as the product of, respectively, the maximal conductance, $\bar{g}_j$, activation (m_j) and inactivation variables (h_j), and the difference between the membrane potential V_i and the reversal potential E_j. As in section 8.3.2, activation and inactivation gates follow the simple two-state kinetic scheme introduced by Hodgkin and Huxley (1952):

$$\text{(closed)} \; \underset{\beta(V)}{\overset{\alpha(V)}{\rightleftharpoons}} \; \text{(open)} \, , \tag{8.53}$$

where α and β are voltage-dependent rate constants.

Synaptic currents are described by

$$I_{ki}^{\text{syn}} = \bar{g}_{ki} m_{ki} (V_i - E_{ki}) \, , \tag{8.54}$$

where ki indicates the synaptic contact from neuron k to neuron i, $\bar{g}_{ki}$ is the maximal conductance of postsynaptic receptors, and E_{ki} is the reversal potential. m_{ki} is the fraction of open receptors according to the simple two-state kinetic scheme [equation (8.33)], rewritten in the form:

$$\text{(closed)} + T(V_k) \; \underset{\beta}{\overset{\alpha}{\rightleftharpoons}} \; \text{(open)} \tag{8.55}$$

where $T(V_k)$ is the concentration of transmitter in the cleft and α, β are the rate constants for binding and unbinding of transmitter to the channel. When a spike occurred in cell k, $T(V_k)$ was set to 1 mM during 1 ms, leading to the transient activation of the current and providing the synaptic interaction between cells k and i.

Two cell types were modeled and contained the minimal set of voltage- and calcium-dependent currents to account for their most salient intrinsic firing properties. Thalamocortical (TC) cells contained the calcium current I_T; the hyperpolarization-activated current I_h; and the I_{Na}, I_{KD} currents responsible for action potentials. Thalamic reticular (RE) cells contained I_T, I_{Na}, and I_{KD}. These currents were modeled using Hodgkin–Huxley types of representations (see above) for which all specific details were given in previous papers (McCormick and Huguenard 1992; Destexhe et al. 1993a, 1996a, b). The intrinsic bursting properties of TC and RE cells are shown in figure 8.8 (insets).

TC and RE cells are interconnected using glutamate AMPA and GABAergic $GABA_A$ and $GABA_B$ receptors (von Krosigk et al. 1993) (see scheme in figure 8.8). These synaptic currents were represented by two-state kinetic models for AMPA and $GABA_A$ receptors (see above) and the model of $GABA_B$ receptors given in equation (8.49). More details on the specific parameters used can be found in Destexhe et al. (1996a).

The behavior of the circuit of TC and RE cells is shown in figure 8.8. Under controlled conditions, the network generated oscillatory behavior owing to the interaction between inhibition and the bursting properties of thalamic neurons (figure 8.8B). These oscillations had a frequency (8–12 Hz) and phase relations consistent with the 8–12-Hz spindle oscillations analyzed experimentally (see von Krosigk et al. 1993). The model exhibited the correct frequency only if intrinsic and synaptic currents had kinetics compatible with voltage-clamp experiments (Destexhe et al. 1996a).

Experiments in thalamic slices (von Krosigk et al. 1993) have also demonstrated that in the same thalamic circuits, the 8–12-Hz spindle oscillations can be transformed into a slower (2–4 Hz) and more synchronized oscillation following addition of bicuculline, a $GABA_A$ receptor antagonist. Models could account for these observations only if $GABA_B$ receptors had a nonlinear dependence on the number of presynaptic spikes (see section 8.6). This transformation is shown in figure 8.8B–C. These models of small circuits were also extended to larger networks of thalamic neurons, and the propagating properties found in ferret thalamic slices (Kim et al. 1995) could be reproduced by the model (Destexhe et al. 1996a). Similar conclusions were also reached by another modeling investigation (Golomb et al. 1996).

Models based on accurate kinetic representations of intrinsic and synaptic currents therefore reproduce experimental data, both at the single-cell level for the repertoire of firing properties of thalamic cells, and at the network level for the genesis of oscillations by thalamic circuits. Similar models were used to investigate other problems, such as why the reticular nucleus oscillates *in vivo* but not *in vitro* (Destexhe et al. 1994a), why oscillations have propagating properties *in vitro* but are coherent *in vivo* (Contreras et al. 1996, Destexhe et al. 1998a), or to explain paroxysmal behavior in a way that is consistent with different experimental models of epileptic seizures (Destexhe 1998, 1999).

In conclusion, this example shows that simplified kinetic models are useful tools for studying network mechanisms involving large numbers of neurons and synapses. The use of Hodgkin–Huxley-type models for representing voltage-dependent currents and simplified kinetic models for representing synaptic currents led to population behavior consistent with experimental measurements. To yield correct oscillation frequency and phase relations, the modeled ionic currents had to have rise and decay kinetics consistent with experimental data, but other properties were not required, such as AMPA receptor desensitization or the activation–inactivation coupling in sodium channels. This suggests that

A

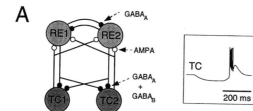

B   Spindle oscillations

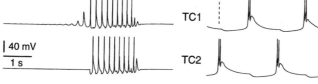

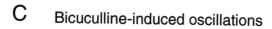

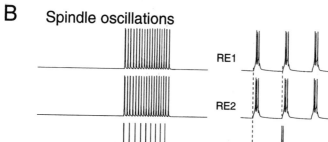

C   Bicuculline-induced oscillations

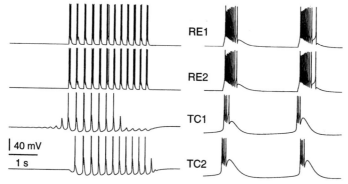

the rise and decay kinetics of currents are important determinants for this type of network behavior.

8.8 Discussion

This chapter has reviewed kinetic models for ion channels underlying membrane excitability, synaptic transmission, and second-messenger actions. We discuss here the assumptions and limits of kinetic models and how to include them in a more general framework that includes the metabolism of the cell.

8.8.1 Assumptions of Kinetic Models

Markov kinetic models assume that (1) the gating of ion channels can be described by transition diagrams involving a finite number of states and (2) the transition probability between these states is independent of time. Because the flux of ions though single channels can be measured directly, it has been possible to observe directly the rapid and stochastic transitions between conducting and nonconducting states (Neher and Sakmann 1976). Such rapid transitions are predicted by finite-state diagrams by opposition to a continuum of states that would predict smooth variations in current. A finite number of states is also justified thermodynamically by the existence of local energy minimas, with the consequence that the configuration of the channel protein in the membrane can be approximated by a set of distinct conformational states separated by large energy barriers (Hille 1992).

Several alternative formalisms have been suggested for modeling ion channels. Diffusional (Millhauser et al. 1988) or continuum gating models (Levitt 1989) are Markovian but posit an infinite number of states. Fractal (Liebovitch and Sullivan 1987) or deterministically chaotic (Liebovitch and Toth 1991) models assume a finite number of states, but allow time-dependent transition rates. Differentiation between discrete multistate Markov models and any of these alternatives hinges on high time-resolution studies of channel openings. Analysis of single-channel openings and closings has shown

Figure 8.8
Modeling the interplay of intrinsic and synaptic currents in networks of thalamic neurons. (A) Scheme of a circuit of thalamic cells. Thalamocortical relay (TC) and thalamic reticular (RE) neurons are connected with AMPA, $GABA_A$ and $GABA_B$ receptors. Each cell type also generates bursts of action potentials in response to current injection (insets) due to the presence of a low-threshold calcium current. (B) 8–12-Hz spindle oscillations arising from the mutual excitatory and inhibitory loop between TC and RE cells. In this case, RE cells produce few action potentials and the inhibitory postsynaptic potential (IPSP) in TC cells is dominated by $GABA_A$-mediated currents. TC cells fire every two cycles in alternation (left Panel). (C) 2–4-Hz oscillations obtained when $GABA_A$ receptors are suppressed. In this case, all cells oscillate in phase (left panel) and the large bursts produced by RE cells evoke $GABA_B$-mediated IPSPs in TC cells. Modeling the phase relations and the coexistence of these two types of oscillations requires one to model intrinsic and synaptic currents with correct kinetics and the nonlinear dependence of $GABA_B$ responses. Modified from Destexhe et al. (1996a).

that finite-state Markov models are most consistent with experimental data (McManus et al. 1988, Sansom et al. 1989). Moreover, finite-state Markov models so far have been successful in accounting for nearly all types of ion channels recorded with single-channel techniques (see Hille 1992, Sakmann and Neher 1995).

Another assumption implicit when using kinetic models is the fact that the kinetic formalism can be used to describe large populations of molecules. This assumption is clearly justified when describing Na^+ or K^+ currents resulting from several thousands of ion channels, but it may break down in other cases. For example, in central synapses, the release of transmitter is thought to open only about 5–50 postsynaptic receptors (Edwards et al. 1990, Hestrin 1992, Traynelis et al. 1993). Consequently, individual synaptic currents may have significant variability owing to probabilistic channel openings (see e.g., Silver et al. 1992). In situations in which this variability is of interest, stochastic rather than kinetic models would be more appropriate.

8.8.2   Simplified Models

Although a substantial number of states are necessary to account for the single-channel and macroscopic behavior of ion channels, this complexity is not always necessary. The voltage-dependent sodium channel was shown to be best described by a Markov model involving nine states (Vandenberg and Bezanilla 1991), but the simpler model of Hodgkin and Huxley (1952), as well as simplified three-state models (Destexhe et al. 1994c), generated similar action potentials (figure 8.3). For synaptic ion channels, the phenomenon of receptor desensitization can be modeled relatively well using a three-state scheme (figure 8.5). Therefore, models incorporating different levels of complexity can be used according to the type of behavior to be modeled.

In large-scale network simulations, where many thousands of cells and synapses must be simulated, it is clear that the simplest representation for voltage-dependent and synaptic currents is needed for reasons of computational efficiency. Integrate-and-fire models with simplified synaptic couplings are classically used in neural networks (reviewed in Arbib 1995). However, electrophysiological experiments show that in most regions of the brain, neurons exhibit complex intrinsic firing properties (Llinás 1988) and possess multiple types of synaptic receptors (McCormick 1992). Modeling these properties clearly requires more sophisticated representations than integrate-and-fire models. In the example of thalamic oscillations, simplified schemes for synaptic interactions and Hodgkin–Huxley models for voltage-dependent currents provided acceptable representations of model network behavior as observed experimentally (see section 8.7).

This approach therefore lies between the biophysical accuracy of multistate Markov schemes derived from single-channel recordings, and highly simplified integrate-and-fire representations commonly used in neural networks. Computational models allow us to explore the dynamic possibilities of cells possessing multiple voltage-dependent currents and synaptic receptor types. Simplified models of ionic currents will become increasingly useful in exploring these complex systems as new ion channel and receptor subtypes are identified and as new interactions with cellular biochemistry are discovered.

8.8.3   Integration with Molecular Biology

Modeling membrane excitability, synaptic transmission, and second-messenger actions using similar equations provides a full description of a wide range of cellular interactions using the same formalism (Destexhe et al. 1994c). Besides its aesthetic advantage, this approach is also useful because it uses a language that is compatible with molecular and biochemical descriptions. It is therefore a natural description to adopt in order to link electrical activity to biochemistry, which is likely to become of increasing interest in the future.

A wide range of biological phenomena can be addressed by models that consider ion channels in a molecular and biochemical context as well as an electrical one. As a prominent example, the action of second messengers such as G-proteins was considered here in the context of $GABA_B$ receptors. G-proteins may not only act on ion channels, but may also affect various biochemical pathways through adenylate cyclase, protein kinases, phospholipase C, and gene regulation (Berridge and Irvine 1989, Nelson and Alkon 1991, Tang and Gilman 1991, Birnbaumer 1992, Clapham and Neer 1993, Gutkind 1998, Selbie and Hill 1998). It is clear that when the action of synaptic receptors with biochemical pathways is clarified, it will uncover a new dimension of complexity in cellular interactions. It is likely to become increasingly difficult to understand the long-term behavior of networks intuitively, and computational models should play an important role in helping us to understand these complex interactions.

Appropriate models are therefore needed to integrate electrophysiological knowledge with the intricate web of second messengers, protein phosphorylation systems, and the deeper machinery of signal transduction and gene regulation. Kinetic models provide a natural way of integrating electrophysiology with cellular biochemistry, in which ion channels are considered as a special and important class of enzymes rather than as a completely distinct subject. Kinetic models of interactions through second messengers and G-proteins mark only the initial stages of this integration.

Note

1. All models shown here were simulated using NEURON (Hines and Carnevale 1997).

References

Aldrich, R. W., and Stevens, C. F. (1987). Voltage-dependent gating of single sodium channels from mammalian neuroblastoma cells. *J. Neurosci.* 7: 418–431.

Aldrich, R. W., Corey, D. P., and Stevens, C. F. (1983). A reinterpretation of mammalian sodium channel gating based on single channel recording. *Nature* 306: 436–441.

Andersen, O., and Koeppe, III R. E. (1992). Molecular determinants of channel function. *Physiol. Rev.* 72: S89–S158.

Andersen, P., and Rutjord, T. (1964). Simulation of a neuronal network operating rhythmically through recurrent inhibition. *Nature* 204: 289–190.

Andrade, R., Malenka, R. C., and Nicoll, R. A. (1986). A G protein couples serotonin and GABA$_B$ receptors to the same channels in hippocampus. *Science* 234: 1261–1265.

Arbib, M. (ed.) (1995). *The Handbook of Brain Theory and Neural Networks.* MIT Press, Cambridge, Mass.

Armstrong, C. M. (1969). Inactivation of the potassium conductance and related phenomena caused by quaternary ammonium ion injection in squid axons. *J. Gen. Physiol.* 54: 553–575.

Armstrong, C. M. (1981). Sodium channels and gating currents. *Physiol. Rev.* 62: 644–683.

Armstrong, C. M., and Hille, B. (1998). Voltage-gated ion channels and electrical excitability. *Neuron* 20: 371–380.

Augustine, G. J., and Charlton, M. P. (1986). Calcium dependence of presynaptic calcium current and postsynaptic response at the squid giant synapse. *J. Physiol.* 381: 619–640.

Bartol, T. M., Jr., Land, B. R., Salpeter, E. E., and Salpeter, M. M. (1991). Monte Carlo simulation of miniature endplate current generation in the vertebrate neuromuscular junction. *Biophys. J.* 59: 1290–1307.

Berridge, M. J., and Irvine, R. F. (1989). Inositol phosphates and cell signaling. *Nature* 341: 197–205.

Bezanilla, F. (1985). Gating of sodium and potassium channels. *J. Membr. Biol.* 88: 97–111.

Birnbaumer, L. (1992). Receptor-to-effector signaling through G proteins: roles for beta gamma dimers as well as alpha subunits. *Cell* 71: 1069–1072.

Blaustein, M. P. (1988). Calcium transport and buffering in neurons. *Trends Neurosci.* 11: 438–443.

Borg-Graham, L. J. (1991). Modeling the nonlinear conductances of excitable membranes. in W. Wheal and J. Chad, eds. pp. 247–275. *Cellular and Molecular Neurobiology: A Practical Approach,* Oxford University Press, New York.

Breitwieser, G. E. and Szabo, G. (1988). Mechanism of muscarinic receptor-induced K$^+$ channel activation as revealed by hydrolysis-resistant GTP analogues. *J. Gen. Physiol.* 91: 469–493.

Brown, D. A. (1990). G-proteins and potassium currents in neurons. *Annu. Rev. Physiol.* 52: 215–242.

Brown, A. M., and Birnbaumer, L. (1990). Ionic channels and their regulation by G protein subunits. *Annu. Rev. Physiol.* 52: 197–213.

Bush, P., and Sejnowski, T. J. (1991). Simulations of a reconstructed cerebellar Purkinje cell based on simplified channel kinetics. *Neural Computation* 3: 321–332.

Chabala, L. D. (1984). The kinetics of recovery and development of potassium channel inactivation in perfused squid giant axons. *J. Physiol.* 356: 193–220.

Chen, C., and Hess, P. (1990). Mechanisms of gating of T-type calcium channels. *J. Gen. Physiol.* 96: 603–630.

Clapham, D. E., and Neer, E. J. (1993). New roles for G-protein βγ-dimers in transmembrane signaling. *Nature* 365: 403–406.

Clay, J. R. (1989). Slow inactivation and reactivation of the potassium channel in squid axons. *Biophys. J.* 55: 407–414.

Clements, J. D. (1996). Transmitter time course in the synaptic cleft: its role into central synaptic function. *Trends Neurosci.* 19: 163–171.

Clements, J. D., Lester, R. A. J., Tong, J., Jahr, C., and Westbrook, G. L. (1992). The time course of glutamate in the synaptic cleft. *Science* 258: 1498–1501.

Colquhoun, D., and Hawkes, A. G. (1977). Relaxation and fluctuations of membrane currents that flow through drug-operated channels. *Proc. Roy. Soc. Lond.* B 199: 231–262.

Colquhoun, D., and Hawkes, A. G. (1981). On the stochastic properties of single ion channels. *Proc. Roy. Soc. Lond.* B 211: 205–235.

Colquhoun, D., Jonas, P., and Sakmann, B. (1992). Action of brief pulses of glutamate on AMPA/kainate receptors in patches from different neurons of rat hippocampal slices. *J. Physiol.* 458: 261–287.

Contreras, D., Destexhe, A., Sejnowski, T. J., and Steriade, M. (1996). Control of spatiotemporal coherence of a thalamic oscillation by corticothalamic feedback. *Science* 274: 771–774.

Destexhe, A. (1998). Spike-and-weave oscillations based on the properties of GABA$_B$ receptors. *J. Neurosci.* 18: 9099–9111.

Destexhe, A. (1999). Can GABA$_A$ conductances explain the fast frequency of absence seizures in rodents? *Eur. J. Neurosci* 11: 2175–2181.

Destexhe, A., and Sejnowski, T. J. (1995). G-protein activation kinetics and spill-over of GABA may account for differences between inhibitory responses in the hippocampus and thalamus *Proc. Natl. Acad. Sci. U.S.A.* 92: 9515–9519.

Destexhe, A., and Sejnowski, T. J. (2000). *The Thalamocortical Assembly,* Oxford University Press, Oxford (in press).

Destexhe, A., Babloyantz, A., and Sejnowski, T. J. (1993a) Ionic mechanisms for intrinsic slow oscillations in thalamic relay neurons. *Biophys. J.* 65: 1538–1552.

Destexhe, A., McCormick, D. A., and Sejnowski, T. J. (1993b). A model of 8–10 Hz spindling in interconnected thalamic relay and reticularis neurons. *Biophys. J.* 65: 2474–2478.

Destexhe, A., Contreras, D., Sejnowski, T. J., and Steriade, M. (1994a). Modeling the control of reticular thalamic oscillations by neuromodulators. *NeuroReport* 5: 2217–2220.

Destexhe, A., Mainen, Z., and Sejnowski, T. J. (1994b). An efficient method for computing synaptic conductances based on a kinetic model of receptor binding. *Neural Computation* 6: 14–18.

Destexhe, A., Mainen, Z. F., and Sejnowski, T. J. (1994c). Synthesis of models for excitable membranes, synaptic transmission and neuromodulation using a common kinetic formalism. *J. Comput. Neurosci.* 1: 195–230.

Destexhe, A., Bal, T., McCormick, D. A., and Sejnowski, T. J. (1996a) Ionic mechanisms underlying synchronized oscillations and propagating waves in a model of ferret thalamic slices. *J. Neurophysiol.* 76: 2049–2070.

Destexhe, A., Contreras, D., Steriade, M., Sejnowski, T. J., and Huguenard, J. R. (1996b). In vivo, in vitro and computational analysis of dendritic calcium currents in thalamic reticular neurons. *J. Neurosci.* 16: 169–185.

Destexhe, A., Contreras, D., and Steriade, M. (1998a) Mechanisms underlying the synchronizing action of corticothalamic feedback through inhibition of thalamic relay cells. *J. Neurophysiol.* 79: 999–1016.

Destexhe, A., Mainen, Z. F., and Sejnowski, T. J. (1998b). Kinetic models of synaptic transmission. In *Methods in Neuronal Modeling,* 2nd ed. C. Koch and I. Segev, eds., pp. 1–26. MIT Press, Cambridge, Mass.

Dutar, P., and Nicoll, R. A. (1988). A physiological role for GABA$_B$ receptors in the central nervous system. *Nature* 332: 156–158.

Edwards, F. A., Konnerth, A., and Sakmann, B. (1990). Quantal analysis of inhibitory synaptic transmission in the dentate gyrus of rat hippocampal slices: a patch-clamp study. *J. Physiol.* 430: 213–249.

Fitzhugh, R. (1965). A kinetic model of the conductance changes in nerve membrane. *J. Cell. Comp. Physiol.* 66: 111–118.

Golomb, D., Wang, X. J., and Rinzel, J. (1996). Propagation of spindle waves in a thalamic slice model. *J. Neurophysiol.* 75: 750–769.

Gutkind, J. S. (1998). The pathways connecting G protein-coupled receptors to the nucleus through divergent mitogen-activated protein kinase cascades. *J. Biol. Chem.* 273: 1839–1842.

Harris, A. L., Spray, D. C., and Bennett, M. V. L. (1981). Kinetic properties of a voltage-dependent junctional conductance. *J. Gen. Physiol.* 77: 95–117.

Hestrin, S. (1992). Activation and desensitization of glutamate-activated channels mediating fast excitatory synaptic currents in the visual cortex. *Neuron* 9: 991–999.

Hille, B. (1992). *Ionic Channels of Excitable Membranes.* Sinauer Associates, Sunderland, Mass.

Hines, M. L., and Carnevale, N. T. (1997) The NEURON simulation environment. *Neural Computation* 9: 1179–1209.

Hodgkin, A. L., and Huxley, A. F. (1952). A quantitative description of membrane current and its application to conduction and excitation in nerve. *J. Physiol.* 117: 500–544.

Horn, R. J., Patlak, J., and Stevens, C. F. (1981). Sodium channels need not open before they inactivate. *Nature* 291: 426–427.

Johnson, F. H., Eyring, H., and Stover, B. J. (1974). *The Theory of Rate Processes in Biology and Medicine.* Wiley, New York.

Jonas, P., Major, G., and Sakmann, B. (1993). Quantal components of unitary EPSCs at the mossy fibre synapse on CA3 pyramidal cells of rat hippocampus. *J. Physiol.* 472: 615–663.

Jones, E. G. (1985). *The Thalamus.* Plenum Press, New York.

Keller, B. U., Hartshorne, R. P., Talvenheimo, J. A., Catterall, W. A., and Montal, M. (1986). Sodium channels in planar lipid bilayers. Channel gating kinetics of purified sodium channels modified by batrachotoxin. *J. Gen. Physiol.* 88: 1–23.

Kim, U. Ball, T., and McCormick, D. A. (1995). Spindle waves are propagating synchronized oscillations in the ferret LGNd in vitro. *J. Neurophysiol.* 74: 1301–1323.

Kim, U., Sanches-Vives, M. V., and McCormick, D. A. (1997). Functional dynamics of GABAergic inhibition in the thalamus. *Science* 278: 130–134.

Koch, C., and Segev, I. (eds.) (1998). *Methods in Neuronal Modeling.* 2nd ed. MIT Press, Cambridge, Mass.

Labarca, P., Rice, J. A., Fredkin, D. R., and Montal, M. (1985). Kinetic analysis of channel gating. Application to the cholinergic receptor channel and the chloride channel from *Torpedo Californica. Biophys. J.* 47: 469–478.

Lamb, T. D., and Pugh, E. N. (1992). A quantitative account of the activation steps involved in phototransduction in amphibian photoreceptors. *J. Physiol.* 449: 719–758.

Lester, R. A., and Jahr, C. E. (1992). NMDA channel behavior depends on agonist affinity. *J. Neurosci.* 12: 635–643.

Levitt, D. G. (1989). Continuum model of voltage-dependent gating. *Biophys. J.* 55: 489–498.

Liebovitch, L. S., and Sullivan, J. M. (1987). Fractal analysis of a voltage-dependent potassium channel from cultured mouse hippocampal neurons. *Biophys. J.* 52: 979–988.

Liebovitch, L. S., and Toth, T. I. (1991). A model of ion channel kinetics using deterministic chaotic rather than stochastic processes. *J. Theor. Biol.* 148: 243–267.

Llinás, R. R. (1988). The intrinsic electrophysiological properties of mammalian neurons: a new insight into CNS function. *Science* 242: 1654–1664.

Lytton, W. W. (1996). Optimizing synaptic conductance calculation for network simulations. *Neural Computation* 8: 501–509.

Magleby, K. L., and Stevens, C. F. (1972). A quantitative description of end-plate currents. *J. Physiol.* 223: 173–197.

Marom, S., and Abbott, L. F. (1994). Modeling state-dependent inactivation of membrane currents. *Biophys. J.* 67: 515–520.

McCormick, D. A. (1992). Neurotransmitter actions in the thalamus and cerebral cortex and their role in neuromodulation of thalamocortical activity. *Progr. Neurobiol.* 39: 337–388.

McCormick, D. A., and Huguenard, J. R. (1992). A model of the electrophysiological properties of thalamocortical relay neurons. *J. Neurophysiol.* 68: 1384–1400.

McManus, O. B., Weiss, D. S., Spivak, C. E., Blatz, A. L., and Magleby, K. L. (1988). Fractal models are inadequate for the kinetics of four different ion channels. *Biophys. J.* 54: 859–870.

Millhauser, G. L., Saltpeter, E. E., and Oswald, R. E. (1988). Diffusion models of ion-channel gating and the origin of power-law distributions from single-channel recording. *Proc. Natl. Acad. Sci.* U.S.A. 85: 1503–1507.

Neher, E. (1998). Vesicle pools and Ca^{2+} microdomains: new tools for understanding their role in neurotransmitter release. *Neuron* 20: 389–399.

Neher, E., and Sakmann, B. (1976). Single-channel currents recorded from membrane of denervated frog muscle fibers. *Nature* 260: 799–802.

Neher, E., and Stevens, C. F. (1979). Voltage-driven conformational changes in intrinsic membrane proteins. In *The Neurosciences. Fourth Study Program* F. O. Schmitt and F. G. Worden, eds., pp. 623–629. *MIT Press, Cambridge, Mass.*

Nelson, T. J., and Alkon, D. L. (1991). GTP-binding proteins and potassium channels involved in synaptic plasticity and learning. *Molec. Neurobiol.* 5: 315–328.

Otis, T. S., and Mody, I. (1992). Modulation of decay kinetics and frequency of $GABA_A$ receptor-mediated spontaneous inhibitory postsynaptic currents in hippocampal neurons. *Neuroscience* 49: 13–32.

Otis, T. S., Dekoninck, Y., and Mody, I. (1993). Characterization of synaptically elicited $GABA_B$ responses using patch-clamp recordings in rat hippocampal slices. *J. Physiol.* 463: 391–407.

Patneau, D. K., and Mayer, M. L. (1991). Kinetic analysis of interactions between kainate and AMPA: evidence for activation of a single receptor in mouse hippocampal neurons. *Neuron* 6: 785–798.

Perozo, E., and Bezanilla, F. (1990). Phosphorylation affects voltage gating of the delayed rectifier K^+ channels by electrostatic interactions. *Neuron* 5: 685–690.

Rall, W. (1967). Distinguishing theoretical synaptic potentials computed for different somadendritic distributions of synaptic inputs. *J. Neurophysiol.* 30: 1138–1168.

Rall, W. (1995). *The Theoretical Foundation of Dendritic Function,* I. Segev, J. Rinzel, and G. M. Shepherd, eds. MIT Press, Cambridge, Mass.

Raman, I. M., and Trussell, L. O. (1992). The kinetics of the response to glutamate and kainate in neurons of the avian cochlear nucleus. *Neuron* 9: 173–186.

Sakmann, B., and Neher, E. (eds.) (1995). *Single-Channel Recording,* 2nd ed. Plenum Press, New York.

Sansom, M. S. P., Ball, F. G., Kerry, C. J., Ramsey, R. L., and Usherwood, P. N. R. (1989). Markov, fractal, diffusion, and related models of ion channel gating. A comparison with experimental data from two ion channels. *Biophys. J.* 56: 1229–1243.

Selbie, L. A. and Hill, S. J. (1998). G protein-coupled-receptor cross-talk: the fine-tuning of multiple receptor-signaling pathways. *Trends Pharmacol. Sci.* 19: 87–93.

Sigworth, F. J., and Neher, E. (1980). Single Na channel currents observed in cultured rat muscle cells. *Nature* 287: 447–449.

Silver, R. A., Traynelis, S. F., and Cull-Candy, S. G. (1992). Rapid time-course miniature and evoked excitatory currents at cerebellar synapses *in situ. Nature* 355: 163–166.

Standley, C., Ramsey, R. L., and Usherwood, P. N. R. (1993). Gating kinetics of the quisqualate-sensitive glutamate receptor of locust muscle studied using agonist concentration jumps and computer simulations. *Biophys. J.* 65: 1379–1386.

Steriade, M., and Llinás, R. R. (1988). The functional states of the thalamus and the associated neuronal interplay. *Physiol. Rev.* 68: 649–742.

Steriade, M., McCormick, D. A., and Sejnowski, T. J. (1993). Thalamocortical oscillations in the sleeping and aroused brain. *Science* 262: 679–685.

Stevens, C. F. (1978). Interactions between intrinsic membrane protein and electric field. *Biophys. J.* 22: 295–306.

Szabo, G., and Otero, A. S. (1989). Muscarinic activation of potassium channels in cardiac myocytes: kinetic aspects of G protein function in vivo. *Trends Pharmacol. Sci.* Dec., Suppl.: 46–49.

Tang, M. J., and Gilman, A. G. (1991). Type-specific regulation of adenylyl cyclase by G protein beta gamma subunits. *Science* 254: 1500–1503.

Thalmann, R. H. (1988). Evidence that guanosine triphosphate (GTP)-binding proteins control a synaptic response in brain: effect of pertussis toxin and GTP gamma S on the late inhibitory postsynaptic potential of hippocampal CA3 neurons. *J. Neurosci.* 8: 4589–4602.

Thompson, S. M., and Gahwiler, B. H. (1992). Effects of the GABA uptake inhibitor tiagabine on inhibitory synaptic potentials in rat hippocampal slice cultures. *J. Neurophysiol.* 67: 1698–1701.

Thomson, A. M., and Destexhe, A. (1999). Dual intracellular recordings and computational models of slow IPSPs in rat neocortical and hippocampal slices. *Neuroscience* 92: 1193–1215.

Traynelis, S. F., Silver, R. A., and Cull-Candy, S. G. (1993). Estimated conductance of glutamate receptor channels activated during EPSCs at the cerebellar mossy fiber-granule cell synapse. *Neuron* 11: 279–289.

Vandenberg, C. A., and Bezanilla, F. (1991). A model of sodium channel gating based on single channel, macroscopic ionic, and gating currents in the squid giant axon. *Biophys. J.* 60: 1511–1533.

VanDongen, A. M. J., Codina, J., Olate, J., Mattera, R., Joho, R., Birnbaumer, L., and Brown, A. M. (1988). Newly identified brain potassium channels gated by the guanine nucleotide binding protein G_o. *Science* 242: 1433–1437.

von Krosigk, M., Bal, T., and McCormick, D. A. (1993). Cellular mechanisms of a synchronized oscillation in the thalamus. *Science* 261: 361–364.

Wang, X. J., and Rinzel, J. (1992). Alternating and synchronous rhythms in reciprocally inhibitory model neurons. *Neural Computation* 4: 84–97.

Wang, X. J., Golomb, D., and Rinzel, J. (1995). Emergent spindle oscillations and intermittent burst firing in a thalamic model: specific neuronal mechanisms. *Proc. Natl. Acad. Sci. U.S.A.* 92: 5577–5581.

Yamada, W. M., and Zucker, R. S. (1992). Time course of transmitter release calculated from simulations of a calcium diffusion model. *Biophys. J.* 61: 671–682.

Yamada, W. M., Koch, C., and Adams, P. R. (1989). Multiple channels and calcium dynamics. In *Methods in Neuronal Modeling,* C. Koch and I. Segev, eds. pp. 97–134. MIT Press, Cambridge, Mass.

9 Stochastic Simulation of Cell Signaling Pathways

Carl A. J. M. Firth and Dennis Bray

This chapter begins by examining some of the difficulties encountered when conventional deterministic programming methods are applied to cell signaling pathways. These include the combinatorial explosion of large numbers of different species and the instability associated with reactions between small numbers of molecules. The advantages of individual-based, stochastic modeling are then reviewed and a novel stochastic program, called *StochSim,* is described.

The application of stochastic models to signaling pathways is examined with specific reference to the pathway employed by coliform bacteria in the detection of attractants and repellents. Key conceptual advances underlying this approach are the recognition that many individual proteins in a pathway operate in functional units, known as receptor complexes, and that thermally driven flipping of proteins and protein complexes from one conformation to another underlies all signaling events inside the cell.

The bacterial chemotaxis pathway is controlled by two large protein assemblies associated with the plasma membrane—the receptor complex and the flagellar motor. Stochastic modeling of these two complexes and the conformational changes they undergo allows us to integrate biochemical and thermodynamic data into a coherent and manageable account.

The more we learn about cell signaling pathways, the less, paradoxically, we seem to understand. As the numbers of receptors, kinases, and G-proteins carrying signals increase and details of their protein structures, post-translational modifications and interconnections multiply, it becomes increasingly difficult to visualize any pathway as a whole. Of course, it can be argued that many, perhaps most, of the molecular details of any specific pathway will turn out to be irrelevant to the meaning of the signals carried. But how are we to discover which features are important and which are not? Recall that the binding of a ligand to a handful of receptors, or a change of less than 10% in the rate constant of individual enzymes, is enough to switch a behavior pattern of a cell on or off, or to direct its differentiation in an entirely different direction.

The way ahead, we suggest, is twofold. First of all we need to employ the power of computers to store the myriad details of cell signaling pathways. Molecular parameters must be stored in an intelligent way that allows their function in the pathway to be simulated and displayed. Second, we need to search for the principles of organization by which a cell controls its behavior, and which will also help us to decipher the biochemical hieroglyphics used by living systems. These two paths are not completely separate and are likely to cross repeatedly. Principles discovered by thought and experiment will guide the development of new computer programs; in reciprocal fashion, the employment of suitable programs will help us discover new principles.

In this chapter we describe some of our own endeavors in this direction. We first describe our efforts to find a suitable programming base for the representation of intracellular signaling reactions. The difficulties and limitations of conventional deterministic models are described and the advantages of individual-based, stochastic modeling are reviewed. A novel stochastic program we have developed, which has many desirable features for modeling signal pathways, is presented. We then outline two essential principles that, in our estimation, will be vital for the interpretation of intercellular signaling pathways—the formation of large signaling complexes, and the thermally driven flipping of proteins and protein complexes from one conformation to another. Finally we show how these principles can be embodied in suitable software and used to analyze a specific signaling pathway—the pathway by which coliform bacteria detect chemical attractants and repellents.

9.1   Limitations of Deterministic Models

Networks of biochemical reactions are normally studied by quantitative analysis of their characteristic continuous, deterministic rate equations. General-purpose simulators, such as GEPASI (Mendes et al. 1990, Mendes 1993), MIST (Ehlde and Zacchi 1995), and SCAMP (Sauro 1986, 1993), predict the behavior of metabolic pathways by constructing differential equations from user-defined stoichiometric chemical reactions, which they then solve using numerical integration. Deterministic models have also been simulated with more specialized programs written to solve a single set of differential equations, using methods similar to the general-purpose simulators (Bray et al. 1993, Tang and Othmer 1994, Valkema and Van Haastert 1994, Bray and Bourret 1995, Kohn 1998).

Under certain circumstances, deterministic models no longer represent the physical system adequately and cannot be used to predict the concentrations of chemical species. If we could focus on the molecular scale of a chemical reaction, we would see individual molecules randomly bumping into others; if the energies of the colliding molecules were sufficient, they might react. To convert this probabilistic system, with its molecular graininess, into a continuous, deterministic model, one must make several assumptions. These hold within a defined regime, where the discrete, stochastic nature of a chemical system is not apparent. However, certain properties of a system may cause these assumptions to break down, with the result that the use of a continuous, deterministic model leads to a loss in predictive power:

• To convert a spatial distribution of discrete molecules into a single, continuous, variable of concentration, one assumes that the system has infinite volume. The greater the difference between the system and this assumption—in other words, the smaller the volume—the less accurate the model becomes (Erdi and Toth 1989). However, intracellular biochemical reactions frequently occur in minute volumes, whether in a prokaryotic cell or a eukaryotic organelle. For example, something like 200 calcium-gated potassium and sodium ion channels are sufficient to control a key step in several neutrophil signaling pathways (Hallett 1989). Hydrogen (hydroxonium) ions can affect almost every cellular system, yet they are present in the cytoplasm only at around 6000 per eukaryotic cell (Hallett 1989). A signaling event may culminate in the transcription of a single copy gene—representing the concentration of an active gene because a continuous variable would have little meaning (Ko 1991, Kingston and Green 1994).

• Fluctuations, which arise from the stochasticity of the physical system, are often negligible and for many purposes can be ignored. However, if the system operates close to points of unstable equilibria, stochastic fluctuations can be amplified and cause observable, macroscopic effects (Turner 1977, Erdi and Toth 1989). Biological pathways frequently employ positive feedback and other, more complex, nonlinear relationships, which lead to instabilities in state space.

• The representation of a chemical species as a continuous variable of concentration further assumes that the system is absolutely uniform. Rate constants carry with them the assumption that every molecule in the system can be represented by a notional average molecule whose changes can be predicted with absolute certainty. But in reality, biochemical systems frequently exhibit spatial heterogeneity due to compartmentalization, protein association, and diffusion. These processes cannot be described adequately by a single, continuous variable. Mathematical analysis has demonstrated the importance of spatial distributions of reaction substrates on the behavior of a reaction network (Zimmerman and Minton 1993). Similarly, constraining a reacting species within a two-dimensional space, such as a membrane, can also have far-reaching effects on its reaction kinetics.

Living cells embody these properties—small volumes, heterogeneity, and points of unstable equilibria—in the extreme. Flagellar motors in bacterial cells switch between two rotational states primarily under the influence of thermal fluctuations (Scharf et al. 1998). The quantized release of vesicles from nerve endings is probabilistic; different numbers of vesicles are discharged in response to identical signals (Fatt and Katz 1952, Del Castillo and Katz 1954). The output characteristics of sensory detectors, such as retinal rod outer segments (Lamb 1994, Van Steveninck and Laughlin 1996) and the firing of individual nerve cells (Smetters and Zador 1996), are intrinsically stochastic.

The first application of deterministic modeling in biochemistry was in the study of metabolic pathways. Metabolites are modified in a series of chemical reactions in which chemical bonds are broken and formed. The substrate molecules are present in large numbers and the system rarely contains points of unstable equilibria. Under these conditions, deterministic models can yield an accurate representation of metabolism (Sorribas et al. 1995).

By contrast, signal transduction pathways frequently contain small numbers of particles, more complex dynamics, such as positive feedback, and rely heavily on spatial localization to ensure the ordered transmission of a signal from one place to another. This makes a deterministic modeling strategy unsuitable for studying these pathways. There is compelling experimental evidence that regulatory events in organisms do not operate in a deterministic fashion (McAdams and Arkin 1997; see also chapter 2 in this volume). Stochastic characteristics have been found in the regulation of cell cycle progression in Swiss 3T3 cells (Brooks and Riddle 1988), the commitment decision in human hematopoiesis (Mayani et al. 1993), and the control of human CD2 expression in transgenic mice (Elliott et al. 1995).

9.2   Stochastic Models

Stochastic models do not suffer from the same problems as their deterministic counterparts. By treating concentration as a discrete variable, usually the number of molecules of a particular species, there is no longer an assumption that the volume of the system is infinite. Spatial localization can be considered, and fluctuations will be apparent when the stochastic model is simulated (figure 9.1; see also plate 27). Physically, a chemical reaction is a stochastic event involving a discrete number of particles; therefore it is more realistically described using a stochastic reaction model.

In a system with a large volume, away from points of instability, the mean behavior predicted by a stochastic model is in agreement with the corresponding deterministic model (Bartholomay 1962, Erdi and Toth 1989). However, in other systems, the predictions made by the two models are incompatible. Analytical solution of the stochastic model and the equivalent deterministic model demonstrates that the stationary points may differ in their position in state space or in their stability; what is stable in a deterministic model could be unstable in the stochastic model (Erdi and Toth 1989, Zheng and Ross 1991, Mansour and Baras 1992). Often the stationary points in the two models coincide, but thermodynamic fluctuations in the stochastic model may drive the system between stationary points, especially near bifurcations (Baras and Geysermans 1995).

It is tempting to favor the deterministic solutions because they appear to have less "noise" than the stochastic solutions. However, these stochastic fluctuations are often of great importance:

Figure 9.1
Predicted time course of an enzyme-catalyzed cyclic reaction showing the progressive emergence of random behavior as the reaction volume becomes smaller. The reaction is the simple interconversion of substrate (green), S, and product (red), P, by two enzymes. The top graph shows the time course of the reaction simulated by MIST, a conventional continuous, deterministic simulator (Ehlde and Zacchi 1995). The next four graphs show changes in S and P predicted by StochSim for different volumes of the reaction system. Note that at the smallest volume considered, each step in concentration corresponds to a change of one molecule. (See plate 27.)

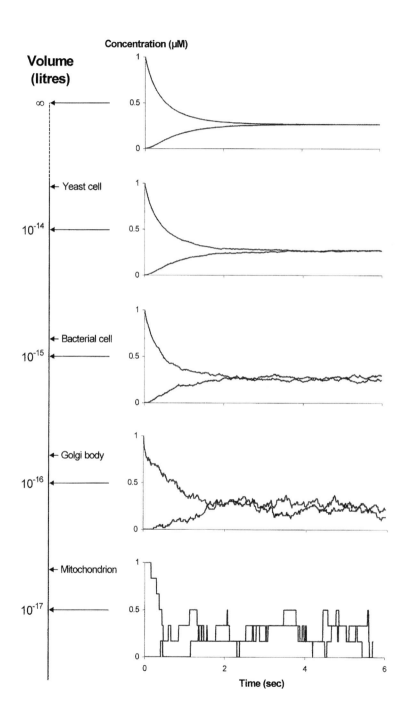

• According to the fluctuation–dissipation theorem, the processes leading to equilibrium are interconnected with the fluctuations around the equilibrium point (Erdi and Toth 1989). For example, in a biochemical system at equilibrium, the kinetic rate constants can be calculated by analyzing the fluctuations. A deterministic model cannot yield such kinetic data once equilibrium has been reached.

• Fluctuations typically obey a Poisson distribution, which becomes Gaussian in the continuous limit. Close to points of instability, however, the fluctuation distribution changes, often in an irregular manner (Turner 1977). In a system close to a chaotic regime, for instance, the effect of intrinsic noise is to anticipate the chaos, smearing the boundary between the prechaotic and chaotic behaviors (Matias 1995).

• Stochastic fluctuations are crucial to the phenomenon of stochastic resonance. The noise enhances the response of a system to weak signals, sometimes boosting only those within a specific range of frequencies (Collins et al. 1995).

9.2.1 Previous Stochastic Simulators

In 1976, Gillespie developed a discrete, stochastic algorithm to simulate reaction models (Gillespie 1976). This method was subsequently used by other groups to analyze biochemical kinetics (Gillespie 1977, 1979; Hanusse and Blanche 1981; Stark et al. 1993; Breuer and Petruccione 1995; Matias 1995; McAdams and Arkin 1997; Arkin et al. 1998). The Gillespie algorithm makes timesteps of variable length; in each timestep, based on the rate constants and population size of each chemical species, one random number is used to choose which reaction will occur, and another random number determines how long the step will last. The chemical populations are altered according to the stoichiometry of the reaction, and the process is repeated (see chapter 2 for further details).

To determine which chemical reaction will be taking place in a given timestep, the Gillespie algorithm calculates the probability of each reaction occurring relative to another by multiplying the rate constant of each reaction with the concentration of its substrates. A random number is then used to determine which reaction will occur according to this set of weighted probabilities. The speed of the algorithm will therefore be on the order of n^{-1}, where n is the number of chemical reactions in the model. If the number of reactions in the model is doubled, the time taken to simulate the reactions doubles.

Though the Gillespie algorithm is suitable for simulating many systems, it is sufficiently far from an accurate representation of the physical events underlying chemical reactions that it suffers from several problems of particular relevance to signaling pathways:

• It is becoming increasingly apparent that even biochemical systems with relatively few species can contain a large number of possible reactions owing to their complexity. Intracellular signaling pathways, for example, often utilize large complexes of proteins to integrate and process information (Williams et al. 1993). Each component of these complexes is usually modified at various sites to control its activity and chemical properties, typically by the covalent binding of a phosphate or methyl group. Overall, the multiprotein complex may contain upward of 20 sites, each of which can often be modified independently. The properties of the complex will depend on each of these sites, determining how it will participate in chemical reactions. With 20 sites, the complex can exist in a total of 2^{20}, or one million, unique states, each of which could react in a slightly different way. Therefore, if our multiprotein complex interacted with only 10 other chemical species, a detailed model may contain as many as ten million distinct chemical reactions—a combinatorial explosion. With a speed on the order n^{-1}, the Gillespie algorithm could not reasonably simulate this number of reactions.

• The Gillespie algorithm maintains the total number of molecules of each chemical species, but does not represent individual particles. Therefore, it is not possible to associate physical quantities with each, nor trace the fate of particular molecules over a period of time. This means that it is not possible to extend this algorithm to a more thermodynamically realistic model in which energies and conformational states are associated with each molecule. Similarly, without the ability to associate information on position and speed of travel with each particle, the algorithm cannot be adapted to simulate diffusion, localization, or spatial heterogeneity.

9.3   A Novel Stochastic Simulator

We have developed a simple, novel algorithm (Morton-Firth 1998, Morton-Firth and Bray 1998) that is more efficient at simulating small-volume systems, with large numbers of reactions, than conventional simulators and that also allows the study of stochastic phenomena. Molecules are represented as individual software objects, which interact according to probabilities derived from rate constants and other physical parameters. A number of dummy molecules, or "pseudo-molecules," are used to simulate unimolecular reactions. If a molecule reacts with a pseudo-molecule, the former may undergo a unimolecular reaction.

The stochastic simulator, called *StochSim,* consists of a platform-independent core simulation engine encapsulating the algorithm described below and a separate graphical user interface (figure 9.2). StochSim is available for download from http://www.zoo.cam.ac.uk/zoostaff/morton/stochsim.htm.

Execution follows a simple algorithm, illustrated in figure 9.3. Time is quantized into a series of discrete, independent time slices. In each time slice, one molecule (not a pseudo-molecule) is selected at random. Then another object, in this case either a molecule or a pseudo-molecule, is selected at random. If two molecules are selected, any reaction that

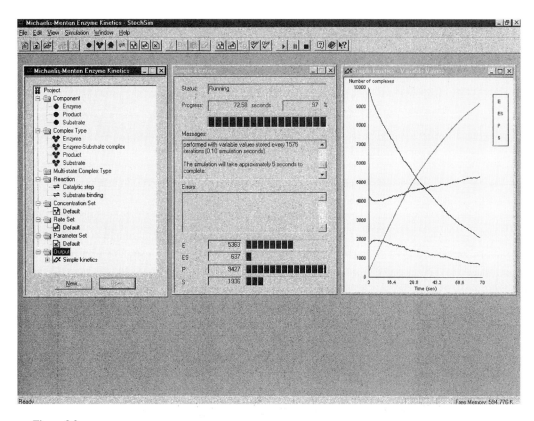

Figure 9.2
Screen shot from StochSim during a running simulation.

occurs will be bimolecular; if one molecule and a pseudo-molecule are selected, it will be unimolecular. Another random number is then generated and used to see if a reaction occurs. The probability of a reaction is retrieved from a lookup table: if the probability exceeds the random number, the particles do not react; on the other hand, if the probability is less than the random number, the particles react, and the system is updated accordingly. The next time slice then begins with another pair of molecules being selected. Although it sounds complicated, the sequence of events can be executed in a very short time because each iteration consists of a small number of elementary operations; even a relatively slow Pentium computer can carry out over five hundred thousand iterations every second.

Enzymes and other proteins are frequently modified in biochemical pathways to alter their intrinsic properties, such as activity, binding affinity, and localization. The representation of each molecule as a distinct object in the algorithm provides a unique opportunity

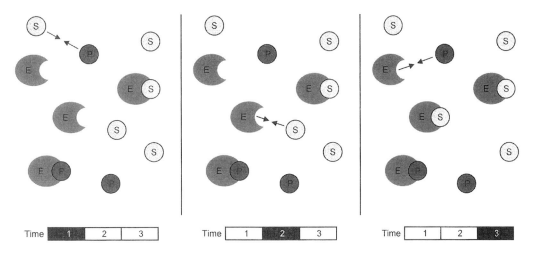

Figure 9.3
Series of three slices illustrating the basic algorithm used by StochSim. Each symbol represents a different molecular object in the program: E represents an enzyme, S the substrate, and P the product. The number of these objects is smaller than in a typical simulation. (Left) In the first time slice, the computer randomly selects two objects, in this case a substrate and product molecule. When the reaction lookup table is accessed, it indicates that substrate molecules do not react with product molecules; the first time slice is now complete. (Middle) An enzyme and substrate molecule are selected in the second time slice. Again, the program goes to the reaction lookup table, but this time it finds that the two molecules selected react with a certain probability, p. A random number between 0 and 1 is generated and if it is less than p, as in this example, the reaction occurs. The program updates the system accordingly at the end of the time slice by binding the substrate to the enzyme to form the enzyme–substrate complex. (Right) The newly made enzyme–substrate complex is now one of the objects in the simulation, and the program continues to select another pair of molecules for reaction.

for storing the state—a representation of the modifications made—with each molecule. In contrast, previous simulators treated each variant as a new chemical species, often resulting in a combinatorial explosion in the total number of species when molecules had multiple sites that could be modified.

9.4 Multistate Molecules

"Multistate molecules" are used in the algorithm whenever a molecule can exist in more than one state. They contain a series of binary flags, each of which represents a site or property of the molecule; typically, these are used to represent covalent modification, such as phosphorylation, binding to another molecule, or conformational change. The flags determine how the multistate molecule will react, so that, for instance, it may participate in a reaction at an increased rate as a result of acetylation. The flags themselves can be modified by a reaction, perhaps in a binding reaction with another molecule, or they can be instantaneously equilibrated according to a fixed probability. The latter tactic is used with processes such as ligand binding and conformational change, which occur several orders of magnitude faster than other chemical reactions in the system.

If a multistate molecule is selected for the next iteration, the algorithm proceeds in the following manner: First, any instantaneously equilibrated flags are assigned to be on or off according to a probability dependent on the state of the molecule. The probability that the protein conformation flag is set, for example, might vary according to the state of the ligand occupancy and protein phosphorylation flags. When these flags have been set, the program inspects the other reactions in which the selected molecule can take part, with reference to its current state. Random numbers are used to decide whether any reaction will take place in the current step, and, if so, *which* reaction will occur. The reaction will then be performed, if appropriate, and the relevant flags sets.

9.5   Simulation of Cell Signaling Pathways

9.5.1   Signaling Complexes and Allostery

The ability to represent multistate molecules, and in particular to codify conformational states, has important benefits for the simulation of cell signaling pathways. Receptors, protein kinases, protein phosphatases, and other components of cell signaling pathways are often physically associated with each other, forming compact clusters of molecules, called *signaling complexes,* which may be attached to the cell membrane or cytoskeleton (Mochly-Rosen 1995, Bray 1998, Zuker and Ranganathan 1999). Rather than allowing signals to diffuse haphazardly in the cell to the next molecule in the pathway, these signaling complexes operate as computational units, each receiving one or more inputs, processing the signals internally, then generating one or more specific outputs. Signaling complexes provide an intermediate level of organization analogous to the integrated circuits and microprocessors used in the construction of large electronic circuits. They could help us make sense of the seemingly impenetrable jungle of molecular interactions that characterize even the simplest forms of cellular communication.

The mechanism of integration within signaling complexes has not been identified with certainty, but from what we know of protein molecules, it seems likely that conformational changes will be a major force. The ability of protein molecules to exist in different three-dimensional shapes, or conformations, each with a different enzymatic activity, is one of the cornerstones of contemporary molecular biology. Originally recognized as a device to allow biochemical pathways and gene expression to be regulated by low molecular weight metabolites, allostery is now seen to underlie every molecular process carried out by a living cell. A large proportion of proteins in a cell are capable of undergoing conformational transitions. Changes of protein shape are thermally driven, with the probability of transition determined by the free energy of the two states. These free energies are themselves influenced by modifications to the protein, such as phosphorylation or methylation, and by the large or small molecules with which the protein associates. Furthermore, as shown in the pioneering work of Changeux and his associates on the nicotinic acetylcholine receptor, allosteric transitions are deeply involved in the transmission of signals inside a cell (Edelstein and Changeux 1998). Signaling pathways are like miniature circuits composed of multiple biochemical switches, each with its own probability of being either on or off (Bray 1995).

Returning to signaling complexes, we see that the most probable mode of operation of even large clusters of proteins is through the cooperative spread of allosteric conformations. These can propagate rapidly through a complex, allowing the complex as a whole to switch rapidly from one state to another. Each conformational state of a complex can have an associated characteristic enzymatic activity through which it influences other molecules downstream. What we normally refer to as the *activity* of the complex (in a deterministic context) will in reality be the fraction of time spent in the active conformation. This physical process is captured in a natural way by the representation of multistate molecules in StochSim.

9.5.2 The Chemotactic Response

The cluster of proteins associated with the chemotactic receptor of coliform bacteria, here called the Tar complex, is particularly well understood and illustrates many features found also in larger eukaryotic complexes. The Tar complex is built around a dimeric transmembrane receptor (Tar) that monitors many different factors, including pH, temperature, and the concentration of several amino acids, sugars, and metal ions.

Escherichia coli swim by means of 5 to 10 long flagella anchored to motor structures positioned randomly over the cell surface. Since each flagellum is a left-handed helical structure, anticlockwise rotation of the flagellum (viewed from the free end of the flagellum, looking toward the cell) tends to push against the cell body. The flagella form a bundle behind the cell as a result of the viscous drag of the medium, propelling it forward in a smooth swimming or running motion at a speed of approximately 20 μm s-1 (Liu and Papadopoulos 1996). When rotating clockwise, the flagella fail to form a coherent bundle and pull the cell in different directions, causing it to tumble erratically with no net motion in any direction. In the absence of any stimuli, the cells alternate between these two behaviors as if sniffing out food, with runs lasting approximately 0.8 s and tumbles 0.2 s (Stock et al. 1991).

Chemical attractants, such as aspartate, suppress tumbling, thereby extending run length. If an *E. coli* cell placed near a source of attractant is, by chance, swimming in the direction of the attractant, the probability of tumbling will be reduced and the cell will continue to swim closer; if the cell happens to be swimming away, it will soon tumble, re-orient, and start swimming in a new direction—possibly bringing it nearer to the attractant. The resulting biased random walk therefore leads to a net migration toward the source of attractant. To avoid the cell becoming trapped in a series of extended runs in the general vicinity of an attractant at low concentrations, *E. coli* quickly adapt to a given concentration, allowing it to respond positively to further increases and negatively to any decreases. Coupled with an extraordinary sensitivity, whereby the binding of less than 10 attractant molecules can produce a measurable change in swimming behavior, adaptation allows the cell to respond to concentrations of aspartate between 2 nM and 1 mM (Mesibov et al. 1973, Jasuja et al. 1999).

The direction of flagellar rotation is controlled by a series of coupled phosphorylation and methylation reactions that transmit information on the external chemical environment from cell surface receptors to the flagellar motor complex (figure 9.4). The relationship between the direction of rotation of multiple flagella on a cell and its swimming behavior is complicated and not fully understood, so the chemotactic response is often characterized by the motion of individual flagella. Typically, cells are tethered to glass coverslips via a single flagellum so that the motor complex turns the cell body rather than the flagellum. This can be directly observed by using a microscope to look down through the coverslip.

9.6   The Model

We extend the work of Barkai and Leibler (1997) to build a detailed and physically realistic computer model of the Tar complex that incorporates all of the chemical steps known to participate in its response to aspartate. Using StochSim, individual molecules and molecular complexes are represented as separate units that interact stochastically, according to experimentally determined rate constants. Each complex in the population is represented in the computer as a discrete object that flips rapidly between two alternative conformations. Ligand binding, methylation, and association with intracellular signaling molecules are represented internally by binary flags (figure 9.5) that, in combination, alter the probability of occupation of the two conformations (which is equivalent to changing their free energies). The resulting model provides an accurate and physically realistic simulation of the chemotactic response, reproducing most *in vivo* and *in vitro* observations, and allows us to highlight problems that were not apparent in earlier, less molecularly detailed models.

9.7   Simulation Results

Incorporation of the free energy values from table 9.1 into our stochastic model of the Tar complex allows us to predict the response of a bacterium to a range of aspartate concentra-

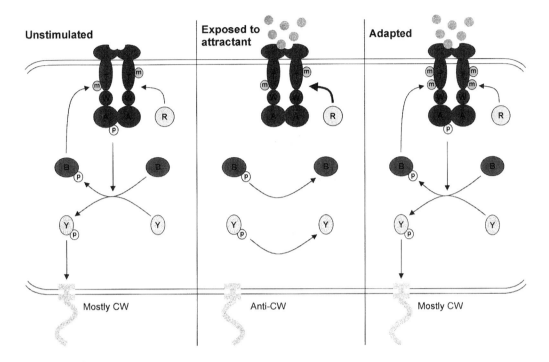

Figure 9.4
Signaling pathway responsible for bacterial chemotaxis. In the absence of chemoattractants, the receptor complex, containing two receptors, two molecules of CheW, and two molecules of CheA, is active in promoting the autophosphorylation of CheA. [Recent studeis suggest that the receptor complex may exist in even higher-order structures with a different stoichiometry (Liu et al. 1997, Bray et al. 1998).] Phosphate groups are transferred from CheA to CheY, forming phosphorylated CheY, which binds to FliM in the flagellar switch complex, increasing the probability that the motor will rotate in a clockwise direction. Attractants, such as the amino acid aspartate, bind to chemotactic receptors and reduce the activity of the ternary complex. The rate of production of phosphorylated CheY is thereby decreased, causing the motor to revert to its native anticlockwise rotation state. In the continued presence of attractants, the chemotaxis receptors become more highly methylated as a result of: (1) the increased local activity of the methyltransferase CheR, (2) the reduced local activity of the methylesterase CheBp, and (3) the reduced global concentration of CheBp, caused by a reduction in phosphate flux from CheA to CheB.

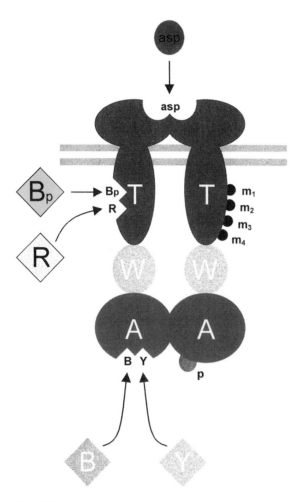

Figure 9.5
Illustration of binding sites and methylation states for the chemotactic receptor complex, which contains two receptor molecules, two CheW molecules, and two CheA molecules. Receptor dimers can be in one of five methylation states [with zero, one, two, three, or four methyl (m) groups], and can also bind ligand and phosphorylated CheB and CheR; CheA dimers can be phosphorylated and bind CheB and CheY. In addition, the receptor can be aspartate sensitive (representing Tar) and the complex is either in an inactive or active conformation (these are not shown). This diagram is not meant to represent the actual positions of binding sites on the complex (Morton-Firth et al. 1999).

Table 9.1
Free energies of Tar complexes based on thermodynamic considerations

Species	p	ΔG (kJ/mol)	Species	p	ΔG (kJ/mol)
Nonligand bound			Ligand bound		
E_0	0.020	9.64	E_0a	0.003	14.46
E_1	0.125	4.82	E_1a	0.020	9.64
E_2	0.500	0.00	E_2a	0.125	4.82
E_3	0.875	−4.82	E_3a	0.500	0.00
E_4	0.997	−14.46	E_4a	0.980	−9.64

Note: E symbols represent the Tar complex in different states of methylation and ligand occupancy. ΔG represents the free energy of the active conformation relative to the inactive conformation, which is directly related to the probability that a complex will be in an active conformation, p, by $\Delta G = -RT \ln(p/\{1-p\})$. The values are calculated as follows: from experiment, the probability of E_2 activity is 0.5, giving $\Delta G = 0.0$; the probability for E_2a is 0.125, giving $\Delta G = 4.82$ kJ mol^{-1}. To ensure consistency with the experimental observation that each receptor has approximately three methyl groups on average in the presence of saturating aspartate, ΔG for E_3a must also be 0.0. The difference between the relative free energies of ligand-bound and nonligand-bound receptor must be the same for each methylation state since we are assuming that the dissociation constant is independent of methylation. Therefore, the ΔG of E_3 is −4.82 kJ mol^{-1}. The difference in free energies between the other methylation states is largely a matter of choice; for consistency, we have assumed that the step in free energy from E_2 to E_3 is the same as in E_1 to E_2 and also E_0 to E_1. However, to ensure that the probability of E_4a activity approaches 1.0 (a requirement of robust adaptation), we set the step in free energy between E_3 and E_4 at twice that of lower methylation states.

tions (Morton-Firth et al. 1999). The change in swimming behavior in response to the addition of saturating aspartate predicted by the stochastic simulation accurately matches experimental data within the limits of experimental and stochastic error. As shown in figure 9.6, the simulated bias rises to a maximum immediately upon exposure to aspartate and then falls asymptotically to its prestimulus value (Stock 1994). Note that the experimental values represent measurements averaged over a large number of bacteria, whereas the simulation shows the statistical fluctuations expected in the swimming of individual bacteria.

The duration of the response to aspartate predicted by StochSim is proportional to the amount of receptor bound, as in real bacteria. Experiments performed using methylaspartate, a nonmetabolized derivative of aspartate, demonstrate this relationship (Spudich and Koshland 1975). Taking into account the difference in receptor binding between aspartate and methylaspartate, the duration of the response measured in these experiments agrees with that predicted by StochSim over 4 orders of magnitude in aspartate concentration—an impressively wide range (figure 9.7).

Each simulation generates an extremely large body of data relating to the receptor complexes in the bacterium over many thousands of timesteps. We examined, for example, the detailed changes in the methylation of receptors as they adapted to different levels of aspartate. This revealed that although there was a net shift toward higher methylation levels, the population was highly heterogeneous (figures 9.8A and B). There was also a

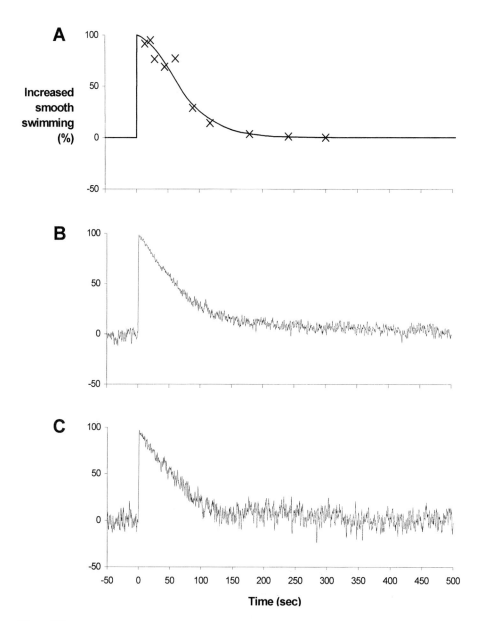

Figure 9.6
Time course of adaptation, experimental and simulated. The increase in smooth swimming behavior (bias) is measured as a percentage of maximum. 1mM aspartate is added at 0. (A) Experimental data taken from Stock (1994), in which 10 measurements (x) are fitted with a smooth curve (−). (B) The mean swimming behavior predicted by five simulations. (C) Swimming behavior predicted in a single simulation. In both (B) and (C), the data have been smoothed by calculating a 1.0-s running average.

transient reduction in the effective dissociation constant (K_D) for aspartate following stimulation due to a temporary decrease in the proportion of active receptors (figure 9.8C). Many other parameters could in principle be followed. The amount of quantitative information in these simulations far exceeds anything yet measured experimentally, although where measurements have been made, there is excellent quantitative agreement.

A similar approach can be used to examine the other multiprotein complex in the chemotactic pathway—the flagellar motor. Thus we were able to use StochSim to simulate the interaction between CheYp and the motor, and to evaluate different possible models for the switching of the motor from clockwise to anticlockwise rotation (Morton-Firth 1998). By combining the model of the motor complex with the receptor complex, we can achieve a complete model of chemotaxis within an *E. coli* cell (figure 9.9). Signals processed by the receptor complex are transmitted to 10 flagellar motors via CheYp; the decision to run or tumble is made democratically, according to the voting hypothesis (Ishihara et al. 1983). The resultant average run length was 0.89 s, average tumble length was 0.33 s, and swimming bias 0.71, which are in close agreement with experimental measurements of 0.8 s, 0.2 s (Stock et al. 1991), and 0.7, respectively (Liu and Parkinson 1989).

9.8   Free Energy Considerations

Our experience with this program led us to appreciate the central importance of conformational changes in the function of the receptor complex, and the usefulness of free energy values in cataloging these changes. Methylation, ligand occupancy, and association with modifying enzymes all result in changes in the free energy values of the two conformational states. The performance of the complex—its rate of phosphorylation, methylation, and binding affinity for ligand—can all be derived from these free energy values. Although direct measurements of free energies have not been made, we were able to estimate the free energy differences between cognate active and inactive conformations for a range of states. Free energy changes provide a convenient "currency" with which to relate the different actions of the complex. We believe that free energy values will become increasingly useful as data on this and other signal complexes accumulate.

Our use of free energy values allowed us to correlate the change in activity associated with ligand binding to the binding affinity of that ligand. Furthermore, we found that we could make the simplifying assumption that the two conformations each had a fixed affinity constant for aspartate regardless of their level of methylation or association with internal signaling proteins. The effective dissociation constant for aspartate of an individual receptor then depended solely on the occupancy of the two states (that is, the free energy changes between the two) and agreed with such measurements as have been made.

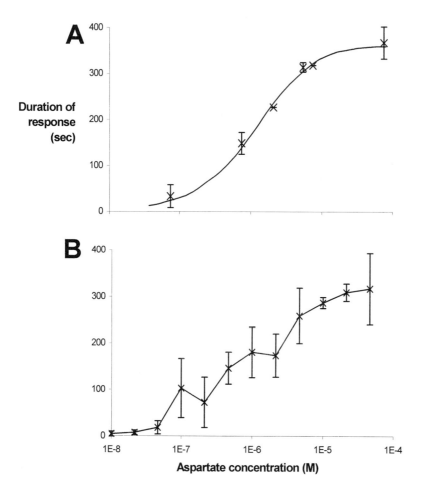

Figure 9.7
Duration of the adaptation response to different concentrations of aspartate. (A) Experimental data taken from Berg and Tedesco (1975) in which they measured the time between exposure of 17 bacteria to methylaspartate and the restoration of anticlockwise rotation (x) and then fit the data with a smooth curve (–). For comparison with aspartate data, the equivalent concentration of aspartate has been calculated to achieve the same level of Tar occupancy based on dissociation constants of 160 μM and 1.2 μM for methylaspartate and aspartate, respectively (Mesibov et al. 1973, Biemann and Koshland, 1994). (B) Duration of the response predicted by StochSim. Each value represents the time taken for the bias to fall to within 0.001 of the prestimulus value, averaged over three simulations.

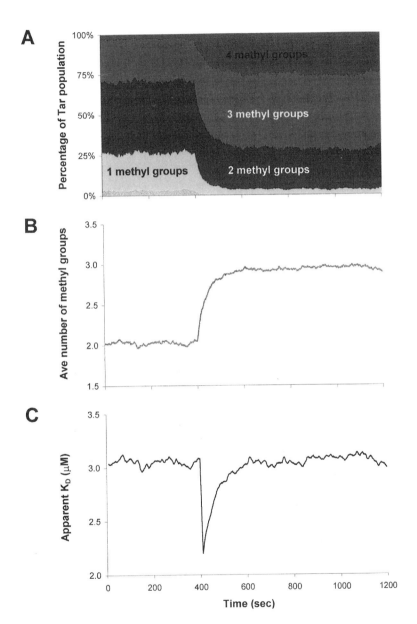

Figure 9.8
Changes in methylation of Tar following exposure to aspartate. Simulations show the response to a saturating concentration of aspartate (added at 400 s). (A) Distribution of different methylation levels, expressed as a percentage of the total population of Tar receptors. (B) Average number of methyl groups per Tar receptor. (C) The apparent dissociation constant for aspartate.

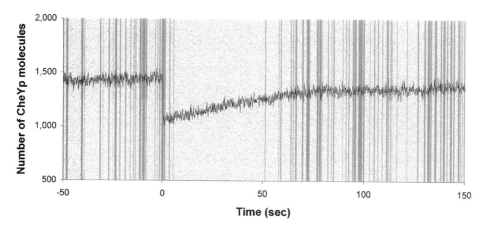

Figure 9.9
Simulated swimming behavior of an *E. coli* cell in response to saturating aspartate. The solid bars represent runs of varying lengths, interspersed with tumbles; the line indicates the concentration of cytoplasmic CheYp.

9.9   Conclusions

The simulation strategy used in our program was strongly influenced by our view of how signal complexes work. Software objects representing individual receptor complexes were equipped with sets of binary flags representing covalent modifications or binding interactions. Each complex in the program flipped between one of two conformations with a probability determined by its current combination of binary flags. During execution of the program, discrete time slices were simulated in each of which two species, drawn from the pool of molecules and complexes involved in chemotaxis, were selected at random. Interactions between the selected species (such as binding or catalysis) were determined using a random number generator and a lookup table containing the probability of their reaction. Any individual complex could, if required, be traced throughout the course of a simulation as it passed through multiple different states.

Our model proved to be an effective way to examine the behavior of the aspartate detection system. The program was fast enough to simulate the behavior of all of the chemotactic signaling molecules in an individual bacterium over a period of several minutes on a modest desktop computer. Furthermore, because of its modular structure, the program was also transparent, allowing every possible state of the complex to be represented and all interactions to be included. For example, competition between CheB and CheR for binding to the Tar receptor and the conformational dependence of methylation and phosphorylation, were included explicitly in the model. Although some of the detailed interactions have been modeled previously, conventional simulations using multiple differential equations have difficulty dealing with all of the manifold possibilities in combination.

9.9.1   Predicting the Unpredictable

Many biochemical systems, particularly those with few chemical reactions, are less expensive to simulate using a deterministic rather than a stochastic model. To what extent are we justified in sacrificing this economy in our pursuit of realism? In other words, what do we gain by employing a stochastic simulation strategy?

Deterministic modeling can be likened to playing a game with dice that have three and a half spots—the mean score—on each face. Take the average of a thousand dice rolls and you might not notice whether you are using deterministic or stochastic dice; roll just once, and the difference is clear. Intracellular signaling pathways use few dice, relying instead on handfuls of particles interacting in small volumes. Eukaryotic cells, several orders of magnitude larger than prokaryotes, restrict sections of the pathway to specific compartments or heterogeneous volumes, such as those found within polarized cells. The resultant stochasticity is not hidden away internally, but is readily apparent from experimental measurements (Hallett and Pettit 1997).

Our discrete, stochastic approach has also allowed us to develop a model in keeping with experimental evidence; by contrast, conventional simulations have used gross simplifications to reduce the pathway to something more manageable. Although the removal of extraneous detail and complexity is often invaluable in the study of isolated pathways, previous models of bacterial chemotaxis have been forced to exclude important features, such as multiple methylation sites (Levin et al. 1999) and phosphotransfer reactions (Barkai and Leibler 1997). To extend a deterministic model further, for example, to test the effects of CheZ activation by CheA binding, is not a trivial exercise. As many as 1000 chemical species may have to be added (the receptor complex can bind CheZ in any methylation state, whether bound to CheR or CheB, and so on); each of these new species may undergo perhaps 10 reactions each, requiring the addition of a further 10,000 reactions, most unrelated to the CheZ activation mechanism being tested!

However, stochastic simulation does not hold all the answers. Often, when a complex system is simulated, the results are equally difficult to interpret, depending on what question we are trying to answer. We may be able to demonstrate that a given model reproduces the experimentally observed behavior, but we may not understand why—in other words, what features of the model are responsible for the behavior of the system. For this reason, this and other methods of simulation should be performed in conjunction with conventional methods of mathematical analysis.

Acknowledgments

We wish to thank Dr. Matthew Levin and Tom Shimizu for helpful discussions. This work was funded by Trinity College, Cambridge, and the Medical Research Council.

References

Arkin, A., Ross, J., and McAdams, H. H. (1998). Stochastic kinetic analysis of developmental pathway bifurcation in phage lambda-infected *E. coli* cells. *Genetics* 149: 1633–1648.

Baras, F., and Geysermans, P. (1995). Sensitivity of non-linear dynamical systems to fluctuations: Hopf bifurcation and chaos. *Nuovo Cim. D* 17: 709–723.

Barkai, N., and Leibler, S. (1997). Robustness in simple biochemical networks. *Nature* 387: 913–917.

Bartholomay, A. F. (1962). Enzymatic reaction-rate theory: a stochastic approach. *Ann. N.Y. Acad. Sci.* 96: 897–912.

Berg, H. C., and Tedesco, P. M. (1975). Transient response to chemotactic stimuli in *Escherichia coli. Proc. Natl. Acad. Sci. USA* 72: 3235–3239.

Biemann, H. P., and Koshland, D. E. Jr. (1994). Aspartate receptors of *Escherichia coli* and *Salmonella typhimurium* bind ligand with negative and half-of-the-sites cooperativity. *Biochemistry* 33: 629–634.

Bray, D. (1995). Protein molecules as computational elements in living cells. *Nature* 376: 307–312.

Bray, D. (1998). Signaling complexes: Biophysical constraints on intracellular communication. *Annu. Rev. Biophys. and Biomolec. Structure* 27: 59–75.

Bray, D., and Bourret, R. B. (1995). Computer analysis of the binding reactions leading to a transmembrane receptor-linked multiprotein complex involved in bacterial chemotaxis. *Mol. Biol. Cell* 6: 1367–1380.

Bray, D., Bourret, R. B., and Simon, M. I. (1993). Computer simulation of the phosphorylation cascade controlling bacterial chemotaxis. *Mol. Biol. Cell* 4: 469–482.

Bray, D., Levin M. D., and Morton-Firth, C. J. (1998). Receptor clustering as a cellular mechanism to control sensitivity. *Nature* 393: 85–88.

Breuer, H. P., and Petruccione, F. (1995). How to build master equations for complex systems. *Continuum Mech. Thermodyn.* 7: 439–473.

Brooks, R. F., and Riddle, P. N. (1988). Differences in growth factor sensitivity between individual 3T3 cells arise at high frequency. *Exp. Res.* 174: 378–387.

Collins, J. J., Chow, C. C., and Imhoff, T. T. (1995). Stochastic resonance without turning. *Nature* 376: 236–238.

Del Castillo, J., and Katz, B. (1954). Quantal components of the end-plate potential. *J. Physiol.* 124: 560–573.

Edelstein, S. J., and Changeux, J-P. (1998). Allosteric transitions of the acetylcholine receptor. *Adv. Prot. Chem.* 51: 121–184.

Ehlde, M., and Zacchi, G. (1995). MIST: a user-friendly metabolic simulator. *Comput. Appl. Biosci.* 11: 201–207.

Elliott, J. I., Festenstein, R., Tolaini, M., and Kioussis, D. (1995). Random activation of a transgene under the control of a hybrid HCD2 locus control region Ig enhance regulatory element. *EMBO J.* 14: 575–584.

Erdi, P., and Toth, J. (1989). *Mathematical Models of Chemical Reactions.* Manchester University Press, Manchester.

Fatt, P., and Katz, B. (1952). Spontaneous subthreshold activity at motor nerve endings. *J. Physiol.* 117: 109–128.

Gillespie, D. T. (1976). A general method for numerically simulating the stochastic time evolution of coupled chemical reactions. *J. Comput. Phys.* 22: 403–434.

Gillespie, D. T. (1977). Exact stochastic simulation of coupled chemical reactions. *J. Phys. Chem.* 81: 2340–2361.

Gillespie, D. T. (1979). A pedestrian approach to transitions and fluctuations in simple non-equilibrium chemical systems. *Physica A* 95: 69–103.

Hallett, M. B. (1989). The unpredictability of cellular behavior: trivial or fundamental importance to cell biology? *Perspect. Biol. Med.* 33: 110–119.

Hallett, M. B., and Pettit, E. J. (1997). Stochastic events underlie Ca^{2+} signaling in neutrophils. *J. Theoret. Biol.* 186: 1–6.

Hanusse, P., and Blanche, A. (1981). A Monte Carlo method for large reaction-diffusion systems. *J. Chem. Phys.* 74: 6148–6153.

Ishihara, A., Segall, J. E., Block, S. M., and Berg, H. C. (1983). Coordination of flagella on filamentous cells of *Escherichia coli. J. Bacteriol.* 155: 228–237.

Jasuja, R., Keyoung, J., Reid G. P., Trentham, D. R., and Khan S. (1999). Chemotatic responses of *Escherichia coli* to small jumps of photo released L-aspartate. *Biophys. J.* 76: 1706–1719.

Kingston, R. E., and Green, M. R. (1994). Modeling eukaryotic transcriptional activation. *Curr. Biol.* 4: 325–332.

Ko, M. S. H. (1991). A stochastic model for gene induction. *J. Theoret. Biol.* 153: 181–194.

Kohn, K. W. (1998). Functional capabilities of molecular network components controlling the mammalian G1/S cell cycle phase transition. *Oncogene* 16: 1065–1075.

Lamb, T. D. (1994). Stochastic simulation of activation in the G-protein cascade of phototransduction. *Biophys. J.* 67: 1439–1454.

Liu, J. D., and Parkinson, J. S. (1989). Role of CheW protein in coupling membrane receptors to the intracellular signaling system of bacterial chemotaxis. *Proc. Natl. Acad. Sci. USA* 86: 8703–8707.

Liu, Y., Levit, M., Lurz, R., Surette, M. G., and Stock, J. B. (1997). Receptor-mediated protein kinase activation and the mechanism of transmembrane signaling in bacterial chemotaxis. *EMBO J.* 16: 7231–7240.

Liu, Z., and Papadopoulos, K. D. (1996). A method for measuring bacterial chemotaxis parameters in a microcapillary. *Biotechnol. Bioeng.* 51: 120–125.

Mansour, M. M., and Baras, F. (1992). Microscopic simulation of chemical systems. *Physica A* 188: 253–276.

Matias, M. A. (1995). On the effects of molecular fluctuations on models of chemical chaos. *J. Chem. Phys.* 102: 1597–1606.

Mayani, H., Dragowska, W., and Lansdorp, P. M. (1993). Lineage commitment in human hematopoiesis involves asymmetric cell division of multipotent progenitors and does not appear to be influenced by cytokines. *J. Cell. Physiol.* 157: 579–586.

McAdams, H. H., and Arkin, A. (1997). Stochastic mechanisms in gene expression. *Proc. Natl. Acad. Sci. USA* 94: 814–819.

Mendes, P. (1993). GEPASI: a software package for modeling the dynamics, steady states and control of biochemical and other systems. *Comput. Appl. Biosci.* 9: 563–571.

Mendes, P., Barreto, J. M., Gomes, A. V., and Freire, A. P. (1990). Methodology for simulation of metabolic pathways and calculation of control coefficients. In *Control of Metabolic Processes,* A. Cornish-Bowden and M. L. Cardenas (eds.) pp. 221–224. Plenum Press, New York.

Mesibov, R., Ordal, G. W., and Adler, J. (1973). The range of attractant concentration for bacterial chemotaxis and the threshold and size of response over this range. *J. Gen. Physiol.* 62: 203–223.

Mochly-Rosen, D. (1995). Localization of protein kinases by anchoring proteins: a theme in signal transduction. *Science* 268: 247–251.

Morton-Firth, C. J. (1998). Stochastic Simulation of Cell Signaling Pathways. Ph.D. thesis, Cambridge University, Cambridge. Available for download from http://www.zoo.cam.ac.uk/zoostaff/morton.

Morton-Firth, C. J., and Bray, D. (1998). Predicting temporal fluctuations in an intracellular signaling pathway. *J. Theoret. Biol.* 192: 117–128.

Morton-Firth C. J., Shimizu, T. S., and Bray, D. (1999). A free-energy-based stochastic simulation of the Tar receptor complex. *J. Mol. Biol.* 286: 1059–1074.

Sauro, H. M. (1986). Control Analysis and Simulation of Metabolism. Ph.D. thesis, Oxford Polytechnic, Oxford.

Sauro, H. M. (1993). SCAMP: a general-purpose simulator and metabolic control analysis program. *Comput. Appl. Biosci.* 9: 441–450.

Scharf, B. E., Fahrner, K. A., Turner, L., and Berg, H. C. (1998). Control of direction of flagellar rotation in bacterial chemotaxis. *Proc. Natl. Acad. Sci. USA* 95: 201–206.

Smetters, D. K., and Zador, A. (1996). Synaptic transmission—noisy synapses and noisy neurons. *Curr. Biol* 6: 1217–1218.

Sorribas, A., Curto, R., and Cascante, M. (1995). Comparative characterization of the fermentation pathway of *Saccharomyces cerevisiae* using biochemical systems-theory and metabolic control analysis—model validation and dynamic behavior. *Math. Biosci.* 130: 71–84.

Spudich, J. L., and Koshland, J. E. Jr. (1975). Quantitation of the sensory response in bacterial chemotaxis. *Proc. Natl. Acad. Sci. USA* 72: 710–713.

Stark, S. M., Neurock, M., and Klein, M. T. (1993). Strategies for modeling kinetic interactions in complex mixtures: Monte Carlo algorithms for MIMD parallel architectures. *Chem. Eng. Sci.* 48: 4081–4096.

Stock, J. B. (1994). Adaptive responses in bacterial chemotaxis. In *Regulation of Cellular Signal Transduction Pathways by Desensitization and Amplification,* D. R. Sibley and M. D. Houslay, eds. Wiley, Chichester.

Tang, Y., and Othmer, H. G. (1994). A G protein-based model of adaptation in *Dictyostelium discoideum. Math. Biosci.* 120: 25–76.

Turner, J. S. (1977). Discrete simulation methods for chemical kinetics. *J. Phys. Chem* 81: 2379–2408.

Valkema, R., and Van Haastert, P. J. M. (1994). A model for cAMP-mediated cGMP response in *Dictyostelium discoideum. Mol. Biol. Cell* 5: 575–585.

Van Steveninck, R. D. R., and Laughlin, S. B. (1996). Light adaptation and reliability in blowfly photoreceptors. *Int. J. Neural Systems* 7: 437–444.

Williams, N. G., Paradis, H., Agarwal, S., Charest, D. L., Pelech, S. L., and Roberts, T. M. (1993). Raf-1 and p21(v-ras). cooperate in the activation of mitogen-activated protein kinase. *Proc. Natl. Acad. Sci. USA* 90: 5772–5776.

Zheng, Q., and Ross, J. (1991). Comparison of deterministic and stochastic kinetics for nonlinear systems. *J. Chem. Phys.* 94: 3644–3648.

Zimmerman, S. B., and Minton, A. P. (1993). Macromolecular crowding: biochemical, biophysical and physiological consequences. *Annu. Rev. Biophys. Biomol. Struct.* 22: 27–65.

Zuker, C. S., and Ranganathan, R. (1999). The path to specificity. *Science* 283: 650–561.

10 Analysis of Complex Dynamics in Cell Cycle Regulation

John J. Tyson, Mark T. Borisuk, Kathy Chen, and Bela Novak

During its division cycle, a cell replicates all its components and divides them more or less evenly between two daughter cells, so that each daughter inherits the machinery and information necessary to repeat the process. In eukaryotes, the genome is accurately replicated during a distinct phase of the cell cycle (called the S phase, for DNA synthesis), and the sister DNA molecules are carefully separated in a later phase of the cycle (M phase, for mitosis). The timing of S and M phases is controlled by a complex network of biochemical reactions, based on cyclin-dependent kinases and their associated proteins. In this chapter we analyze the simplest case of cell cycle control, the meiotic and mitotic cycles of frog eggs, to illustrate how computational models can make connections between molecular mechanisms and the physiological properties of a cell. The mechanism of this "simple" case is already so complex that informal biochemical intuition cannot reliably deduce how it works. Our intuition must be supplemented by precise mathematical and computational tools. To this end, we convert the mechanism of cell cycle control into differential equations and show how the dynamic properties of the control system (e.g., stable steady-state or oscillation) can be manipulated by changing parameter values. In this way, cell cycle control can be switched from an embryonic pattern (S-M) to a somatic pattern (G1-S-G2-M), by changing levels of gene expression. This theme (gene expression → system dynamics → cell physiology) is a new paradigm for molecular cell biology, and this sort of mathematical modeling will become increasingly important as we try to bridge the gap between the behavior of intact cells and the molecular circuitry of disassembled control systems.

10.1   The "Last Step" of Computational Molecular biology

The fundamental goal of every molecular cell biologist is to understand a certain aspect of cell physiology in terms of its underlying molecular mechanism and ultimately in terms of the genes that encode the molecules. In making this connection, several levels of explanation can be identified: gene sequence → protein sequence → protein structure → protein function → protein–protein interactions. Once the molecular network has been worked out, there is one final step that is often neglected: from mechanism to physiology. This is the domain of biochemical kinetics: not the classical lore of Michaelis–Menten rate laws and competitive versus noncompetitive inhibitors, but the modern theory of dynamic systems, bifurcations, phase-space portraits, and numerical simulations.

   In this chapter we use a specific problem—control of meiotic and mitotic division cycles in frog egg development—to illustrate how we make connections from mechanism to physiology. We describe the problems involved, the analytical and computational tools available, and the kind of information sought. Other fine examples of the "last step" include the work of Bray et al. (1993) on bacterial chemotaxis, Oster and colleagues on molecular motors (Peskin 1993, Elston and Oster 1997, Elston et al. 1998), Goldbeter et al. (1990) and Sneyd et al. (1995) on calcium oscillations and waves, McAdams and Shapiro (1995) on λ-phage infection, and Sherman (1997) on pulsatile insulin secretion.

In a recent issue of *Trends in Biochemical Science,* Dennis Bray (1997) described the crisis in biological reductionism, "What are we to do with the enormous cornucopia of genes and molecules we have found in living cells? . . . [One] road to salvation is through the keyboard of a computer. . . . If we can use computer-based graphical elements to understand the world of protein structure, why should we not do the same for the universe of cells? The data are accumulating and the computers are humming. What we lack are the words, the grammar and the syntax of the new language." In fact, the language exists. Its "words" are rate laws, its "sentences" are differential equations, its "meanings" can be deduced by dynamic systems theory, as we will show.

10.2 The Cell Division Cycle in Frog Eggs

The basic physiology of frog egg development is summarized in figure 10.1. In the ovary, immature, diploid egg cells grow very large and stop in the G2 phase, with replicated DNA. In response to a hormone, progesterone, a batch of immature oocytes are stimulated to mature. They enter meiosis I, when homologous chromosomes are separated to the daughter nuclei, and one of the nuclei is discarded as a polar body. Then they go immediately into meiosis II, when all the replicated chromosomes are lined up by the spindle with sister chromatids attached to opposite poles. Here the mature egg stops, at metaphase of meiosis II, awaiting fertilization. Fertilization triggers a large calcium wave that propagates through the egg and, among other things, releases the maternal nucleus from its metaphase arrest. The sister chromatids are separated; one set is discarded, and the other set fuses with the sperm nucleus to create a diploid zygote. The zygote now enters a series of twelve rapid, synchronous, mitotic division cycles (alternating phases of DNA synthesis and mitosis) to create a hollow ball of 4096 cells called a *blastula.* At this point in time, called the *midblastula transition* (MBT), dramatic changes occur: gaps (G1 and G2) are introduced between the phases of DNA synthesis and mitosis, subsequent cell cycles are no longer synchronous, and mRNA synthesis from the zygotic nucleus becomes essential for further development. (Before MBT, the fertilized egg was directing protein synthesis from its store of maternal mRNA.)

The synchronous mitotic cycles of the early embryo are driven by cytoplasmic reactions that periodically activate a substance called *M-phase-promoting factor* (MPF). Even in the

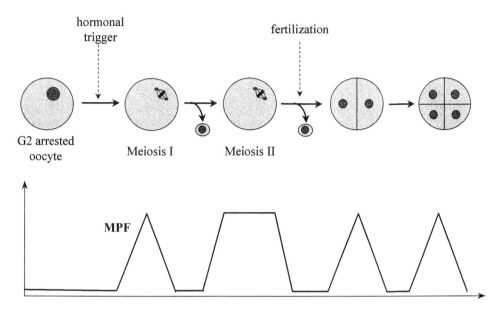

Figure 10.1
MPF fluctuations in the meiotic and mitotic cycles of frog eggs. See text for explanation. Adapted from Murray and Hunt (1993).

presence of drugs that block DNA synthesis (Kimelman et al. 1987) or spindle formation (Gerhart et al. 1984), the cytoplasmic clock continues driving periodic MPF activity. The clock continues ticking even in enucleated embryos (Hara et al. 1980) and in cell-free cytoplasmic extracts of frog eggs (Murray and Kirschner 1989).

This brief description of frog egg development, from immature oocyte to MBT, identifies several characteristic states of the division control system:

• G2 arrest (immature oocyte)—blocked before the G2 to M transition, with low MPF activity

• metaphase arrest (mature oocyte)—blocked before the meta to anaphase transition, with high MPF activity

• spontaneous cycles (early embryo)—rapid, autonomous oscillations of MPF activity, driving alternating phases of DNA synthesis and mitosis

• regulated cycles (after MBT)—slow, asynchronous division cycles containing all four phases (G1, S, G2, M) and coordinated to the chromosome cycle at three checkpoints (in G1, G2, and metaphase)

10.3   The Molecular Machinery of MPF Regulation

In 1988 it was discovered that MPF is a cyclin-dependent kinase (Lohka et al. 1988), that is, a heterodimer of Cdk1 (the kinase subunit) and cyclin B (the regulatory subunit). Active MPF phosphorylates several protein targets, thereby initiating the events of mitosis. MPF activity in frog eggs is controlled in two ways: by the availability of cyclin B, and by phosphorylation of the Cdk1 subunit. During spontaneous cycling, cyclin B is synthesized steadily and degraded in bursts at the end of mitosis (Evans et al. 1983). Its degradation is regulated by the anaphase-promoting complex (APC) (Glotzer et al. 1991, King et al. 1996, Murray 1995), a group of enzymes that attach ubiquitin moieties to cyclins. Polyubiquitinated cyclins are then rapidly degraded by proteasomes. APC activity, in turn, is activated by MPF (Felix et al. 1990), so the regulatory system contains a negative feed-back loop: Cyclin activates MPF, MPF activates the APC, and the APC destroys cyclin. MPF activation of APC has a significant time delay, suggesting that the signal is indirect. The intermediary enzyme may be Plx1, a pololike kinase in *Xenopus* (Descombes and Nigg 1998).

In addition, Cdk1 activity is affected by activatory phosphorylation at a threonine residue (T167) and inhibitory phosphorylation at a tyrosine residue (Y15) (Solomon 1993). T167 is rapidly phosphorylated by an unregulated kinase (CAK). More important, the enzymes that phosphorylate and dephosphorylate Y15 (Wee1 and Cdc25, respectively) are themselves regulatory enzymes (Izumi et al. 1992, Kumagai and Dunphy 1992, Smythe and Newport 1992). MPF activates Cdc25 and inhibits Wee1, so the control system also contains positive feedback loops: MPF helps its friend, Cdc25, and hinders its enemy, Wee1.

Figure 10.2, which summarizes the molecular details just described, represents a hypothesis for the cell division control system in frog eggs. If correct, it should account for the basic physiological properties of egg cell cycles, and it should make testable predictions. To study the mechanism, we use standard principles of biochemical kinetics to convert the box-and-arrow diagram into a set of rate equations (table 10.1) (Novak and Tyson 1993b). Our model consists of nine differential equations for free cyclin, the four phosphorylation states of MPF, and the four regulatory enzymes: Wee1, Cdc25, APC, and IE (the "intermediary enzyme" between MPF and APC). To describe the rates of the component steps, we need 26 parameters (rate constants and Michaelis constants). Many of the rate constants can be estimated (Marlovits et al. 1998) from ingenious kinetic experiments in frog egg extracts, as indicated in table 10.2. The other parameters in the model are assigned values consistent with certain experimental observations (e.g., the 15–20-min time lag between MPF activation and maximal degradation of cyclin B by APC). Given the parameter values in table 10.2, the rate equations in table 10.1 can be solved numerically, and

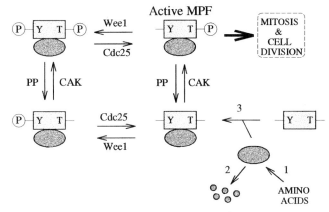

A. The Phosphorylation States of MPF

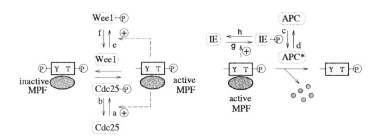

B. Positive Feedback Loops   **C.** Negative Feedback Loop

Figure 10.2
A molecular mechanism for M-phase control in frog eggs. See text for explanation. Key: rectangle, Cdk1 subunit; oval, cyclin subunit; left phosphorylation site ("Y" for tyrosine) is inhibitory, right phosphorylation site ("T" for threonine) is activatory. From Borisuk and Tyson (1998) and adapted from Novak and Tyson (1993b).

they exhibit stable limit cycle oscillations with a period of about 60 min (figure 10.3), which is consistent with observations of oscillating extracts (Murray and Kirschner 1989).

This model has been studied numerically by Novak and colleagues (Marlovits et al. 1998; Novak and Tyson 1993b), and the results compared carefully with many sorts of experiments on extracts and intact embryos. Surprisingly, nearly the same model is also successful in accounting for many physiological and genetic properties of cell division in fission yeast (Novak and Tyson 1995). Such detailed comparisons are necessary to make the connection between mechanism and physiology. They determine the scope and reliability of the model and suggest new ways to test the mechanism. Where they fail, they uncover weaknesses or oversights, and suggest ways to improve and expand our understanding of control systems.

Table 10.1
Novak–Tyson model of MPF regulation in frog eggs

$$\frac{d[\text{Cyclin}]}{dt} = k_1 - k_2[\text{Cyclin}] - k_3[\text{Cdk1}][\text{Cyclin}]$$

$$\frac{d[\text{YT}]}{dt} = k_{\text{pp}}[\text{MPF}] - (k_{\text{wee}} + k_{\text{cak}} + k_2)[\text{YT}] + k_{25}[\text{PYT}] + k_3[\text{Cdk1}][\text{Cyclin}]$$

$$\frac{d[\text{PYT}]}{dt} = k_{\text{wee}}[\text{YT}] - (k_{25} + k_{\text{cak}} + k_2)[\text{PYT}] + k_{\text{pp}}[\text{PYTP}]$$

$$\frac{d[\text{PYTP}]}{dt} = k_{\text{wee}}[\text{MPF}] - (k_{\text{pp}} + k_{25} + k_2)[\text{PYTP}] + k_{\text{cak}}[\text{PYT}]$$

$$\frac{d[\text{MPF}]}{dt} = k_{\text{cak}}[\text{YT}] - (k_{\text{pp}} + k_{\text{wee}} + k_2)[\text{MPF}] + k_{25}[\text{PYTP}]$$

$$\frac{d[\text{Cdc25P}]}{dt} = \frac{k_a[\text{MPF}]([\text{total Cdc25}] - [\text{Cdc25P}])}{K_a + [\text{total Cdc25}] - [\text{Cdc25P}]} - \frac{k_b[\text{PPase}][\text{Cdc25P}]}{K_b + [\text{Cdc25P}]}$$

$$\frac{d[\text{Wee1P}]}{dt} = \frac{k_e[\text{MPF}]([\text{total Wee1}] - [\text{Wee1P}])}{K_e + [\text{total Wee1}] - [\text{Wee1P}]} - \frac{k_f[\text{PPase}][\text{Wee1P}]}{K_f + [\text{Wee1P}]}$$

$$\frac{d[\text{IEP}]}{dt} = \frac{k_g[\text{MPF}]([\text{total IE}] - [\text{IEP}])}{K_g + [\text{total IE}] - [\text{IEP}]} - \frac{k_h[\text{PPase}][\text{IEP}]}{K_h + [\text{IEP}]}$$

$$\frac{d[\text{APC*}]}{dt} = \frac{k_c[\text{IEP}]([\text{total APC}] - [\text{APC*}])}{K_c + [\text{total APC}] - [\text{APC*}]} - \frac{k_d[\text{AntiIE}][\text{APC*}]}{K_d + [\text{APC*}]}$$

$$k_{25} = V'_{25}([\text{total Cdc25}] - [\text{Cdc25P}]) + V''_{25}[\text{Cdc25P}]$$

$$k_{\text{wee}} = V'_{\text{wee}}[\text{Wee1P}] + V''_{\text{wee}}([\text{total Wee1}] - [\text{Wee1P}])$$

$$k_2 = V'_2([\text{total APC}] - [\text{APC*}]) + V''_2[\text{APC*}]$$

Source: Borisuk and Tyson (1998).
Notes: We write a differential equation for the concentration or relative activity of each of the nine regulatory proteins in figure 10.2 The k_i's and V_i's are rate constants and the K_j's are Michaelis constants. The total concentrations of Cdk1, Cdc25, Wee1, IE, and APC are all taken to be constant.

Although such comparisons are crucial elements of the last step of computational molecular biology, they are best left to the original literature, lest they bog us down in too many technical details. Instead, we want to show how dynamic systems theory can be used to analyze the model and show how it works.

10.4   Isolating the Positive and Negative Feedback Loops

In much the same way that a molecular biologist can simplify the full MPF machinery of frog egg extracts by genetic manipulations that isolate one or two reactions, a theoretician can simplify the mathematical model by isolating a subset of equations. For instance, we can concentrate on the positive feedback loops by studying the simplified mechanism in

Table 10.2
Parameter values used in the Novak–Tyson model

These parameters are dimensionless:			
K_a / [total Cdc25]	0.1	K_e / [total Wee1]	0.1
K_b / [total Cdc25]	1.0	K_f / [total Wee1]	1.0
K_c / [total APC]	0.01	K_g / [total IE]	0.01
K_d / [total APC]	1.0	K_h / [total IE]	0.01
These paramenters have units of min^{-1}:			
k_1	0.01	k_{pp}	0.004
k_3 [Cdk1]	0.5	k_a [Cdk1] / [total Cdc25]	2.0
V_2' [total APC]	0.005	k_b [PPase] / [total Cdc25]	0.1
V_2'' [total APC]	0.25	k_c [total IE] / [total APC]	0.13
V_{25}' [total Cdc25]	0.017	k_d [Anti IE] / [total APC]	0.13
V_{25}'' [total Cdc25]	0.17	k_e [Cdk1] / [total Wee1]	2.0
V_{wee}' [total Wee1]	0.01	k_f [PPase] / [total Wee1]	0.1
V_{wee}'' [total Wee1]	1.0	k_g [Cdk1] / [total IE]	2.0
k_{cak}	0.64	k_h [PPase] / [total IE]	0.15

Source: Marlovits et al. (1998).

figure 10.4 described by three differential equations. Because the association of Cdk1 and cyclin B subunits is a fast reaction, the concentration of free cyclin B will be very small, as long as Cdk1 is in excess. This eliminates the differential equation for free cyclin B, and we are left with two nonlinear ordinary differential equations for active MPF (M) and total cyclin (Y) (Novak and Tyson 1993a):

$$\frac{dM}{dt} = k_1 - (k_2' + k_2''M^2)M + (k_{25}' + k_{25}''M^2)(Y - M) - k_{wee}M$$

$$\frac{dY}{dt} = k_1 - (k_2' + Kk_2''M^2)Y.$$

Such equations are usually studied by graphical methods in the phase plane (figure 10.5). The two curves plotted there are: (1) the cyclin balance curve (sigmoidal shape), where cyclin synthesis is exactly balanced by degradation:

$$Y = \frac{k_1}{k_2' + k_2''M^2}$$

and (2) the MPF balance curve (N-shaped), where MPF synthesis and activation are exactly balanced by degradation and inactivation

$$Y = M\left[1 + \frac{k_{wee} + k_2' + k_2''M^2}{k_{25}' + k_{25}''M^2}\right] - \frac{k_1}{k_{25}' + k_{25}''M^2}$$

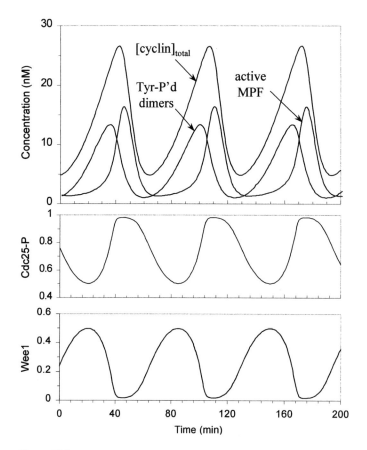

Figure 10.3
Numerical solution of the rate equations in table 10.1, using the parameter values in table 10.2 (Marlovits et al. 1998).

Wherever these balance curves intersect, the control system is in a time-independent steady state. A steady state may be stable or unstable with respect to small perturbations. These ideas are best understood from the figures (figure 10.5). As the parameters of the model change, the balance curves move on the phase plane, and the number and stability of steady states change. Several characteristic types of phase plane portraits appear:

• a G2-arrested state with lots of cyclin but little MPF activity, because most of the dimers are inhibited by tyrosine phosphorylation (figure 10.5A)

• an M-arrested state with lots of active MPF (figure 10.5B)

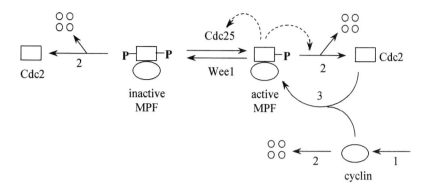

L = [cyclin B monomer] Q = [Cdk1 monomer]
M = [active MPF] P = [inactive MPF]
C = [total Cdk1] = $M+P+Q$ = constant
Y = [total cyclin] = $M+P+L$ = variable

$$F_2(M) = k_2' + k_2'' M^2$$
$$F_{25}(M) = k_{25}' + k_{25}'' M^2$$

$$\frac{dL}{dt} = k_1 - F_2(M) \cdot L - k_3 \cdot L \cdot (C + L - Y)$$
$$\frac{dM}{dt} = k_3 \cdot L \cdot (C + L - Y) - F_2(M) \cdot M$$
$$+ F_{25}(M) \cdot (Y - L - M) - k_{\text{wee}} \cdot M$$
$$\frac{dY}{dt} = k_1 - F_2(M) \cdot Y$$

Figure 10.4
A simplified mechanism emphasizing the role of the positive feedback loops. Symbols are the same as in figure 10.2. From Tyson et al. (1997).

• an oscillatory state exhibiting sustained, periodic fluctuations of MPF activity and total cyclin protein (figure 10.5C)

• coexisting stable states of G2 arrest and M arrest (figure 10.5D)

Portraits (5A), (5B), and (5C) are reminiscent of the major stages of frog egg development: immature oocyte, mature oocyte, and spontaneous oscillations prior to MBT, respectively. Is portrait (5D) observable in frog eggs or extracts? Are other distinct portraits possible? How many?

Next, for the sake of argument, let us ignore the positive feedback effects and concentrate on the negative feedback loop (figure 10.6), described by three differential equations:

$$\frac{dM}{dt} = k_1 - [k_2'(1 - A) + k_2''A]M \ ,$$

$$\frac{dE}{dt} = \frac{k_3 M(1 - E)}{J_3 + 1 - E} - \frac{k_4 E}{J_4 + E} \ ,$$

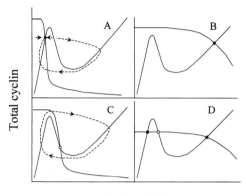

Active MPF

Figure 10.5
Phase-plane portraits for the simplified mechanism in figure 10.4. The sigmoidal curve is the cyclin-balance curve, and the N-shaped curve is the MPF-balance curve. (A) Stable steady state with most MPF inactive. The dashed trajectories indicate that the steady state is "excitable," i.e., a small addition of active MPF will drive the cell into mitosis. (B) Stable steady state with most MPF active. (C) Unstable steady state surrounded by a limit cycle oscillation (dashed trajectory). (D) Bistability: two stable steady states separated by an unstable saddle point. From Tyson et al. (1997).

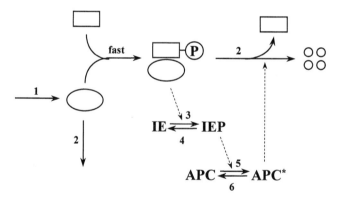

Figure 10.6
A simplified mechanism emphasizing the time-delayed negative feedback loop. Symbols are the same as in figure 10.2. Adapted from Goldbeter (1991).

$$\frac{dA}{dt} = \frac{k_5 E(1-A)}{J_5 + 1 - A} - \frac{k_6 A}{J_4 + A}.$$

A similar system of equations was first proposed by Goldbeter (1991) as a "minimal cascade" model of the cell cycle. The three equations cannot be reduced to two without destroying the possibility of oscillations (Griffith 1968), so phase plane analysis is impossible. Instead, we will characterize this model by a "one-parameter bifurcation diagram." Holding all other parameters fixed, we vary a single parameter, k_1 (the rate constant for cyclin synthesis), and plot the steady-state MPF activity as a function of k_1 (figure 10.7A). We find that the steady-state MPF activity varies smoothly from low to high levels, but there is a region of k_1 values where the steady state is unstable and is surrounded by stable limit cycle oscillations. The limit cycles connect smoothly to the locus of steady states at the points where the steady state changes stability. This sort of connection between steady states and limit cycles is called a *Hopf bifurcation,* and is a common feature of dynamic systems. (For a simple description of these bifurcations, see the appendix in Borisuk and Tyson, 1998.) Figure 10.7A illustrates that at a point of Hopf bifurcation, the amplitude of oscillation goes to zero, but the period of oscillation is finite.

Next we can vary a second parameter, k_2'' (the rate constant for cyclin degradation when APC is active), to produce a "two-parameter" bifurcation diagram (figure 10.7B), which shows how a locus of Hopf bifurcations delineates in parameter space the region of spontaneous oscillations from the region of stable steady states.

10.5   Bifurcation Diagrams for the Positive Feedback Subsystem and the Full Model

The positive feedback model (figure 10.4) shows a variety of different bifurcations: saddle-node (SN), Hopf (H), and saddle-loop (SL). Figure 10.8 shows schematically how these bifurcation points are related to each other. The SN loci come together at a cusp point, and the SN, H, and SL come together at Takens–Bogdanov (TB) bifurcation points. The basic structure of the diagram in figure 10.8 is well known for dynamic systems with positive feedback (autocatalysis) (Guckenheimer 1986). It is distinctly different from the diagram in figure 10.7B for the negative feedback loop.

When we consider the full model, the bifurcation diagram (figure 10.9A; see also plate 28) clearly contains elements from figures 10.7B and 10.8. In addition, because of the way the feedback loops couple together, the bifurcation diagram contains some new elements. Consider the one-parameter bifurcation diagram (figure 10.9B), starting with limit cycles issuing from the upper Hopf bifurcation associated with the negative feedback loop. As k_1 decreases, this family of limit cycles disappears, not at the other Hopf bifurcation (as in figure 10.7A), but at a novel bifurcation point, called a *saddle-node-loop* (SNL). In con-

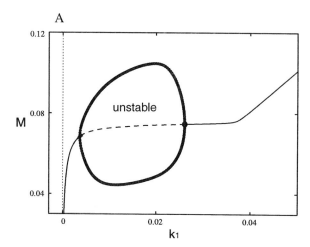

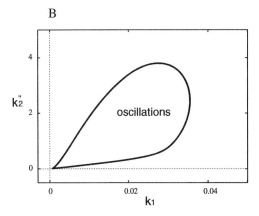

Figure 10.7
Bifurcation diagrams for the negative feedback model in figure 10.6. Basal parameter values: $k_1 = 0.01$, $k_2' = 0.005$, $k_2'' = 0.5$, $k_3 = 2$, $k_4 = 0.15$, $k_5 = 0.13$, $k_6 = 0.13$, $J_3 = J_4 = J_5 = 0.01$, $J_6 = 1$. (A) One-parameter bifurcation diagram. As a function of the rate constant for cyclin synthesis (k_1), we plot the steady state activity of MPF (thin line, solid and dashed) and the maximum and minimum MPF activities (thick lines) during limit cycle oscillations. The steady state loses stability at points of Hopf bifurcation. (B) Two-parameter bifurcation diagram. The locus of Hopf bifurcation points is traced out as the rate constants for cyclin synthesis (k_1) and degradation (k_2'') change.

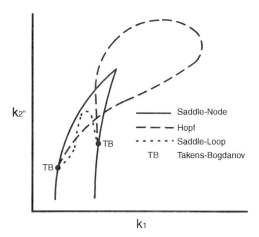

Figure 10.8
Schematic bifurcation diagram for the positive feedback model in figure 10.4. The true bifurcation diagram is much more complicated; see Borisuk and Tyson (1998).

trast to the characteristics of a Hopf bifurcation, at an SNL the amplitude of oscillation remains finite while the frequency of oscillation goes to zero (period goes to infinity). Figure 10.9A has many other curious features, which are discussed in detail in Borisuk and Tyson (1998).

10.6 Biological Significance of Bifurcation Analysis

Dynamic systems, like the MPF control system in frog eggs, consist of nonlinear differential equations, one for each time-dependent variable in the molecular mechanism. These equations contain many parameters necessary to describe the rates of the chemical reactions comprising the mechanism. Once the parameters are specified, the differential equations can be solved and the solutions characterized as to the number and stability of steady states, existence of limit cycles, etc. Such characteristics of the solution set can change only at points of bifurcation, which are special combinations of parameter values where (for instance) steady states change stability, and limit cycles come or go. To get a global view of the behavioral complexity of the control system, we would like to subdivide parameter space into regions where the dynamic characteristics of the system are unchanging. Each region would then be bounded by a bifurcation set, which separates regions of different characteristics. How can we envision this global structure in a parameter space of 10–100 dimensions?

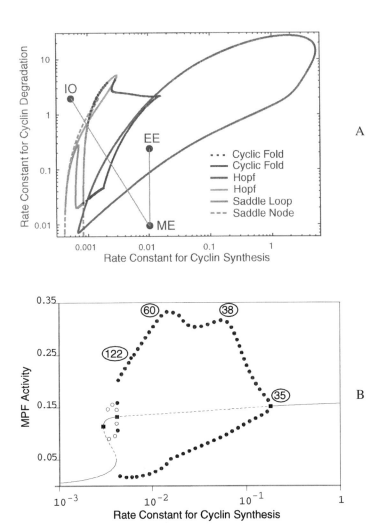

Figure 10.9
Bifurcation diagrams for the full model in figure 10.1. (A) Two-parameter bifurcation diagram. See text for explanation. (B) One-parameter bifucation diagram. A stable limit-cycle solution (period = 35 min) appears at a super-
critical Hopf bifurcation at large k_1 (rate constant for cyclin synthesis = 0.02). The closed circles indicate the
maximum and minimum activities of MPF during an oscillation. As k_1 decreases, the limit cycle grows in amplitude and period (circled numbers), until at $k_1 = 0.004$, this branch of oscillations ends at a saddel-node-loop bifurcation (finite amplitude, infinite perod). For $0.003 < k_1 < 0.004$, there exist several branches of small-amplitude
limit cycles (both stable and unstable) associated with the Takens–Bogdanov bifurcations of the system. From
Borisuk and Tyson (1998). (See plate 28.)

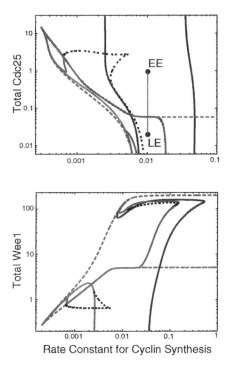

Figure 10.10
Alternative slices through the "avocado." See text for explanation. From Borisuk and Tyson (1998). (See plate 29.)

Suppose parameter space were only three dimensions, and imagine that the area of interest resembles an avocado. That is, the skin of the fruit represents the most extreme parameter values possible. As we poke inside the skin, we find only a smooth, uniform, rather boring fleshy region corresponding to a unique, stable, steady-state solution of the differential equations. But were we to slice through the fruit with a sharp knife, we would find a hard kernel inside, where something more interesting is happening. Slicing through this kernel in several directions, we can begin to make out the structure of the seed and the embryo it contains.

Now, if the avocado were 26-dimensional, we would have to slice it up in many different directions and look at the cross-sections of the kernel inside. Figure 10.9A is one such cross-section. Figure 10.10 (see also plate 29) shows two more. Superficially all three cross-sections look different, but when examined closely, they present the same basic dynamic structures from different perspectives. They are no more or less different than portraits of a human face in profile and frontal views. The kernel of our model consists of a cusp-shaped region of multiple steady states, criss-crossed by a locus of Hopf and SL bifurcations (generated by the positive feedback loops in the equations), and abutting a second region of oscillatory solutions bounded by Hopf bifurcations (generated by the time-delayed negative feedback loop).

The parameters in our model include rate constants and total enzyme concentrations that are under genetic control. Changing the expression of certain genes will change the parameter values of our model and move the control system across bifurcation boundaries into regions of qualitatively different behavior. For instance, in figure 10.9A we see three large regions of parameter space (low, steady MPF activity; oscillatory MPF activity; and high, steady MPF activity) that correspond to the three basic physiological states of frog eggs (immature oocyte, IO; autonomous mitotic cycles of the early embryo, EE; and mature egg, ME). Moving from one physiological state to another requires a change in parameter values, presumably driven by changes in gene expression. For instance, oocyte maturation, induced by progesterone, could easily be driven by upregulation of cyclin synthesis and downregulation of cyclin degradation, which would carry the control system along the developmental path IO→ME in figure 10.9A. If fertilization then upregulates the APC, path ME→EE, the control system will move into the region of spontaneous oscillations driven by the negative feedback loop.

At MBT, spontaneous oscillations cease, implying that the mitotic control system moves out of the region of spontaneous limit cycle oscillations. This might occur, for instance, if Cdc25 were downregulated: see path between EE and LE (late embryo) in figure 10.10A. The system crosses an SNL bifurcation. The period of the spontaneous cycles increases abruptly, and then the oscillations are replaced by a stable steady state of low MPF activity (i.e., G2 arrest). Such cells could be induced to enter the M phase and divide if Cdc25 were temporarily upregulated to move the control system across the SNL bifurcation into the oscillatory region. Then, if the Cdc25 level were to come back down after division, the daughter cells would halt at the G2 checkpoint, waiting for another signal to divide.

10.7   Experimental Predictions

The bifurcation diagram in figure 10.9B makes several unexpected predictions that can be tested experimentally. The bifurcation parameter, k_1 (rate constant for cyclin synthesis), can be controlled in frog egg extracts by a protocol developed by Andrew Murray. Endogenous mRNAs are eliminated, and exogenously synthesized cyclin mRNA is added to the extract. By changing the concentration of cyclin mRNA in the extract, we can change k_1. Figure 10.9B predicts for this preparation:

- There will be a stable steady state of low MPF activity at low k_1.
- There will be a stable steady state of high MPF activity at high k_1.

• The control system will undergo a Hopf bifurcation, creating stable, small-amplitude MPF oscillations, as k_1 decreases from high values.

• As k_1 decreases further, the amplitude of MPF oscillations will increase rapidly (less mRNA generates more protein), while its period will not change much (a slower rate of accumulation of cyclin does not lengthen the cycle proportionally).

• When k_1 gets small, the period of MPF oscillations will become progressively longer, although its amplitude won't change much, until the oscillatory state is replaced by a stable steady state of low MPF activity (SNL bifurcation).

• Just below the SNL bifurcation, when we are still inside the cusp, the system should exhibit bistable behavior (coexistence of two stable steady states).

10.8   Conclusion

The reproduction of eukaryotic cells is controlled by a complex network of biochemical interactions among cyclin-dependent kinases and their associated proteins (cyclins, phosphatases, inhibitors, and proteolytic enzymes). This network, we claim, is a dynamic system with many qualitatively different modes of behavior. The control system can be switched between different modes by changes in parameter values that drive the system across bifurcations. This bifurcations create and remove steady states and limit cycles, which correlate (in ways yet to be determined) with the insertion and removal of checkpoints in the cell cycle. If this point of view is correct, then to understand the physiology of cell division, we must first understand the bifurcation properties of the differential equations describing the underlying molecular control mechanism. Making these connections and understanding not only the cell cycle but any other aspect of cellular regulation will require new analytical and computational tools, and new ways of thinking about molecular biology. Dynamic systems theory provides a general framework for taking the last step of computational biology, from molecular mechanisms to the physiology of living cells.

Acknowledgments

Our work on cell cycle control is supported by the National Science Foundation of the United States (DBI-9724085) and Hungary (T-022182), and by the Howard Hughes Medical Institute (75195-512302).

References

Borisuk, M. T., and Tyson, J. J. (1998). Bifurcation analysis of a model of mitotic control in frog eggs. *J. Theor. Biol.* 195: 69–85.

Bray, D. (1997). Reductionism for biochemists: how to survive the protein jungle. *Trends Biochem. Sci.* 22: 325–326.

Bray, D., Bourret, R. B., and Simon, M. I. (1993). Computer simulation of the phosphorylation cascade controlling bacterial chemotaxis. *Mol. Biol. Cell* 4: 469–482.

Descombes, P., and Nigg, E. A. (1998). The polo-like kinase Plx1 is required for M phase exit and destruction of mitotic regulators in *Xenopus* egg extracts. *EMBO J.* 17: 1328–1335.

Elston, T., and Oster, G. (1997). Protein turbines I: the bacterial flagellar motor. *Biophys. J.* 73: 703–721.

Elston, T., Wang, H., and Oster, E. (1998). Energy transduction in ATP synthase. *Nature* 391: 510–514.

Evans, T., Rosenthal, E. T., Youngbloom, J., Distel, D., and Hunt, T. (1983). Cyclin: a protein specified by maternal mRNA in sea urchin eggs that is destroyed at each cleavage division. *Cell* 33: 389–396.

Felix, M.-A., Labbe, J.-C., Doree, M., Hunt, T., and Karsenti, E. (1990). Triggering of cyclin degradation in interphase extracts of amphibian eggs by cdc2 kinase. *Nature* 346: 379–382.

Gerhart, J., Wu, M., and Kirschner, M. W. (1984). Cell cycle dynamics of an M-phase-specific cytoplasmic factor in *Xenopus laevis* oocytes and eggs. *J. Cell Biol.* 98: 1247–1255.

Glotzer, M., Murray, A. W., and Kirschner, M. W. (1991). Cyclin is degraded by the ubiquitin pathway. *Nature* 349: 132–138.

Goldbeter, A. (1991). A minimal cascade model for the mitotic oscillator involving cyclin and cdc2 kinase. *Proc. Natl. Acad. Sci. U.S.A.* 88: 9107–9111.

Goldbeter, A., Dupont, G., and Berridge, M. J. (1990). Minimal model for signal-induced Ca^{2+} oscillations and for their frequency encoding through protein phosphorylation. *Proc. Natl. Acad. Sci. U.S.A.* 87: 1461–1465.

Griffith, J. S. (1968). Mathematics of cellular control processes. I. Negative feedback to one gene. *J. Theor. Biol.* 20: 202–208.

Guckenheimer, J. (1986). Multiple bifurcation problems for chemical reactors. *Physica* 20D: 1–20.

Hara, K., Tydeman, P., and Kirschner, M. W. (1980). A cytoplasmic clock with the same period as the division cycle in *Xenopus* eggs. *Proc. Natl. Acad. Sci. U.S.A.* 77: 462–466.

Izumi, T., Walker, D. H., and Maller, J. L. (1992). Periodic changes in phosphorylation of the *Xenopus* cdc25 phosphatase regulate its activity. *Mol. Biol. Cell* 3: 927–939.

Kimelman, D., Kirschner, M., and Scherson, T. (1987). The events of the midblastula transition in *Xenopus* are regulated by changes in the cell cycle. *Cell* 48: 399–407.

King, R. W., Deshaies, R. J., Peters, J. M., and Kirschner, M. W. (1996). How proteolysis drives the cell cycle. *Science* 274: 1652–1659.

Kumagai, A., and Dunphy, W. G. (1992). Regulation of the cdc25 protein during the cell cycle in *Xenopus* extracts. *Cell* 70: 139–151.

Lohka, M. J., Hayes, M. K., and Maller, J. L. (1988). Purification of maturation-promoting factor, an intracellular regulator of early mitotic events. *Proc. Natl. Acad. Sci. U.S.A.* 85: 3009–3013.

Marlovits, G., Tyson, C. J., Novak, B., and Tyson, J. J. (1998). Modeling M-phase control in *Xenopus* oocyte extracts: the surveillance mechanism for unreplicated DNA. *Biophys. Chem.* 72: 169–184.

McAdams, H. H., and Shapiro, L. (1995). Circuit simulation of genetic networks. *Science* 269: 650–656.

Murray, A. (1995). Cyclin ubiquitination: the destructive end of mitosis. *Cell* 81: 149–152.

Murray, A., and Hunt, T. (1993). *The Cell Cycle. An Introduction.* W. H. Freeman, New York.

Murray, A. W., and Kirschner, M. W. (1989). Cyclin synthesis drives the early embryonic cell cycle. *Nature* 339: 275–280.

Novak, B., and Tyson, J. J. (1993a). Modeling the cell division cycle: M-phase trigger, oscillations and size control. *J. Theor. Biol.* 165: 101–134.

Novak, B., and Tyson, J. J. (1993b). Numerical analysis of a comprehensive model of M-phase control in *Xenopus* oocyte extracts and intact embryos. *J. Cell Sci.* 106: 1153–1168.

Novak, B., and Tyson, J. J. (1995). Quantitative analysis of a molecular model of mitotic control in fission yeast. *J. Theor. Biol.* 173: 283–305.

Peskin, C. S., Odell, G. M., and Oster, G. F. (1993). Cellular motions and thermal fluctuations— the Brownian ratchet. *Biophys. J.* 65: 316–324.

Sherman, A. (1997). Calcium and membrane potential oscillation in pancreatic β cells. In *Case Studies in Mathematical Modeling: Ecology, Physiology and Cell Biology,* H. G. Othmer, F. R. Adler, M. A. Lewis, and J. C. Dallon, eds., pp. 199–217. Prentice-Hall, Upper Saddle River, N.J.

Smythe, C., and Newport, J. W. (1992). Coupling of mitosis to the completion of S phase in *Xenopus* occurs via modulation of the tyrosine kinase that phosphorylates p34^{cdc2}. *Cell* 68: 787–797.

Sneyd, J., Keizer, J., and Sanderson, M. (1995). Mechanisms of calcium oscillations and waves: a quantitative analysis. *FASEB J.* **9**: 1463–1472.

Solomon, M. J. (1993). Activation of the various cyclin/cdc2 protein kinases. *Curr. Opin. Cell Biol.* 5: 180–186.

Tyson, J. J., Chen, K. C., and Novak, B. (1997). The eukaryotic cell cycle: molecules, mechanisms and mathematical models. In *Case Studies in Mathematical Modeling: Ecology, Physiology and Cell Biology,* H. G. Othmer, F. R. Adler, M. A. Lewis, and J. C. Dallon, eds., pp. 127–147. Prentice-Hall, Upper Saddle River, N.J.

11 Simplifying and Reducing Complex Models

Bard Ermentrout

11.1 Introduction

The skill and innovation of experimental biologists has enabled them to obtain more and more information about their preparations. This presents a challenge to anyone who wishes to create a mathematical model or simulation of a given system. At what point does the model cease to have explanatory value, having become too complex to do anything more than simulate a variety of parameter values and initial conditions for a system? Often the models that are proposed have dozens of parameters, many of which may not be known for the particular system studied. Furthermore, the complexity of the models makes it difficult to study sensitivity to parameters and initial conditions, even on fast computers. This difficulty is magnified when the systems that are simulated are inherently stochastic, for then one can ask: How many sample paths are enough? In addition to the computational difficulties and the incomplete knowledge of parameters, there is also the issue interpreting the output of the model. Large simulations produce a tremendous amount of output, much of which is likely to be useless for the particulars of a given experiment. Finally, for many biological systems, one can only guess at the mechanism. A simulation does not tell you how dependent the behavior of a system is on the particular form of the mechanism that you have chosen. Only a detailed analysis can tell you that, and for complex models and simulations this is difficult at the very least and usually is impossible.

These concerns lead many modelers to propose simplified models. The multiple time and space scales in biological systems force one to generally focus on some particular level, often using heuristic approximations for the finer details, which are neglected. The hope is that the knowledge of the finer levels will suggest the correct heuristics for modeling and understanding the higher levels of the system. This principle of reductionism has served the sciences well.

In physics, where there are well-defined laws, it is often possible to use a microscopic description to derive a macroscopic model. One does not usually treat every water molecule separately when studying flow through a tube; instead, one uses the Navier–Stokes equations that describe the macroscopic behavior of fluids. There are several reasons for this. One is the obvious computational complexity. The second and equally important reason is that the behavior of individual molecules cannot be studied nor is it of interest.

Modeling a biological system should be viewed in the same manner. If one wants to study the behavior of large-scale electrical activity in a region of the cortex, then is it necessary to include dozens of channels into each nerve cell? Since the properties of individual neurons are not known, the choice of parameters for each of these channels is at best an average of similar systems. Thus, one approach has been to use simplified models. The difficulty with simplifying is knowing how to choose the simple model, how to connect it to the measured phenomena, and how to determine its relationship to the details that lurk beneath. Is it possible to construct simplified models that are *quantitative* rather than just qualitative?

In this chapter we describe some methods that allow one to derive quantitatively correct models from more complex systems. We will attempt to show that the assumptions of certain dynamic properties and extreme differences in time and space scales can be exploited to produce simple and often analyzable models.

11.1.1 Averaging

In many models, there are diverse time scales and space scales. Extreme time differences often allow one to assume that during the changes in a fast quantity, a slow quantity can be considered constant. On the other hand, if the slow quantity is of interest, then the fluctuations or changes in the fast quantity occur so fast that the slow quantity only "sees" the mean or average of them. This intuitively appealing idea can be made rigorous by a procedure called *averaging*. Related to this is the idea of "spatial averaging" in which one assumes that individual influences of one system on another are small but manifold. Thus, one averages over these; such an average or approximation is often called the *mean-field approximation*.

Both averaging and mean-field approaches provide a method for making complex models much simpler without losing quantitative details. I will describe a number of these ideas through a variety of examples.

11.2 Master Equations

Many biological problems can be cast as continuous time jump processes in which a system switches from one state to the next. The simplest example would be the random opening and closing of a channel. Another example is the growth of an actin polymer in the presence of some cell signal where the states are the length of the polymer. Any system involving rates, such as chemical reactions, can be viewed as a continuous time process with jumps. If the probability of jumping from one state to the next depends only on the current state (that is, it has no dependence on history), then the process is called a *Markov process*.

In the simplest case, a quantity of interest takes on n values and the rate of jumping between states k and j is given by $M_{kj}dt$ where dt is the time interval (figure 11.1). This is easily simulated, and from the simulation it is possible to obtain the probabilities of being in any given state. The key to analyzing this is to assume that one does the simulation over a long period of time or that we are interested in the average properties of the process. Let

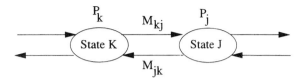

Figure 11.1
Diagram showing the probability of jumping between state k and j in a Markov chain.

$P_j(t)$ be the probability of being in state j at time t. From the figure, the change in probability P_j is the net influx from P_k minus the outflux from P_j, i.e.,

$$\Delta P_j = M_{kj} P_k \Delta t - M_{jk} P_j \Delta t.$$

Taking the limit leads to the differential equation:

$$\frac{dP_j}{dt} = \sum_{k=1}^{n} (MkjP_k - M_{jk} P_j) \equiv \sum_{k=1}^{n} A_{kj} P_k . \tag{11.1}$$

This equation is called the *master equation.* (It is also called the *Chapman–Kolmogorov equation.*) The initial probabilities must sum to one and then it is a simple matter of solving this equation. The steady-state probabilities arc the desired result. The matrix A has a zero eigenvalue since columns sum to zero. The eigenvector (normalized, of course) corresponding to this eigenvalue gives the steady-state probabilities. Thus, by looking at this mean or averaged system, the steady-state behavior is found.

A more interesting situation arises when one of the variables involved itself obeys a differential equation. Consider, for example, the simple model:

$$\frac{dx}{dt} = z - x , \tag{11.2}$$

where z randomly switches between two states, 0,1 at some constant rate, r. Since x cannot instantly follow the variable, z, we expect the probability $P(x = X,t)$ to be a continuous function of X. Figure 11.2 shows histograms for the distribution of x for this system when z randomly switches back and forth between the two states 0 and 1 at a rate r. For fast changes, we expect x to hover around the mean value of z, which is $\frac{1}{2}$. This is seen in figure 11.2 (upper left). However, for slow rates, $r = 0.2$ for example, x has enough time to move to $z = 0$ or $z = 1$ so that the distribution is strongly bimodal, as seen in the figure. There is a transition from a unimodal to a bimodal distribution as the rate is decreased. Once again, we can appeal to the master equation to understand this transition.

Suppose that z has n states and that

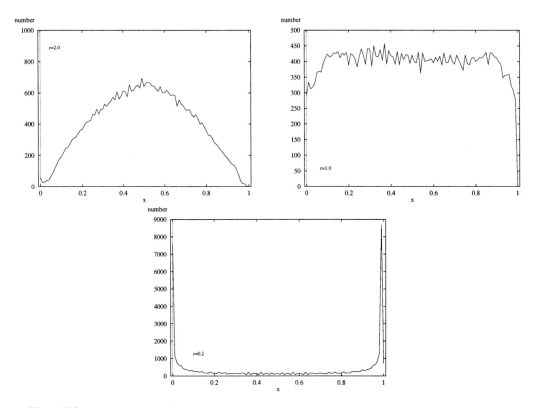

Figure 11.2
Histograms for the state variable x in the two-state Markov model as the rate varies from fast $r = 2$ to intermediate, $r = 1$ to slow $r = 0.2$.

$$\frac{dx}{dt} = f(x, z).$$

Furthermore, suppose that the rates of transition from state j to state k may be x-dependent, $M_{jk}(x)$. Then we need to define the probabilities of being in state j with $x = X$, $P_j(X,t)$. Averaging once again over many sample paths, one finds that

$$\frac{\partial P_j(X,t)}{\partial t} = -\frac{\partial}{\partial X}[f(X, j)P_j(X,t)]$$

$$+ \sum_{k=1}^{n} [M_{kj}(X)P_k(X,t) - M_{jk}(X)P_j(X,t)].$$

Applying this to our system, we obtain:

$$\frac{\partial P_0}{\partial t} = -\frac{\partial}{\partial X}(-XP_0) - rP_0 + rP_1$$

$$\frac{\partial P_1}{\partial t} = -\frac{\partial}{\partial X}[(1 - X)P_1] - rP_1 + rP_0.$$

Let $P(x,t) = P_0(x,t) + P_1(x,t)$ be the probability of $X = x$ at time t and in either state 0 or state 1. The steady-state distribution for this turns out to be:

$$P(x = X) = CX^{r-1}(1 - X)^{r-1} ,$$

where C is a normalization constant. From this it is clear that if $r < 1$, that is, the rates are slow, then the distribution is singular, with asymptotes at 0 and 1. On the other hand, for fast rates, $r > 1$, the distribution is continuous, vanishes at the ends, and is peaked at $X = \frac{1}{2}$. As the transitions become infinitely fast, the distribution becomes infinitely narrow and centered at $X = \frac{1}{2}$, the limit obtained by solving equation (11.2) using the average $z = \frac{1}{2}$. Averaging over sample paths has provided a quantitative method for completely characterizing this simple two-state model coupled to a differential equation.

11.2.1   Application to a Model for Fibroblast Orientation

I will next consider an example from cell biology. Here we consider the effect of density on the behavior of fibroblasts in culture (Edelstein-Keshet and Ermentrout 1991). It is known that fibroblasts will align with each other with a probability that depends on their relative angles of motion. In dense cultures, many patches of parallel cells can be found. In sparse cultures, there are few aligned patches. Thus, we would like to understand how the density affects the appearance of parallel patches. We can treat the formation of arrays as a stochastic process in which cells will switch from one angle to another when they interact with each other. Spatial distribution is obviously quite important, but here we will neglect it and simply study the alignment problem. We distinguish cells that are moving (free) from cells that are attached (bound) to the culture dish surface. The following types of interactions are allowed:

1. There will be random shifts in alignment from an angle θ to a neighboring angle $\theta \pm \phi$ at rate r for free cells.

2. Cells that bump into other bound cells of orientation θ' will reorient to that angle and stick at a rate that is linearly related to the fraction of cells with angle θ' and the difference between the angles, $K(\theta - \theta')$.

3. Bound cells will free up at a rate γ and free cells become bound at a rate η.

Since the area of the culture dish does not change over time and only the number of cells does, we will treat this number N as the parameter. For simplicity assume m discrete angles. Let $F_j(t)$ denote the probability of a cell being free and having an orientation θ_j and let $B_j(t)$ be the probability of a cell being bound and having an orientation θ_j. The master equation yields:

$$\frac{dF_j}{dt} = r(F_{j-1} - 2F_j + F_{j+1}) - N\sum_i K(i-j)B_i F_j$$

$$+ \gamma B_j - \eta F_j$$

$$\frac{dB_j}{dt} = \eta F_j - \gamma B_j + N\sum_i K(i-j)B_j F_i \ .$$

The number of cells, N, appears in the equations because the larger the number of cells, the greater the rate of collision between bound and free cells. One solution to this equation is $F_j = f$ and $B_j = b$ where both f and b are constant. The stability of the uniform state can be analyzed as a function of N, and the critical value of N can be found for a spatial instability. This can then be compared with the stochastic simulations and the biological system. Related reductions of discrete probabilistic simulations are given in Ermentrout and Edelstein-Keshet (1993).

11.2.2 Mean-Field Reduction of a Neural System

The master equation essentially averages over many sample paths in a system and leads to a set of equations for the probability of any given state of the system. Another way to average a system that has intrinsic randomness is the so-called mean-field approximation. Here one uses that idea that there are many interacting subunits that are tightly coupled and so the effect is one of the average of all of them (e.g., Cowan 1968 or van Vreeswijk and Sompolinsky 1998). In this section, we apply the ideas of averaging over units to reduce a random system of many units to a deterministic system with just two equations. The original model attempts to understand the effect that cortical processing has on thalamic input in the somatosensory whisker barrel area of the rat (Pinto et al. 1996). The model barrel contains N_e excitatory cells and N_i inhibitory cells. Each cell is coupled to the other cells with a randomized weight (positive for excitatory and negative for inhibitory) and each cell receives excitatory input from N_T thalamic neurons, again with randomized weights. The model is cast as an integral equation. If V is the voltage, then the probability of firing an action potential is $P(V)$. a typical example is

$$P(V) = 1/(1 + e^{-\beta(V - V_{thr})}).$$

The parameter β determines the sharpness of the probability and V_{thr} is the voltage at which there is a 50% chance of firing a spike.

The potential of the kth excitatory neuron is:

$$V_k^e(t) = \sum_j w_{ee}^{kj} \int_0^t P_e[V_j^e(s)]e^{-(t-s)/\tau_e}\, ds$$

$$+ \sum_j w_{te}^{kj} \int_0^t P_e[V_j^T(s)]e^{-(t-s)/\tau_e}\, ds$$

$$- \sum_j w_{ie}^{kj} \int_0^t P_i[V_j^i(s)]e^{-(t-s)/\tau_i}\, ds,$$

where the probabilities of thalamic cell spikes are given as inputs. A similar equation holds for the inhibitory neurons. Since the cells fire asynchronously and the weights are randomly distributed about some mean, an obvious approximation is to sum up all the excitatory (inhibitory) cells and then divide by their number to arrive at a mean potential. The question is, what equations does this mean satisfy? A preliminary transformation to a differential equation eases the analysis. Let

$$S_j^e(t) = \int_0^t P_e[V_j^e(s)]e^{-(t-s)/\tau_e}\, ds$$

This is the weighted average of the firing probabilities, taking into account the past history. Then

$$\tau_e \frac{dS_k^e}{dt} + S_k^e = P_e(V_k^e)$$

$$V_k^e = \sum_j \left(w_{ee}^{kj} S_j^e + w_{te}^{kj} S_j^T - w_{ie}^{kj} S_j^i \right).$$

We define the mean field

$$S_e = \frac{1}{N_e} \sum_k S_k^e ,$$

and find that we have to deal with sums of the form

$$\frac{1}{N_e} \sum_k P_e \left(\sum_j w_{ee}^{kj} S_j^e + ... \right).$$

The only approximation that is made is to interchange the nonlinearity and the sum. That is, we approximate

$$\frac{1}{N_e} \sum_k P_e(V_k^e) \approx P_e \left(\frac{1}{N_e} \sum_k V_k^e \right).$$

Since P is a sigmoid curve and the most sensitive behavior occurs near the linear portion, this is not a bad approximation. In fact, one can even compensate for this approximation by modifying the averaged sigmoid. The final result is a set of equations for the mean of the activity in a whisker barrel:

$$\tau_e S'_e = P_e (w_{ee} S^e + w_{te} S^T - w_{ie} S^i),$$

with analogous equations for the inhibition. The thalamic drive, S^T, is given as input to the model and

$$w_{ee} = \frac{1}{N_e} \sum_{jk} w_{ee}^{jk}.$$

The other weights have similar definitions. The reduced mean-field model turns out to mimic the behavior of the full system extremely well and in fact was used to make experimental predictions to some novel forms of stimulus in later work (Pinto et al. 1996).

11.3 Deterministic Systems

There are many ways to reduce the dimension and complexity of deterministic systems. Again, all of these exploit the differences in time scales or space scales. Here we will concentrate on techniques for reduction by exploiting time scales. The idea is very simple—if some process occurs over a much faster time scale than the one of interest, look at the mean of the fast process and eliminate it from the system or conversely, hold the slow processes as constant parameters and study the fast process. This approach has many advantages:

• A smaller system arises.

• There is a direct quantitative connection to the detailed system.

• Computations are difficult when there are drastically different time scales; eliminating the fast or slow variables makes the computations easier.

• Separation of time scales enables one to understand the mechanisms underlying the behavior.

11.3.1 Reduction of Dimension by Using the Pseudo-Steady State

Most biochemists are aware of this method since it is the idea that is used to derive the Michaelis–Menten equations (Murray 1989). Recall that one wants to model the following reaction:

$$S + E \underset{k_{-1}}{\overset{k_1}{\longleftrightarrow}} SE, \quad SE \overset{k_2}{\longrightarrow} P + E.$$

Letting $s = [S]$, $e = [E]$, $c = [SE]$, $p = [P]$, we get

$$s' = -k_1 es + k_{-1} c, \quad e' = -k_1 es + (k_{-1} + k_2)c$$

$$c' = k_1 es - (k_{-1} + k_2)c, \quad p' = k_2 c.$$

Clearly, p is obtained by integrating c. Summing the equations for c and e implies that $c' + e' = 0$ so that $c + e = e_0$. We have immediately eliminated two of the four equations and must only look at a two-dimensional system. By introducing dimensionless variables and parameters, we can eliminate one more equation. As an aside, I want to point out the incredible usefulness of rendering a model dimensionless; this allows one to compare parameters that prior to scaling had different units and so could not be compared. This type of comparison enables one to see where small and large parameters lie and thus direct the reduction of dimension. Murray (and many others) introduce the following scaled parameters and variables:

$$\tau = k_1 e_0 t, \quad u(\tau) = s(t)/s_0, \quad v(\tau) = c(t)/e_0$$

$$\lambda = \frac{k_2}{k_1 s_0}, \quad K = \frac{k_{-1} + K_2}{k_1 s_0}, \quad \varepsilon = \frac{e_0}{s_0},$$

where s_0 is the initial substrate. With this scaling, the equations are

$$\frac{du}{dt} = -u + (u + K - \lambda)v, \quad \varepsilon \frac{dv}{d\tau} = u - (u + K)v.$$

Since the amount of enzyme compared with the initial substrate is generally very small, the parameter ε is quite small. The approximation is to set this to zero. This implies $v = u/(u+K)$ so that we finally end up with

$$\frac{du}{d\tau} = -u + (u + K - \lambda)\frac{u}{u + K} = -\lambda \frac{u}{u + K}.$$

There is only one differential equation to solve!

In general, if the equations have the form:

$$\varepsilon \frac{dX}{dt} = F(X,Y), \quad \frac{dY}{dt} = G(X,Y), \, X \in R^m, Y \in R^n,$$

then one sets $\varepsilon = 0$, solves for $X(Y)$, and plugs this back into the equation for Y. We call X the fast variable(s) and Y the slow variable(s). This approach has been used by Rinzel (1985) to reduce the four-dimensional Hodgkin–Huxley equations to a two-dimensional model. The advantages of a two-dimensional model are that the powerful methods of phase-plane analysis can be used to understand the equations. Abbott and Kepler (1990)

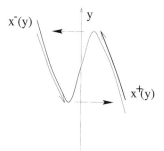

Figure 11.3
Illustrations of a relaxation oscillator. As the slow variable creeps up the right branch, it encounters the knee and must jump to the left branch. Then the slow variable moves down until the lower knee is reached. A jump to the right branch occurs.

generalize Rinzel's ideas and describe a way to simplify arbitrary neural models. In fact, they reduce the six-dimensional Connor–Stevens model to just two dimensions using the differences in time scales.

In many interesting cases, the equation $F(X,Y) = 0$ may have several branches of solutions for a given value of Y. This can lead to oscillations and other phenomena. For example, the classic van der Pol oscillator can be written as:

$$\varepsilon \frac{dx}{dt} = x(1 - x^2) - y, \quad \frac{dy}{dt} = x.$$

It is easiest to visualize what is going on by looking in the (x,y) plane. Figure 11.3 shows a plot of the curve, $y = x(1 - x^2)$ and $x = 0$, which are, respectively, where $dx/dt = 0$ and $dy/dt = 0$. There is one equilibrium point at the origin and it is unstable. Setting $\varepsilon = 0$, we must solve $x - x^3 = y$. For $|y| < 2/3\sqrt{3}$, this equation has three roots. Suppose we take the most positive root, $x^+(y) > 0$ (figure 11.3). Since $x^+(y) > 0$, this means that y will increase. As long as y is below the maximum of the cubic, we can continue to define $x^+(y)$. However, once y exceeds this maximum, the only possible root is $x^-(y)$, and the system "jumps" to the left branch. However, $x^-(y)$ is negative, so y begins to decrease until it reaches the minimum of the cubic and x must jump to the right branch. Thus, an oscillation is formed. This type of model is called a *relaxation oscillator* and is the basis for many models of excitable and oscillatory media. (See Murray 1989 for many examples applied to biology.) The approximate period of the oscillation is given by the amount of time it takes to traverse the two outer branches of the cubic. If we rescale time, introducing $T = t\varepsilon$, then the van der Pol model becomes:

$$\frac{dx}{dt} = x(1 - x^2) - y \quad \frac{dy}{dT} = \varepsilon x.$$

Setting $\varepsilon = 0$, we see that y is essentially constant, so that x will tend to one of up to three fixed points of the equation $x(1 - x^2) = y$. As we saw above, for values of y near zero, there are up to three roots. The middle fixed point is unstable, so only the outer ones are possible steady states. This multiplicity of stable states coupled with the slow movement of another variable (in this case y) has led to an oscillation that jumps between two fixed points. What if the "x" system is two-dimensional or higher? What if the two stable states are not just fixed points? In fact, in many neural models, there are slow conductances. Holding these constant often leads to subsystems that have as stable behavior both a fixed point and a rhythmic solution. Thus, jumping between states produces behavior that is alternately nearly constant and rapidly oscillating. Such behavior is called *bursting* and is important in many physiological and biochemical systems. Wang and Rinzel (1995) use this decomposition of slow and fast systems to characterize and dissect a variety of bursting neurons. While the details are more complicated, the basic ideas are exactly as described in the example illustrated in figure 11.3.

11.3.2 Averaged Equations

A related way to reduce complexity is to again exploit time scales and average, keeping only the slow variables. This approach is valid if the fast variables have a unique solution when the slow variables are held fixed. (In the previous section, the nonuniqueness of the roots of the fast equation was exploited in order to produce an oscillation.) That is, suppose that as the slow variables move around, at any given value of them there is only a single stable behavior of the fast system. Then we can apply the methods of this section.

The basic idea can be gleaned by using the scaled time equations:

$$\frac{dX}{dT} = F(X, Y), \quad \frac{dY}{dT} = \varepsilon G(X, Y).$$

For small ε the slow variables Y do not change too much, so we can treat them as constant in the fast system. Then the fast system will evolve in time until it reaches a (possibly time-dependent) steady state, $X_{ss}(t, Y)$, which depends on the "parameters" Y. We plug this into the Y equation and obtain the slow system:

$$\frac{dY}{dT} = \varepsilon G[X_{ss}(t, Y), Y] \tag{11.3}$$

There are two typical situations. Either the fast system tends to a fixed point so that X_{ss} is independent of time; or the fast system is periodic, in which case X_{ss} is periodic in time. In the former, the slow equation is reduced to:

$$\frac{dY}{dT} = \varepsilon G[X_{xx}(Y), Y] ,$$

which is an equation for the variables Y. We have eliminated X altogether. If the fast system tends to a time-dependent solution, such as a periodic solution, then we can use the "averaging theorem," which states that the behavior of

$$\frac{dY}{dT} = \varepsilon G(Y, T) \text{ with } G(Y, T + P) = G(Y, T)$$

is close to the behavior of the averaged system

$$\frac{d\bar{Y}}{dT} = \varepsilon \frac{1}{P} \int_0^P G(\bar{Y}, T) \, dT.$$

Formally, we can perform the same averaging even if the stimulus is not periodic but varies rapidly compared with ε. In any case, the resulting averaged equation depends only on the slow variable(s), Y. There are many useful applications of this method; we present three of them.

Hebbian Learning   The main mechanism for unsupervised learning is called Hebb's rule. This states that the strength of a connection between two neurons depends on the coincidence of activity between them. We will use the concepts of this subsection to derive a standard model for the growth of weights to a single neuron. Let $V(t)$ be the potential of the neuron and let there be n inputs, $I^1(t),\ldots,I_n(t)$ with corresponding weights, $w^1,\ldots,w_n$. Thus at any given time, the potential satisfies

$$\tau \frac{dV}{dt} = -V + \sum_j w_j I_j(t).$$

The inputs can change with time, but we assume that the change is not very fast, so that

$$V(t) \approx \sum_j w_j I_j(t).$$

Hebb's rule in this case simply says that the weight, w_j, will grow, depending on the correlation between the input and the output, $V(t)I_j(t)$. Thus,

$$\frac{dw_j}{dt} = \varepsilon F[w_j, V(t)I_j(t)]$$

would represent a change in the weights. The simplest case is linear growth:

$$\frac{dw_j}{dt} = \varepsilon V(t)I_j(t) = \varepsilon \sum_k I_k(t)I_j(t)w_k.$$

Since the growth rate is small, $\varepsilon \ll 1$, we average over the inputs and obtain:

$$\frac{dw_j}{dt} = \varepsilon \sum_k C_{kj} w_k ,$$

where C is the zero time correlation matrix of the inputs. The solution to a linear constant coefficient differential equation has the following form:

$$\mathbf{w}(t) = \sum_k \phi_k e^{\varepsilon \lambda_k t} ,$$

where λ_k are the eigenvalues of C and ϕ_k the corresponding eigenvectors. If C is a positive matrix, the Perron–Frobenius theorem implies that the maximal eigenvalue of C has non-negative components. Thus the eigenvector corresponding to the largest eigenvalue will grow fastest and the weights will tend to be proportional to it. This is the basis of "principal component analysis" used in statistics. By using averaging, we have reduced a complex problem to a simple exercise in linear algebra.

Neural Network from Biophysics   The typical biophysical model of a network of neurons has the form

$$C \frac{dV_j}{dt} = -I_{\mathrm{ion},j} + I_{\mathrm{appl},j} - \sum_i g_{ij}^{\mathrm{ex}} s_{ij}^{\mathrm{ex}}(t)(V_j - E_{\mathrm{ex}}) - \sum_i g_{ij}^{\mathrm{in}} s_{ij}^{\mathrm{in}}(t)(V_j - E_{\mathrm{in}}) ,$$

where the synapses satisfy

$$\frac{ds_{ij}^l}{dt} = \alpha_{ij}^l(V_i)(1 - s_{ij}^l) - \beta_{ij}^l s_{ij}^l .$$

Here we have divided the synapses into excitatory and inhibitory although there can be more populations. Furthermore, the model cells are only single compartments but could easily be extended to more. How can we connect this to the simple notion of a neural network or so-called firing rate model in a quantitative manner?

Suppose that the synapses are *slow* relative to the dynamics of the membrane. (This is not in general a good assumption, but for N-methyl-d-aspartate and γ-aminobutyric acid-B synapses, it may not be unreasonable.) Then the synapses can be held "frozen" as parameters and the fast dynamics of the neurons will reach a steady state of either a fixed point or periodic behavior. Let G_j^{ex} (G_j^{in}) be the total excitatory (inhibitory) conductance into cell j. Then the behavior of the cell is determined by only these two parameters, so that $V(t) = V_{ss}(t; Ge^{\mathrm{ex}}, G^{\mathrm{in}})$. Suppose that the time course of a synapse depends only on the presynaptic neuron. That is, $s_{ij}(t) = s_i(t)$. Then,

$$\frac{ds}{dt} = \alpha[V_{ss}(t; G^{ex}, G^{in})](1-2) - \beta s.$$

We now average this and get equations for s:

$$\frac{ds}{dt} = \overline{\alpha}(G^{ex}, G^{in})(1-s) - \beta s.$$

To close the system, we note that G^{ex}, G^{in} are linear sums of the synaptic gating functions, s. In Ermentrout (1994) we applied this technique to a specific model and obtained a neural netlike set of equations. More recently, Chen et al. (1998) derived a greatly simplified model for wave propagation in a thalamic network that produces behavior nearly identical to a large-scale biophysically based computational model and *which can be solved in closed form.*

Weakly Coupled Oscillators   As a final example, I describe a technique that has been used in a variety of biological systems that are made up of many coupled rhythmic elements. The idea is that if the elements are rhythmic, then one of the most important quantities is the relative timing of the rhythms. For example, in quadruped locomotion, the relative phases of the four limbs are precisely what define the gait of the animal. Similarly, in the swimming behavior of the lamprey, the key question is how the metachronal traveling wave is formed. This wave controls the timing of the muscular contractions required for locomotion. Since phase is the relevant experimentally measured quantity, models that consider only the phases as variables make sense. The problem is how to connect an abstract phase model with a concrete mechanistic model for the oscillators. In this section, we show how this is done by using averaging.

The model equations take the form

$$\frac{dX_j}{dt} = F_j(X_j) + \varepsilon G_j(X_1, \dots X_N).$$

We assume that each subsystem (which can represent many variables) has a periodic solution with all the periods identical. (They need not be identical, but we will absorb the differences in the G_j.) Furthermore, in order to make this mathematically rigorous, the strength of interaction between the oscillators must be "weak." A natural question to ask is: What does "weak" mean? There is no simple answer to this; heuristically it means that the coupling is not so strong as to distort the oscillation other than to shift its phase. That is, the wave forms of the oscillating components should not be changed much by coupling; however, they can be shifted.

Let $X_j^0(t)$ be the oscillation in the absence of coupling. Then it can be proven (Kuramoto 1984) that if ε is sufficiently small, the solution to the coupled system is $X_j^0(t + \theta_j)$ where θ_j is a phase shift. The phase shift satisfies a set of equations of the form

$$\frac{d\theta_j}{dt} = \omega_0 + \varepsilon H_j(\theta_1 - \theta_j, \ldots, \theta_N - \theta_j).$$

The parameter ω_0 is the uncoupled frequency and the functions, H_j, are $2\pi\omega$ periodic in each of their arguments. The beauty of this reduction is that even if the basic oscillators lie in a high-dimensional space, there is only one equation for each oscillator. Furthermore, the experimentally important variable, the phase or timing, is the only state variable in the reduced system. Computing H_j from a given model or experimental preparation is difficult, but there is a nice heuristic way to do it. Suppose for simplicity that each oscillator is coupled only to a single component of the other oscillators. For example, if these represent neurons, then the coupling appears only in the somatic voltage equation. An easily computable (both experimentally and numerically) quantity for an oscillator is the phase response curve (PRC). Choose a visible event in your oscillator, e.g., the appearance of a spike. Since the system oscillates, the spike occurs at $t = 0, P, 2P, \ldots$ where P is the period of the oscillator. At $t = t_o < P$, give the oscillator a brief stimulus. This will change the time of the next spike to, say, P_0. The PRC, χ, is defined as

$$\chi(t_0) = \frac{P - P_0}{P}.$$

Thus the PRC measures the fraction of the period lost or gained as a function of the timing of the perturbation. The function H_j is now easily computed:

$$H_j(\theta_1 - \theta_j, \ldots) = \frac{1}{P} \int_0^P \chi(t)[G_j(X_1^0(t + \theta_1 - \theta_j, \ldots))]_1 \, dt \,,$$

where $[G]_1$ is the first component (and by assumption, the only nonzero one) of G. That is, we just average the effects of the coupling against the effect of a perturbation.

One of the easiest applications is to look at a pair of mutually coupled identical oscillators and ask whether they synchronize:

$$\frac{d\theta_1}{dt} = \omega + \varepsilon H(\theta_2 - \theta_1)$$

$$\frac{d\theta_2}{dt} = \omega + \varepsilon H(\theta_1 - \theta_2).$$

Let $\phi = \theta_2 - \theta_1$ and we get

$$\frac{d\phi}{dt} = \varepsilon[H(-\phi) - H(\phi)].$$

The zeros of the right-hand side are the allowable phase shifts between the two oscillators. In particular $\phi = 0$ is always a root. It is stable if $H'(0) > 0$. Thus, one can ask what kinds of interactions lead to stable synchrony. This is a research topic currently of much interest and there are many papers on the subject.

This approach has been used to study the swim generator of the lamprey, where the circuitry is not known in detail. In spite of this lack of detailed knowledge, with very few assumptions, it is possible to suggest experiments and determine mechanisms by simply analyzing the general structure of chains of phase models. A review of this approach can be found in Williams and Sigvardt (1995).

11.4   Discussion and Caveats

Averaging is an intuitively appealing method for reducing the complexity of biological systems that operate on many different time and space scales. There are mathematical techniques that can be brought to bear on complex models that enable us to reduce the dimensionality of the system and to connect one level of detail with another. There are additional mathematical methods that can be used to reduce models to lower dimensions. For example, normal form analysis (Hoppensteadt and Izhikevich 1997) has been used to reduce complex neural networks to simple "canonical models," low-dimensional equations that have all of the properties of their higher-dimensional relatives.

We have presented a number of examples illustrating the application of these simplifying techniques to specific biological systems. The advantages of the simplified models are quite obvious. At the very least, the simulations involve far fewer equations and parameters; at best, a complete analysis of the behavior becomes possible. In many cases the information that they give is quantitative and thus the simplified models can and have been used to suggest specific experiments.

However, there are several questions that arise in this approach to modeling. When are important details being neglected? This is a very difficult question to answer since there will always be aspects that a simplified model cannot tell you, but the more detailed model will. This is an obvious consequence of simplifying. The more dangerous problem is that sometimes these neglected details have consequences for the reduced model, and then the behavior of the reduced model will be misleading. How can you know when important details are missing? The only way to know is to do simulations. A negative answer can never be definitive since there may always be a different instance that you haven't tried which produces a qualitatively different answer to your simplification.

Finally, in the extreme case, the models can become so simplified and far from the original experimental system as to be phenomenological or metaphorical. This is a trap into which many modelers fall, particularly those who neglect to study any of the underlying biology. An entire decade of "catastrophe theory" models illustrates this phenomenon. The best way to avoid the pitfalls inherent in simplification is to continue to maintain contact with the experimental results and at every step of the procedure attempt to justify and, if possible, quantify the assumptions made in going from the details to the simplification.

References

Abbott, L. F., and Kepler, T. B. (1990). Model neurons: from Hodgkin–Huxley to Hopfield. In *Statistical Mechanics of Neural Networks,* L. Garrida, ed. pp. 5–18. Springer-Verlag, Berlin.

Chen, Z., Ermentrout, B., and Wang, X. J. (1998). Wave propagation mediated by GABA-B synapse and rebound excitation in an inhibitory network: a reduced model approach. *J. Computat. Neuro.* 5: 53–60.

Cowan, J. D. (1968). Statistical mechanics of neural nets. In *Neural Networks.* E. R. Caiianello, ed. pp. 181–188. Springer-Verlag, Berlin.

Edelstein-Keshet, L., and Ermentrout, G. B. (1991). Models for contact-mediated pattern formation: cells that form parallel array. *Differentiation* 29: 33–58.

Ermentrout, G. B., (1994). Reduction of conductance-based models with slow synapses to neural nets. *Neural Comp.* 6: 679–695.

Ermentrout, G. B., and Edelstein-Keshet, L. (1993). Cellular automata approaches to biological modeling. *J. Theoret. Biol.* 160: 97–133.

Hoppensteadt, F., and Izhikevich, E. (1997). *Weakly Connected Neural Nets.* Springer-Verlag, Berlin.

Kuramoto, Y. (1984). *Chemical Oscillations, Waves, and Turbulence.* Springer-Verlag, New York.

Murray, J.D. (1989). *Mathematical Biology.* Springer-Verlag, New York.

Pinto, D. J., Brumberg, J. C., Simons, D. J., and Ermentrout, G. B. (1996). A quantitative population model of whisker barrels: re-examining the Wilson-Cowan equations. *J. Comput. Neurosci.* 3: 247–264.

Rinzel, J. (1985). Excitation dynamics: insights from simplified membrane models. *Fed. Proc.* 44, 2944–2946.

van Vreeswijk, C., and Sompolinsky, H. (1998). Chaotic balanced states in a model of cortical circuits. *Neural Comput.* 10: 1321–1373.

Wang, X. J., and Rinzel, J. (1995). Oscillatory and bursting properties of neurons. In *The Handbook of Brain Theory and Neural Networks,* M. A. Arbib, ed. pp. 686–691. MIT Press, Cambridge, Mass.

Williams, T. L. and Sigvardt, K. A. (1995). Spinal cord of lamprey:generation of locomotor patterns. In *The Handbook of Brain Theory and Neural Networks,* M. A. Arbib, ed. pp. 918–921. MIT Press, Cambridge, Mass.

Contributors

Hamid Bolouri
Science and Technology Research Centre
University of Hertfordshire
Hatfield, UK

Mark T. Borisuk
Control and Dynamical Systems
California Institute of Technology
Pasadena, California

Guy Bormann
Born-Bunge Foundation
University of Antwerp
Antwerp, Belgium

James M. Bower
Division of Biology
California Institute of Technology
Pasadena, California

Dennis Bray
Department of Zoology
University of Cambridge
Cambridge, UK

Fons Brosens
Department of Physics
University of Antwerp
Antwerp, Belgium

Jehoshua Bruck
Division of Engineering and Applied
Science
California Institute of Technology
Pasadena, California

Kathy Chen
Department of Biology
Virginia Polytechnic Institute
Blacksburg, Virginia

Eric H. Davidson
Division of Biology
California Institute of Technology
Pasadena, California

Alain Destexhe
UNIC-UPR
CNRS
Paris, France

Bard Ermentrout
Department of Mathematics
University of Pittsburgh
Pittsburgh, Pennsylvania

Carl Firth
Department of Molecular Biology
Astra Pharmaceuticals
Loughborough, UK

Stefanie Fuhrman
Incyte Pharmaceuticals, Inc.
Palo Alto, California

Michael A. Gibson
Computation and Neural Systems Program
California Institute of Technology
Pasadena, California

William A. Goddard III
Beckman Institute
California Institute of Technology
Pasadena, California

Eric Mjolsness
Jet Propulsion Laboratory
California Institute of Technology
Pasadena, California

Bela Novak
Department of Agricultural Chemical
Technology
Technical University of Budapest
Budapest, Hungary

Erik de Schutter
Born-Bunge Foundation
University of Antwerp
Antwerp, Belgium

Roland Somogyi
Molecular Mining Corporation
Kingston, Ontario, Canada

John J. Tyson
Department of Biology
Virginia Polytechnic Institute
Blacksburg, Virginia

Nagarajan Vaidehi
Beckman Institute
California Institute of Technology
Pasadena, California

Xiling Wen
Incyte Pharmaceuticals, Inc.
Palo Alto, California

Chiou-Hwa Yuh
Division of Biology
California Institute of Technology
Pasadena, California

Index

A-amino-3-hydroxy-5-methyl-4-isoxazolepropionic
acid (AMPA) receptor, 242–243, 245, 253
A-P axis, 102–104, 107
Abbott, L. F., 315–316
Ackers, G. K., 27
Adenosine deaminase, 182
Adenosine diphosphate (ADP), 171
Adenosine triphosphate (ATP), 171, 199, 204
ADI method, 197, 207, 216
ADP, 171
Akutsu, T., 44
Alberts, B., 103
Albritton, M. L., 201
Algorithms, 43–44, 61–62, 137, 169, 174, 212. *See also* Reverse engineering (REVEAL) algorithm
Allostery, 273
Alternating direction implicit (ADI) method, 197, 207, 216
AMI method, 163
AMBER force field, 166, 177
Amino acids, 7–8, 166, 175
AMPA receptor, 242–243, 245, 253
Anaphase-promoting complex (APC), 290
Anderson, O., 236
ANN, 41
Anterior-posterior (A-P) axis, 102–104, 107
APC, 290
Apparent diffusion, 191, 205
Approximating dynamics, 40
Arkin, A., 49, 68
Artificial neural networks (ANN), 41
Asynchronicity, 33
Atomic-level simulation and biomacromolecules modeling
  biological problems, 174–184
    binding energy calculation, 179–182
    enzyme reaction mechanisms, 175–177
    human rhinovirus drug reaction, 177–179
    overview, 174–175
    quantitative structure-activity relationship, 182–184
  future research, 183–184
  molecular dynamics, 163–174
    force field, 161, 164–168
    methods, 168–174
    overview, 163–166
    quantum mechanics and, 161, 163–164
    solvents and, effect of, 167–168
  overview, 161–163
ATP, 171, 199, 204
Averaged equations
  Hebbian learning, 318–319
  neural network from biophysics, 319–320
  oscillators, weakly coupled, 320–322
  overview, 317–318

Averaging, 308
Avogadro's number, 192

Backpropagation, 43
Backpropagation through time (BPTT), 43
Barkai, N., 274
Bessel walk process, 217–218
Bezanilla, F., 233–234
Bicoid (*bcd*) transcription factor, 11, 103–104
Bifurcation analysis
  biological significance of, 299–302
  diagrams, 297–298
  Hopf, 297
  predictions, experimental, 302–303
  saddle-loop, 297
  saddle-node, 297
  saddle-node-loop, 299
  Takens-Bogdanov, 297
Binding energy calculation, 179–182
Biochemical models, predictions from
  hybrid methods and, 32–39
    between Boolean networks and differential equations, 32–35
    between differential and stochastic equations, 35–39
    continuous logic models, 34–35
  overview of models, 21–23
  "pure" methods and, 23–32
    Boolean networks, 23–25
    differential equations models, 25–28
    stochastic models, 28–32
  questions regarding, 20–21
Biochemical networks modeling. *See specific models*
Biochemical processes, 5–9
Biological information flow, 127–128
Biological problems
  binding energy calculation, 179–182
  enzyme reaction mechanisms, 175–177
  human rhinovirus drug reaction, 177–179
  overview, 174–175
  quantitative structure-activity relationship, 182–184
Biological significance of bifurcation analysis, 299–302
Biology, understanding
  biochemical processes, 5–9
  biophysical processes, 9
  feedback and gene circuits, 12–13
  prokaryotes vs. eukaryotes, 9–12
Biomacromolecules modeling. *See* Atomic-level simulation and biomacromolecules modeling
Biomolecular network, conceptualizing distributed, 121
Biophysical processes, 9
Blake, C. F., 171

Blastula, 288
Boltzmann factor, 174
Boolean networks
 dynamics of, wiring and rules determining, 122–125
 gene regulation and
  hybrid methods, 32–35
  "pure" methods, 23–25
 inference, 44
 many states converging on one attractor, 125
 overview, 21–22
 parameter estimation, data fitting, and phenomenology, 40
 reverse engineering and, 44, 119, 130–134
 state in, 15
 as theoretical model, 119, 121
 time series, 125
Boundary conditions, 196–197
BPTT, 43
Brameld, K. A., 176
Bray, D., 288
Brownian motion, 189, 191
Bruck, J., 69
Bursting, 317

Caenorhabditis elegans, gene regulation and, 3
Calcium diffusion, 199–205, 208
Cartesian coordinates, 216–217
CAT enzyme measurements, 81
Cell cycle regulation, complex dynamics in
 cell division in frog eggs and, 288–289, 292
 computational molecular biology and, 287–288, 303
 implications of, 303
 M-phase-promoting factor regulation and, 290–303
  bifurcation analysis, 297–303
  feedback loops, isolating positive and negative, 292–297
  molecular machinery of, 290–292
  Novak-Tyson model of, 291–292
 overview, 287
Cell division in frog eggs, 288–289, 292
Cell multiple method (CMM), 165, 171, 174
Cell signaling pathways simulation
 allostery, 273
 chemotactic response, 273–274
 free energy considerations, 279–282
 model description, 274–277
 results, 277–279
 signaling complexes, 272–273
Cell-cell signaling, 107–108
Central nervous system development of rat, 119–120, 134–136
Changeux, J. P., 273
Chaperones, 8
Chapman-Kolmogorov equation, 309. *See also* Master equation

Charge equilibrium (QEq), 164, 176
CHARMM force field, 166
Chemical reactions, 13–15
Chemotactic response, 273–274
Chen, Z., 320
Chitin, 175–176
CI, 163
cis-regulatory elements, 6, 73. *See also* Logical model of *cis*-regulatory control of eukaryotic system
Cluster analysis
 Euclidean, 137–139, 142, 145
 of model network, 127
 mutual information, 139–142
 pathways suggested by, 142–144
CMM, 165, 171, 174
Complex models, simplifying and reducing
 averaging, 308
 deterministic models, 314–322
  averaged equations and, 317–322
  overview, 314
  pseudo-steady state and, 314–317
 master equation and, 308–314
  defined, 308–311
  fibroblast orientation model and, 311–312
  mean-field reduction of neural system, 312–314
 overview, 307–308
 questions raised from, 322–323
Computational learning theory, 39
Computational models. *See* Genetic network inference in computational models; *specific models*
Computational molecular biology, 287–288, 303
Concurrent reactions, 64–67
Configuration, 29
Configuration interaction (CI), 163
Connor-Stevens model, 316
Constitutive transcription factor, 114
Constrained internal coordinates, 170–171
Continuity, 33
Continuous logic models
 hybrid methods and, 34–35
 overview, 22
Continuous-time Markov chain, 30
Cooperativity, 7
Correspondence principle, 211–212
Coulomb component of nonbond energy, 164
Coulomb interactions, 198–199
Crank, J., 193
Crank-Nicholson method, 207, 213–215
Cro, 49–50, 56, 58–59, 63, 69
Cytoplasm, 200–201

D'Ari, R., 32
Data fitting. *See* Parameter estimation, data fitting, and phenomenology
Davidson, E. H., 78
De Schutter, E., 200, 214

Degradation, 8, 55–57
  second-order reactions and, 68
Density functional theory (DFT), 163
Detailed balance condition, 211–212
Deterministic models
  averaged equations and, 317–322
    Hebbian learning, 318–319
    neural network from biophysics, 319–320
    oscillators, weakly coupled, 320–322
    overview, 317–318
  complex models and, 314–322
    averaged equations, 317–322
    overview, 314
    pseudo-steady-state and, 314–317
  limitations of, 264–266, 283
  overview, 314
  probabilistic model and, 52
  pseudo-steady state and, 314–317
D'haeseleer, P., 152
Differential equations models
  gene regulation and
    hybrid methods, 32–39
    "pure" methods, 25–28
    overview, 22
    state in, 15
Diffusion
  apparent, 191, 205
  biophysical processes and, 9
  boundary conditions and, 196–197
  Brownian motion and, 189, 191
  calcium, 199–205, 208
  case study, 216–221
  defined, 9, 191–194
  dimensionality and, 196–197
  implications of modeling, 190
  interactions among molecules and with external
    fields and, 197–207
  macroscopic description of, 194–196
  numerical methods of solving reaction-diffusion sys-
    tem and, 207–216
    diffusion Monte Carlo methods, 207–208
    discretization in space and time, 213
    Green's function Monte Carlo methods, 197, 208–
      213
    Monte Carlo methods, 197–198, 207
    one-dimensional diffusion, 213–215
    overview, 207
    two-dimensional diffusion, 215–216
  one-dimensional, 213–215
  overview, 190–191
  reasons for modeling, 189
  relay race, 206
  two-dimensional, 215–216
Diffusion Monte Carlo methods, 207–208
Dimensionality, 196–197
Ding, H. Q., 174

Discretization in space and time, 213
DNA
  binding sites, 27–28, 57–59
  equilibrium and, 16
  in eukaryotes, 5, 8–9
  force field and, 166
  function of, 3–4
  lamda phage biology and, 49
  transcription factors and, 73
Dreiding FF, 176
Driever, W., 104
Drosophilia melanogaster. 3, 11. See also Trainable
  gene regulation networks with applications to
  Drosophilia pattern formation
Drug action on human rhinovirus, 177–179

Edelstein-Keshet, L., 312
8R-coformycin, 182
8R-deoxycoformycin, 182
Electrodiffusion equation, 198–199
Electrostatic component of nonbond energy, 164
Endo16 cis-regulatory function model
  experimental procedure and data processing, 79–83
  hierarchical cooperative activation model and, 113
  implications of, 97–98
  modular interactions, 83–97
  overview, 79, 83
Endo16 gene, 11, 77–79
Engrailed transcription factor, 103
Enhancers, 11
Enzyme reaction mechanisms, 175–177
Equilibrium, 16–18
Equilibrium constant, 16
Erb, R. S., 44
Ermentrout, G. B., 312
Escherichia coli, 31, 49, 56, 273–274, 279
Euclidean cluster analysis, 137–139, 142, 145
Euclidean distance, 127, 140, 142
Eukaryotes, 5–6, 8–12. See also Logical model of cis-
  regulatory control of eukaryotic system
Even-skipped (eve) stripe transcription factor, 11,
  102–104, 107
Ewald summation, 172
Exons, 7
Expression constructs, 79–81
Extended Hückel method, 163

Feedback, 12–13, 68
Feedback loops, isolating positive and negative, 292–
  297
FEP theory, 179–182
FF, 161, 164–168, 177
Fibroblast orientation model, 311–312
Fick's law, 194
Fitch-Margoliash clustering algorithm, 137
Fokker-Planck method, 23, 35, 37–38

Force field (FF), 161, 164–168, 177
Fourier equation, 174
Free energy, 18–19, 279–282
Free energy perturbation (FEP) theory, 179–182
Frog eggs, cell division in, 288–289, 292
Fundamental equations, 168–169
Furusawa, C., 43
Fushi tarazu (*ftz*) transcription factor, 103, 107

GABA receptors, 152, 243, 247, 253, 257
GABA$_B$-mediated neurotransmission, 247–250
GABAergic signaling gene family, 121, 123
GAD, 152
Gap gene expression, 103–104
Gating, ion channel
  kinetic description of, 225, 227–231
  voltage-dependent, 230–231
Gaussian distribution, 35, 192–193, 208–209
Gene circuits, 12–13, 101
Gene expression
  to functional networks, 120–127
  biomolecular network, conceptualizing distributed, 121
  dynamics of network, wiring and rules determining, 122–125
  many states converging on one attractor, 125–126
  overview, 120–122
  terminology, 127
  gap, 103–104
  large-scale, 132–153
  causal inference of gene interactions and, 150–153
  functional inference from, 132
  high-precision, high-sensitivity assay, 132, 134
  injury analysis and, 148–150
  level-by-level inference from, 154–155
  rat central nervous system development and, 119–120, 134–136
  rat hippocampus development and, 119–120, 145–148
  rat spinal cord development and, 119–120, 136–144, 147–148
  overlapping control of, 147–148
Gene regulation. *See also* Probabilistic model of prokaryotic gene and its regulation; Trainable gene regulation networks with applications to *Drosophilia* pattern formation
  Boolean networks and, 23–25
  hybrid methods, 32–35
  "pure" methods, 23–38
  Caenorhabditis elegans and, 3
  differential equations models and
  hybrid methods, 32–39
  "pure" methods, 25–28
  Drosophilia melanogaster and, 3
  networks, 101
  stochastic models and, 28–32

  hybrid methods, 35–39
  "pure" methods, 28–32
  understanding, 3–5
Generalized mass action (GMA), 42–43
GENESIS software, 200
Genetic algorithms, 44
Genetic network inference in computational models
  constructing, general strategies for, 154–155
  gene expression to functional network, 120–127
  biomolecular network, conceptualizing distributed, 121
  dynamics of network, wiring and rules determining network, 122–125
  many states converging on one attractor, 125–126
  overview, 120–122
  terminology, 127
  large-scale gene expression data and, 132–153
  causal inference of gene interactions and, 150–153
  functional inference from, 132
  high-precision, high-sensitivity assay, 132, 134
  injury analysis and, 148–150
  level-by-level inference from, 154–155
  rat central nervous system development and, 119–120, 134–136
  rat hippocampus development and, 119–120, 145–148
  rat spinal cord development and, 119–120, 136–144, 147–148
  overview, 119–120
  reverse engineering and, 130–132
  shared control processes, inference of, 127–130
Genetic networks modeling. *See specific models*
Genetic programs, evidence for, 148–150
Genetic regulatory networks, 101
GEPASI simulator, 264
Giant (*gt*) transcription factor, 11, 103
Gibbs ensemble, 169
Gibbs free energy, 18–19. *See also* Free energy
Gibson, M. A., 69
Gibson-Bruck algorithm, 62
Gillespie algorithm, 62
Gillespie, D. T., 62, 67, 69
Glass, L., 32
Glutamate AMPA receptor, 242–243, 245
Glutamic acid decarboxylase (GAD), 152
Glycosyl hydrolases, 175
GMA, 42–43
Goddard, W. A., III, 174, 177
Goldbeter, A., 288
Gradient random walk (GRW), 211
Gray, S., 11
Green's function Monte Carlo methods, 197, 208–213
GRW, 211

H bifurcation, 297
Hamahashi, S., 107

Hartree-Fock (HF) density functional theory (DFT), 163
HB component of nonbond energy, 164
HCA model, 11, 110–115
Hebbian learning, 318–319
Helmholtz canonical ensemble, 169
Hertz, J., 43
Heterodimer transcription factor, 112
HF density functional theory (DFT), 163
Hidden Markov models, 39
Hierarchical cooperative activation (HCA) model, 11, 110–115
Hierarchical model, 42
High-precision, high-sensitivity assay, 132, 134
Hippocampus development of rat, 119–120, 145–148
Hodgkin, A. L., 225, 231–232, 234
Hodgkin-Huxley model of voltage-dependent channels, 231–233, 236, 256
Holmes, W. R., 216
Homodimeric transcription factor, 112–114
Hopf (H) bifurcation, 297
Hörstadius, S., 77
Host cell growth, 56–57
HRV, 177–179
Human rhinovirus (HRV), 177–179
Hunchback (hb) transcription factor, 11, 103–104
Huxley, A. F., 225, 231–232, 234
Hybrid methods
  between Boolean networks and differential equations, 32–35
  between differential and stochastic equations, 35–39
  continuous logic models, 34–35
Hydrogen bond (HB) component of nonbond energy, 164
Hydrolysis, 175

ICAM-1 receptors, 177
Injury analysis, 148–150
Integrated network analysis strategy, 155
Interactions among molecules and with external fields, 197–207
Intercellular adhesion molecule 1 (ICAM-1) receptors, 177
Intermediate methods. See Hybrid methods
Introns, 7
Ion channels
  biophysical properties of, 225
  defined, 225
  formalisms about, 255–256
  kinetic description of gating, 225, 227–229, 230–231
  ligand-gated synaptic, 241–245
    defined, 241
    kinetic models of, 242
    Markov models of, 242–243
  Markov models of, 225, 232–233

  ligand-gated synaptic, 242–243
  voltage-dependent, 232–233, 256
  neuronal interactions modeling and, applications to complex, 250–255
  second-messenger-gated synaptic, 246–250
    defined, 246
    GABA$_B$-mediated neurotransmission and, 247–250
    kinetic models of, 246–247
  voltage-dependent, 225–227, 229–237
    Hodgkin-Huxley model of, 231–233, 236, 256
    kinetics of, 230–231
    Markov models of, 232–233, 256
    overview, 225–227, 229–230
    sodium channels, 234–237
Ionotropic receptor, 241

Jackson, J., 40

Kaneko, K., 43
Karasawa, N., 174
Kauffman, S. A., 32
Kepler, T. B., 315–316
Kinetic logic models
  hybrid methods and, 32–34
  overview of, 22
Kinetic models of excitable membranes and synaptic interactions
  assumptions about, 255–257
  ion channels, 227–256
    biophysical properties of, 225
    defined, 225
    formalisms about, 255–256
    gating, 225, 227–231
    interactions modeling and, applications to complex, 250–255
    kinetic description of gating, 225, 227–229, 230–231
    ligand-gated synaptic, 241–245
    Markov models of, 225, 232–233
    second-messenger-gated synaptic, 246–250
    voltage-dependent, 225–227, 229–237
  limits of, 255–257
  molecular biology and, 257
  overview, 225–227
  simplified, 256–257
  transmitter release, 237–241
Kinetics-changes of state, 16
Kitano, H., 107
Knirps (kni) transcription factor, 103
Koeppe, R. E., III, 236
Kramers-Moyal expansion, 38
Kruppel (Kr) transcription factor, 11, 103

Lambda phage biology, 49–52
Langevin approach, 22–23, 35–37

Large-scale gene expression data
  causal inference of gene interactions and, 150–153
  functional inference from, 132
  high-precision, high-sensitivity assay, 132, 134
  injury analysis and, 148–150
  level-by-level inference from, 154–155
  rat central nervous system development and, 119–120, 134–136
  rat hippocampus development and, 119–120, 145–148
  rat spinal cord development and, 119–120, 136–144, 147–148
Lawrence, P., 103
Leibler, S., 274
Liang, S., 44
Ligand gating, 241
Ligand-gated synaptic ion channels
  defined, 241
  kinetic models of, 242
  Markov models of, 242–243
Linear function, 40–41
Logical model of *cis*-regulatory control of eukaryotic system
  considations in modeling, 73–79
    Endo16 gene, 77–79
    problem, 73–74
    sea urchin development, 74–77
  framework, 79–98
    Endo16 cis-regulatory function model, 83–98
    experimental procedure and data processing, 79–83
  future research, 99
  overview, 73

M-phase-promoting factor (MPF)
  defined, 288–289
  regulation, 290–303
    bifurcation analysis and, 297–303
    feedback loops, isolating positive and negative, 292–297
    molecular machinery of, 290–292
    Novak-Tyson model of, 291–292
McAdams, H. H., 288
Macroscopic chemistry, 54
Macroscopic description of diffusion, 194
Markov chains, 39
Markov models of ion channels, 225, 232–233
  ligand-gated synaptic, 242–243
  voltage-dependent, 232–233, 256
Markov process, 311
Markovian systems, 228
Markram, H., 207
Marnellos, G., 44, 107
Mascagni, M., 211
Master equation
  in complex models, 308–314
    defined, 308–311

fibroblast orientation, 311–312
    mean-field reduction of neural systems, 312–314
    probabilistic model of prokaryotic gene and its regulation, 60–61
Mayer, M. L., 242
MBT, 288
MD. *See* Molecular dynamics
Mean-field approximation, 308, 312
Mean-field reduction of neural network, 312–314
Membrane excitability. *See* Kinetic models of excitable membranes and synaptic interactions
Mesoscopic chemistry, 54
Mestl, T., 34
Metabotropic receptors, 246
MIC, 177
Michaelis-Menten equations, 246, 287, 314–315
Michaels, G. S., 44
Microscopic chemistry, 54. *See also* Molecular dynamics
Midblastula transition (MBT), 288
MINDO, 163
Minimal stripe element (MSE2), 108–109
Minimum inhibitory concentration (MIC), 177
MIST simulator, 264
Mjolsness, E., 101–102, 107
MM3 force field, 166
Modeling basics. *See also specific models*
  chemical reactions, 13–15
  overview, 13
  physical chemistry, 15–20
Modified intermediate neglect of differential overlap (MINDO), 163
Molecular biology, 257
  computational, 287–288, 303
Molecular dynamics (MD)
  atomic-level simulation and biomacromolecules modeling, 163–174
    force field, 161, 164–168
    methods, 168–174
    overview, 163–166
    quantum mechanics and, 161, 163–164
    solvents and, effect of, 167–168
  biological problems and, 174–183
    binding energy calculation, 179–182
    enzyme reaction mechanisms, 175–177
    human rhinovirus drug reaction, 177–179
    overview, 174–175
    quantitative structure-activity relationship, 182–184
  defined, 161
  methods
    constrained internal coordinates, 170–171
    fundamental equations, 168–169
    Monte Carlo, 174
    MPSim program, 171–172, 177
    NPT and NVT dynamics, 169–170
    periodic boundary conditions, 172–174

modeling and, 53–54
quantum mechanics and, 161, 163–164
state in, 15
Monod-Wyman-Changeux (MWC) model, 111–112
Monte Carlo algorithms, 61–62, 174
Monte Carlo (MC) methods
diffusion of, 207–208
molecular dynamics methods, 174
numerical methods of solving reaction-diffusion system and, 197–198, 207
stochastic models and, 29
MPF. *See* M-phase-promoting factor
MPSim program, 171–172, 177
mRNA, 5–8, 11–12, 107, 120
MSC-PolyGraf program, 176
MSE2, 108–109
Multimers, 8
Multistate molecules, 272
Murray, J. D., 315
Mutual information, 130
cluster analysis, 139–142
MWC model, 111–112

N-methyl-D-aspartate (NMDA) receptor, 241, 243
Naraghi, M., 206
Neher, E., 206
NEIMO method, 170–171
Nernst-Planck equation, 199
Network models, general strategies for constructing, 154–155. *See also specific types*
Neural network. *See also specific types*
from biophysics, 319–320
mean-field reduction of, 312–314
Neurogenesis, 107–108
Neuronal interactions modeling, complex, 250–255
Neurotransmission, 247–250
Newton-Euler inverse mass operator (NEIMO) method, 170–171
Newton's equations of motion, 169
NMDA receptor, 241, 243
Non-Fickian fluxes, 194, 212
Nose dynamics, 169
Novak, B., 291–292
Novak-Tyson model of M-phase-promoting factor regulation, 291–292
NPT dynamics, 169–170
Numerical methods of solving reaction-diffusion system
diffusion Monte Carlo methods, 207–208
discretization in space and time, 213
Green's function Monte Carlo methods, 197, 208–213
Monte Carlo methods, 197–198, 207
one-dimensional diffusion, 213–215
overview, 207
two-dimensional diffusion, 215–216

Nucleic acids, 166
Nusslein-Volhard, C., 104
NVT dynamics, 169–170

ODE models. *See* Ordinary differential equations models
$1/\Omega$ expansion, 23, 38–39
One-dimensional diffusion, 213–215
OPLS force field, 166
Ordinary differential equations (ODE) models
interactions among molecules and with external fields, 197
overview, 22
parameter estimation, data fitting, and phenomenology, 40–44
trainable gene regulation networks and, 101
Organic systems (OPLS) force field, 166
Oscillators, weakly coupled, 320–322
Oster, G. F., 288

Parameter estimation, data fitting, and phenomenology
Boolean models, 40
ordinary differential equation models, 40–44
overview, 39–40
stochastic models, 44
Partial differential equation (PDE), 192, 211
Patneau, D. K., 242
PCSM, 177, 179
PDE, 192, 211
Pearlmutter, B. A., 43
Pentamer channel stiffening model (PCSM), 177, 179
Periodic boundary conditions, 172–174
PGK, 171
Phenomenology. *See* Parameter estimation, data fitting, and phenomenology
Phosphoglycerate kinase (PGK), 171
Physical chemistry, 15–20
Poisson process, 67
Poisson-Boltzmann approximation, 164
Post-translational modification, 7–8
Probabilistic model of prokaryotic gene its regulation
degradation, 55–56
details, 57–60
calculations, 59
reaction per unit time, 59–60
TF-DNA binding, 57–59
deterministic models and, 52
future research, 69–70
host cell growth, 56–57
lambda phage biology and, 49–52
master equation and, 60–62
overview, 49, 54
purpose of, 52–54
results, 62–69

Probabalistic model of prokaryotic gene its regulation
    (continued)
  stochastic models and, 52–54
  transcription, 54–55
  translation, 55
Prokaryotes, 9–12. *See also* Probabilistic model of
    prokaryotic gene and its regulation
Promoter, 6, 11, 121
  substructure, 108–115
    hierarchical cooperative activation, 110–115
    overview, 108–110
Proteins, 3–4, 166
Pseudo-first-order degradation, 56–57
Pseudo-steady state, 314–317
"Pure" methods
  Boolean networks, 23–25
  differential equations models, 25–28
  stochastic models, 28–32

QEq, 164, 176
Qian, N., 199
QM, 161, 163–164
QSAR, 170, 182–184
Quantified information, 128–130
Quantitative structure-activity relationship (QSAR),
    170, 182–184
Quantum mechanics (QM), 161, 163–164

Rall, W., 245
Random-walk process, 211, 216–218
Ransick, A., 78
Rat
  central nervous system development of, 119–120,
    134–136
  hippocampus development of, 119–120, 145–148
  spinal cord development of, 119–120, 136–144,
    147–148
Reaction kinetics. *See* Differential equations models
Reaction per unit time, 59–60
Reaction-diffusion system
  apparent diffusion coefficient and, 205
  buffers and, 207
  calcium diffusion and, 205–207
  numerical methods of solving, 207–216
    diffusion Monte Carlo methods, 207–208
    discretization in space and time and, 213
    Green's function Monte Carlo methods, 208–213
    Monte Carlo methods, 207
    one-dimensional diffusion and, 213–214
    overview, 207
    two-dimensional diffusion and, 215–216
  study of, 191
Reactions
  concurrent, 64–67

  second-order, 68–69
  sequential, 67–68
Regulatory modules, 11
Reinitz, J., 103–104, 107
Relaxation oscillator, 316
Relay race diffusion, 206
Repressor, 56, 58–59, 63, 69
Reverse engineering (REVEAL) algorithm, 44, 119,
    130–134, 150, 152, 155
Reverse transcription polymerase chain reaction
    (RT-PCR), 132, 134–136, 145
Rhinovirus-1A, 177–179
Rhinovirus-14 , 177–179
Rinzel, J., 315, 317
Risken, H., 35
RNA, 3–4, 177
RNA polymerase (RNAP), 6, 27–28, 50, 54
RT-PCR, 132, 134–136, 145
Runge-Kutta-Fehlberg with Rosenbrock-type exten-
    sions, 211

Saddle-loop (SL) bifurcation, 297
Saddle-node (SN) bifurcation, 297
Saddle-node-loop (SNL) bifurcation, 299
Savageau, M. A., 42
SCAMP simulator, 264
Schroedinger equation, 161
Schroeter, E. H., 107
Sea urchin, 11, 74–77
Second-messenger-gated synaptic channels
  defined, 246
  $GABA_B$-mediated neurotransmission and, 247–250
  kinetic models of, 246–247
Second-order reactions, 68–69
Sejnowski, T. J., 199
Sequential reactions, 67–68
Shannon, Claude, 128
Shannon entropy (H), 128–130
Shapiro, L., 288
Shared control processes, inference of, 127–130, 136–
    137
Sharp, D. H., 104, 111
Shea, M. A., 27
Sherman, A., 211, 288
Sigma-pi neural networks, 41–42
Signaling complexes, 272–273
Sigvardt, K. A., 322
Simpson-Brose, M., 104
Simulated annealing, 43–44
Single gene activity modeling
  basics of modeling and, 13–20
    chemical reactions, 13–15
    overview, 13
    physical chemistry, 15–20

biochemical models, predictions from, 20–39
  hybrid methods and, 32–39
  overview of models, 21–23
  "pure" methods and, 23–32
  questions regarding, 20–21
biology and, understanding, 5–13
  biochemical processes, 5–9
  biophysical processes, 9
  feedback and gene circuits, 12–13
  prokaryotes vs. eukaryotes, 9–12
future research, 45–46
overview, 3–5
parameter estimation, data fitting, and phemenology,
    39–44
  Boolean models, 40
  ordinary differential equation models, 40–44
  overview, 39–40
  stochastic models, 44
roadmap for understanding, 45
SL bifurcation, 297
Small, S, 109
Smolen, P., 200, 214
SN bifurcation, 297
Sneyd, J., 288
SNL bifurcation, 299
Sodium channel models, voltage-dependent, 234–237
Solvents, effect of, 167–168
Somogyi, R., 40
Spatial averaging, 308
Spatial colocalization of related processes, 197
Spatial localization, 266–267
Spinal cord development of rat, 119–120, 136–144,
    147–148
Splicing, 7
Standard force field, 166–167
Standley, C., 243
State of a system, 15
Stochastic models
  of cell signaling pathways, 272–284
    allostery, 273
    chemotactic response, 273–274
    free energy considerations, 279–282
    model description, 274–277
    results, 277–279
    signaling complexes, 272–273
  deterministic model limitations and, 264–266, 283
  fluctuations, 268
  gene regulation and
    hybrid methods, 35–39
    "pure" methods, 28–32
  limitations of, 284
  Monte Carlo methods and, 29
  multistate molecules and, 272
  overview, 263–264
  parameter estimation, data fitting, and phenomenol-
    ogy, 44

previous simulators and, 268–269
probabilistic model and, 52–54
spatial localization and, 266–267
state in, 15
StochSim and, 263, 269–271, 274
StochSim program, 263, 269–271, 274
Strongylocentrotus purpuratus, 11, 74–77
Synaptic interactions. *See* Kinetic models of excitable
    membranes and synaptic interactions
System identification, 39–41

Tailless (*tll*) transcription factor, 103, 110
Takens-Bogdanov (TB) bifurcation, 297
Temperate phage, 49
TF-DNA binding, 57–59
TFs, 4–6, 10–11, 73, 102–104, 107, 112–114
Thalamic oscillations, 250
Thermodynamics, 18–20, 211–212
Thomas, R., 32, 34
Time series table, 132
Trainable gene regulation networks with applications
    to *Drosophilia* pattern formation
  cell-cell signaling, 107–108
  even-skipped (*eve*) stripe transcription factor, 102–
    104, 107
  examples, 102–108
    anterior-posterior axis, 102–104, 107
    even-skipped (*eve*) stripe, 102–104, 107
    gene circuit model, 103, 107–108
  gap gene expression, 103–104
  neurogenesis, 107–108
  ordinary differential equations models and, 101
  overview, 101–102
  promoter substructure, 108–115
    hierarchical cooperative activation, 110–115
    overview, 108–110
Training parameters, 43–44
trans-regulatory elements, 6, 73, 101
Transcription
  basics of, 5–6
  in eukaryotes, 10
  in probabilistic model, 54–55
  transcription factors and, 4–6, 10–11, 73, 102–104,
    107, 112–114
Transcription factors (TFs), 4–6, 10–11, 73, 102–104,
    107, 112–114
Transition table, 132
Translation, 7–8, 11–12, 55
Transmitter release
  kinetic model of, 238–239
  overview, 237–238
  simplified model of, 239–241
Trypanosoma brucei PGK, 171
Two-dimensional diffusion, 215–216

Universal force field (UFF), 167

Vaidehi, N., 177
van der Waals (VDW) component of nonbond energy,
    164
Van Kampen, N. G., 35, 37, 39
Van Kampen's $1/\Omega$ expansion, 23, 38–39
Vandenberg, C. A., 233–234
Vandenberg-Bezanilla model of sodium channel,
    233–234, 236
VDW component of nonbond energy, 164
Verlet algorithm, 169
Voltage-dependent ion channels
  Hodgkin-Huxley model of, 231–233, 236, 256
  kinetics of, 230–231
  Markov models of, 232–233, 256
  overview, 225–227, 229–230
  sodium channels, 234–237
von Hippel, P. H., 6

Walk processes, 211, 216–218
Wang, X. J., 317
Williams, T. L., 322
Wingless transcription factor, 103

Yamada, W. M., 238
Yuh, C. H., 11, 113

Zucker, R. S., 238